Molecular Symmetry and Group Theory

Molecular Symmetry and Group Theory

Robert L. Carter

Department of Chemistry
University of Massachusetts Boston

John Wiley & Sons, Inc.

ACQUISITIONS EDITOR	NEDAH ROSE
MARKETING MANAGER	KIMBERLY MANZI
PRODUCTION EDITOR	DEBORAH HERBERT
DESIGNER	ANN MARIE RENZI
ILLUSTRATION EDITOR	EDWARD STARR
COVER DESIGN	DAVID LEVY

This book was set in 10/12 Times Ten by York Graphic Services, Inc. and printed and bound by Courier Companies, Inc. (Westford).
The cover was printed by Phoenix Color Corporation.

This book is printed on acid-free paper. ∞

Library of Congress Cataloging-in-Publication Data
Carter, Robert L., 1944–
Molecular symmetry and group theory / Robert L. Carter.
    p.   cm.
    Includes index.
    ISBN 0-471-14955-1 (paper : alk. paper)
    1. Molecular theory.   2. Symmetry (Physics)   3. Group theory.
I. Title.
QD461.C32   1997
541.2'2'015122—dc21   97-25445
                                                                 CIP

Printed in the United States of America

10   9   8   7   6

For my students—
Past, present, and future

PREFACE

For more than 25 years, I have taught the junior–senior advanced inorganic chemistry course and have introduced symmetry and group theory as tools for studying bonding and spectroscopy. For most of this period there were no treatments of group theory in the standard inorganic texts, and my students were forced to rely mainly on the lecture material, occasionally supplemented with a rudimentary self-teaching text. In the 1980s, inorganic texts that included group theory began to appear, but in very limited number. In some cases the text as a whole was not suitable for the level or organization of my course, while in others the treatment of group theory was necessarily so brief as to be of little help to most students. These frustrations, which should be familiar to most professors of inorganic chemistry, compelled me in 1989 to begin to write a text that would suit the needs of my course and most especially my students. Over the past eight years and seven offerings of the course, that original idea has grown through a succession of drafts into the present book. In the process, the manuscript developed beyond being a simple supplement to existing inorganic chemistry texts (although that need persists) to become a book that can also serve the needs of a stand-alone introductory course on group theory at the advanced undergraduate or lower graduate levels. Although it has evolved beyond the limits of its initial intent, this book retains the original purpose that motivated my writing it—to provide students with a thorough but understandable introduction to molecular symmetry and group theory as applied to chemical problems.

In keeping with this goal, I have tried to write in a style that invites the reader to discover by example the power of symmetry arguments for understanding otherwise intimidating theoretical problems in chemistry. To this end, the text emphasizes the meaning and chemical significance of the mathematics of group theory, rather than rigorous derivation. Calling upon my own remembered experiences in learning this material and upon my experience in teaching it for many years, I have tried to anticipate those questions and troublesome points students typically have with the subject. As a practical matter, this book shows very explicitly some of the most effective techniques for applying group theory to chemical problems. Some of these (e.g., the tabular method of reducing representations, the use of group–subgroup relationships for dealing with infinite-order groups) are known to many seasoned practitioners but have somehow escaped presentation in other texts. Other techniques and methods of approach are uniquely my own. In addition to fundamentals of theory and application, I have tried to show how group theory has contributed and continues to contribute to our theoretical understanding of structure and bonding. It is my belief that students gain a greater appreciation for any theoretical topic when it is shown how and when the ideas evolved. I would hope that students will realize from this that symmetry and group theory considerations are not peripheral to the theory of structure and bonding but rather are central to a complete understanding.

When teaching a graduate level course in group theory, I cover all topics in this text, at least in the depth presented and essentially in the order of the chapters. When teaching the junior–senior advanced inorganic chemistry course, owing to the constraints of time and level, the coverage is more selective in both range and depth. For this purpose, I customarily cover all of the material in Chapters 1 through 4. However, since this is most students' first encounter with symmetry and group theory, I do not think it necessary to introduce the more advanced topic of projection operators, the subject of Chapter 5. Therefore, I routinely skip this material at this level. In keeping with this, I have written the succeeding chapters in this text so as not to depend upon knowledge of projection operators. For the undergraduate course, I do cover the use of group theory for deducing spectroscopic selection rules for infrared and Raman activity (Chapter 6) but do not go into the depth of coverage on overtones, combinations, and other spectroscopic complications presented in Section 6.5. Likewise, with transition metal complexes I cover all the topics in Chapter 7 but gloss over the details of splitting of terms and the development of correlation diagrams (Sections 7.4 and 7.5), concentrating more on the use of Tanabe–Sugano and Orgel diagrams for interpreting absorption spectra (Section 7.6). Beyond the confines of any course, this book should serve the needs of advanced undergraduate students, graduate students, and professional chemists seeking to learn or review symmetry and group theory on their own. For this purpose, beyond Chapters 1 through 4, readers should feel free to delve into the topics of the remaining chapters as their interests and needs dictate.

Many individuals have contributed to making this a better book than it would have been without their constructive criticisms. First and foremost, I am most appreciative of the many students who used earlier editions of this material in my courses and beyond, particularly those who were forthcoming in their comments. While it is nice to receive compliments, I must confess I more greatly valued your calling to my attention points of confusion and incidents of errors in the earlier versions of the text. Likewise, I am indebted to the many reviewers solicited by John Wiley & Sons who offered constructive critiques of the manuscript at various stages of development. While I have not incorporated every one of their suggestions (which at times were divergent), I have gladly accepted every idea that seemed to further the goals of the text, consistent with my general approach. I am especially grateful to my colleague Professor Leverett J. Zompa for many useful discussions and his critical review of Chapter 7. Finally, I would be remiss not to acknowledge the much appreciated support and patience of Greg Cloutier throughout the course of this project.

Robert L. Carter

CONTENTS

Molecular Symmetry and Group Theory

CHAPTER 1

Fundamental Concepts

Our focus in this text will be the application of symmetry arguments to solve physical problems of chemical interest. As a first step in any application of this sort, we must identify and catalogue the complete symmetry of the system. Once this is done, we can employ the mathematics of groups to simplify the physical problem and subsequently to obtain chemically useful solutions to it. The advantages of this approach, relative to "brute force" techniques, tend to increase as the symmetry of the system increases. When the system has a high degree of structural regularity, complex problems can have elegantly simple solutions. Even in seemingly simple cases, symmetry arguments may provide insights that are difficult to achieve with other approaches.

We will confine our discussion to physical problems of isolated molecules or complex ions. This means that we will only need to consider the symmetry of the species itself, and not any symmetry that may exist as a result of associations with neighboring molecules. In general, the results we will obtain will be correct for samples of dilute gases, where intermolecular forces and influences are negligible. To a lesser extent, the results may be valid for certain liquid samples and dilute solutions. However, in these cases observed behavior may depart significantly from predictions based on the symmetry of the isolated molecules. Sometimes these departures are, in fact, structurally revealing.

In the case of solids, especially ionic crystals and network solids, associations between individual molecules and ions may be considerable, and results based on individual symmetries are least likely to be correct. Interaction between the oriented molecules or ions, and the regularity of the solid itself, result in new kinds of symmetry relationships not found in isolated species. An introduction to the symmetry of solids (space group symmetry) is beyond the intent of this book, but can be found in introductory texts dealing with crystallography or structure determination by x-ray diffraction. Nonetheless, the principles we will explore in relation to problems of isolated molecular species will provide a foundation for later study of applications based on crystal symmetry.

1.1 Symmetry Operations and Elements

The symmetry of molecules is defined in terms of *symmetry elements* and *symmetry operations*. A symmetry element is an imaginary geometrical entity such as a line, plane, or point about which a symmetry operation can be performed. A symmetry operation is a movement of an object about a symmetry element such that the object's orientation and position before and after the operation are indistinguishable. This means that the operation carries every point of the object into an equivalent point or back into the identical point.

Another way of determining whether a particular symmetry exists for an object is to perform the following test. Observe the object, and then turn away while someone performs the symmetry operation. When you turn around and observe the object again, you should not be able to tell whether or not the symmetry operation was actually performed. Note that the object need not be in the identical position it had before the operation (although it may be). It is only necessary that the position be indistinguishable and therefore equivalent for the object to possess the particular symmetry.

To appreciate the difference between indistinguishable and identical positions, try the following exercise. Make a square cutout from stiff paper (e.g., card stock). Label the corners A, B, C, and D. Turn the labeled side down on a piece of wood or heavy cardboard, and place a thumb tack in the middle, so that the cutout spins freely. Check the identities of the corners in the starting position, and then give the cutout a spin. Square it up so that the blank side appears to be in the same orientation as it was before spinning. Now check the identities of the corners. If they are the same as before, the position is identical to the starting position. Otherwise, the position is simply indistinguishable. Note that without the labels, when viewed from the blank side, you cannot tell an indistinguishable orientation from the identical (original) one.

The symmetry of a molecule or ion can be described in terms of the complete collection of symmetry operations it possesses. The total number of operations may be as few as one or as many as infinity. Regardless of the number of operations, all will be examples of only five types. These are (1) a seemingly trivial operation called *identity,* (2) *rotation* (sometimes called *proper rotation*), (3) *reflection,* (4) *inversion,* and (5) a two-part operation called *rotation–reflection* (or *improper rotation*). The elements about which these operations are performed are, respectively, (1) the object itself, (2) a line (*rotation axis* or *proper axis*), (3) a plane (*reflection plane* or *mirror plane*), (4) a point (*inversion center* or *center of symmetry*), and (5) a line (*improper axis* or *alternating axis*). All the corresponding symmetry elements will pass through a common point at the center of the structure. For this reason, the symmetry of isolated molecules and ions is called *point group symmetry.*

The simplest of all symmetry operations is *identity,* given the symbol E (or I in older texts). Every object possesses identity. If it possesses no other symmetry, the object is said to be *asymmetric.* As an operation, identity does nothing to the molecule. It exists for every object, because the object itself exists. The need for such an operation arises from the mathematical requirements of group theory, as we shall see later. Of more immediate concern, identity is often the result of carrying out a particular operation successively a certain number of times. In other words, if you keep doing the same operation repeatedly, eventually you may bring the object back to the identical (not simply equivalent) orientation from which you started. Normally, we will want to designate the results of successive or compound operations by their most direct single equivalents. Thus, if a series of repeated operations carries the object back to its starting point, the result would be identified simply as identity. This will become clearer as we examine the results of sequential symmetry operations throughout this section.

The operation of *rotation* is designated by the symbol C_n, indicating that rotation about an axis by $2\pi/n$ radians ($360°/n$) brings the object into an equivalent position. The value n of a C_n rotation is the *order of the rotation*. It is common to refer to the operation as an n-fold rotation and to refer to the corresponding element as an n-fold rotational axis. Sometimes the term *proper axis* is used to refer to the element associated with rotation, distinguishing it from an improper axis, discussed later in this section. For example, C_4 indicates a fourfold rotation, by which rotation through $2\pi/4 = 90°$ brings the object into an equivalent position, indistinguishable from the starting configuration.

Figure 1.1 shows the effects of successive fourfold rotations about an axis perpendicular to the plane of a planar MX_4 molecule. The four identical X atoms have been labeled X_A, X_B, X_C, and X_D (i.e., $X_A = X_B = X_C = X_D$) so that we can follow the results of each operation. Of course, without these labels the atoms are indistinguishable, and the result of this (or any) symmetry operation would be indistinguishable from the starting positions. In the figure we have arbitrarily defined the rotations in a clockwise manner, but we could just as well have defined them in the opposite direction. It is only necessary that we be consistent in defining successive rotations about the same axis. Notice that carrying out two successive C_4 rotations about the same axis (which we could designate C_4^2) has the same effect as a single C_2 rotation ($2\pi/2 = 180°$). Normally, the simpler notation, C_2, would be preferred. If we continue with a third C_4 rotation, we arrive at a new equivalent configuration, which is the same as we would have obtained by a single fourfold rotation in the opposite direction (here, counterclockwise). To avoid ambiguity, this is designated C_4^3, meaning three successive fourfold rotations in our chosen direction. In general, any C_n rotation carried out $n-1$ times will have the same effect as a C_n rotation in the opposite direction. If we designate a counterclockwise n-fold rotation as C_n^{-1}, then we may write in general $C_n^{n-1} = C_n^{-1}$; for example, $C_4^3 = C_4^{-1}$.* Finally, if we perform a fourth C_4 rotation, our MX_4 molecule will be brought back into its starting (identical) position. Carrying out four successive C_4 operations about the same axis is equivalent to identity; that is, $C_4^4 = E$. In general, any C_n rotation carried out n times in succession will carry the object back into its original configuration; that is, $C_n^n = E$. Thus, as shown in Fig. 1.1, this single element, a C_4 axis, is associated with three unique symmetry operations: C_4, $C_4^2 = C_2$, $C_4^3 = C_4^{-1}$. The fourth op-

---

*For additional verification of this relationship, try demonstrating that $C_4 = C_4^{-3}$ and $C_2 = C_2^{-1}$ in Fig. 1.1.

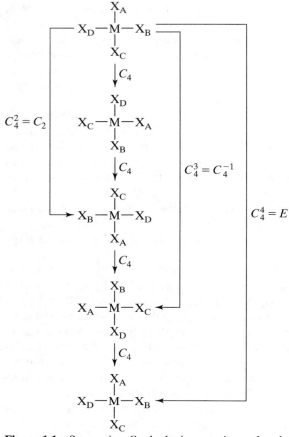

Figure 1.1 Successive C_4 clockwise rotations of a planar MX_4 molecule about an axis perpendicular to the plane of the molecule ($X_A = X_B = X_C = X_D$).

eration, C_4^4, is just a repeat of identity, E. The C_4 and C_4^3 rotations are often described as two C_4 rotations, where it is understood that one rotation is taken in a clockwise sense, and the other is taken in a counterclockwise sense.

As this example shows, the symbol for carrying out m successive n-fold rotations of $360°/n$ about a single axis is C_n^m, where the result is the same as a single rotation by $(m/n)2\pi = (m/n)360°$. To express the compound rotation as its simplest equivalent, m and n are written such that the fraction m/n is in its lowest form; for example, $C_4^2 = C_2$, $C_6^4 = C_3^2$, $C_8^6 = C_4^3$. Furthermore, rotations beyond full circle are expressed as the equivalent single rotation that is less than $2\pi = 360°$. For example, C_4^5 is a rotation by $450°$, equivalent to a $90°$ rotation about the same axis; that is, $C_4^5 = C_4$.

As we have seen, carrying out two C_4 operations in succession about the same axis is the same as C_2. In a formal sense, then, we may say that there also exists a C_2 axis collinear with the C_4 axis. These two axes (C_4 and C_2) are not the only rotational axes of MX_4. There are four other C_2 axes in the plane of the molecule, as shown in Fig. 1.2. These twofold axes are distinguished from the previously defined C_2 axis by adding prime (′) and double prime (″) to their symbols. In general, when two or more axes of the same n-fold order exist, the axis or axes collinear with the highest-order axis or axes in the system are designated without modification. All others of the same n-fold order are distinguished by adding prime or double prime notations. In the present case, the two C_2' axes are defined so as to pass through more atoms than the two C_2'' axes. Only two notations are needed for the four axes, because both C_2' axes belong to the same *class*, while the two C_2'' axes belong to a separate class. We will define the concept of class more generally later, but for now note that the two C_2' axes are geometrically equivalent to each other and distinct from the C_2'' axes, which are likewise geometrically equivalent to each other and distinct from the two C_2' axes. In listing the complete set of symmetry operations for a molecule, operations of the same class are designated by a single notation preceded by a coefficient indicating the number of equivalent operations comprising the class. For square planar MX_4, the rotational operations grouped by class are $2C_4$ (C_4 and C_4^3), C_2 (collinear with C_4), $2C_2'$, and $2C_2''$.

The highest-order rotational axis an object possesses (i.e., the C_n axis for which n is greatest) is called the *principal axis of rotation*. In some highly symmetrical systems (e.g., tetrahedron, octahedron), there may be more than one principal axis, but most less symmetrical systems with rotational symmetry will have only one. For planar MX_4, the principal axis is the C_4 axis, about which the operations of C_4 and C_4^3 are performed.

The operation of *reflection* defines bilateral symmetry about a plane, called a *mirror plane* or *reflection plane*. Accordingly, if a molecule possesses a mirror plane, it will be bisected by that plane. The symbol for the reflection operation and its corresponding element (the mirror plane) is lowercase sigma, σ. If the operation of reflection exists, for any point a distance r along a normal to the mirror plane there will be an equivalent point at a distance $-r$ (cf. Fig. 1.3).

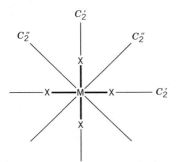

Figure 1.2 The C_2' and C_2'' axes of a planar MX_4 molecule.

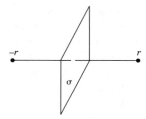

Figure 1.3   Two points, equidistant from a mirror plane σ, related by reflection.

Figure 1.4 shows the five mirror plánes found in a square plane, such as a planar MX_4 molecule. The five planes are grouped into three classes: σ_h, $2\sigma_v$, and $2\sigma_d$. As with the C_2 axes of planar MX_4, the class groupings of the three kinds of mirror planes can be justified on geometrical grounds. Notice that all mirror planes pass through a common point at the center of the molecule.

Figure 1.5 shows the effects of the reflection operations of these mirror planes. As in Fig. 1.1, the four identical X atoms have been labeled X_A, X_B, X_C, and X_D to show the effects of the operations. At first glance, the operation of σ_h appears to do nothing to the molecule, because it lies entirely within the mirror plane. However, if each of the five atoms had a directional property perpendicular to the plane (e.g., p_z orbitals, or vibrational motions of the atoms), the operation of σ_h would transform the property into the negative of itself. The effect of either σ_v operation is to leave two *trans*-related

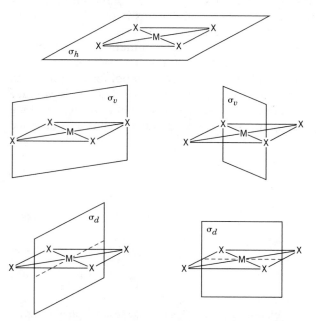

Figure 1.4   Mirror planes of a square planar molecule MX_4.

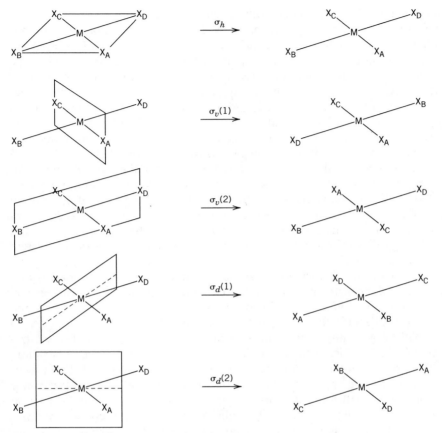

Figure 1.5 Effects of reflections of a planar MX$_4$ molecule (X$_A$ = X$_B$ = X$_C$ = X$_D$).

X atoms and the M atom unaffected and to transpose the other two *trans*-related X atoms, which lie on either side of the vertical plane. The effect of either σ_d operation is to exchange both pairs of X atoms across a dihedral plane, leaving the M atom unaffected.

Note that for any mirror plane, performing two successive reflections about the same plane brings the object back into its original (identical) configuration; that is, $\sigma\sigma = \sigma^2 = E$. Thus, any mirror plane is associated with only one operation, unlike a rotational axis, which may be a common element for a series of operations.

The notation for the planes in planar MX$_4$ is typical of the notation in other systems with several mirror planes in various orientations. A σ_h plane (horizontal mirror plane) is defined as perpendicular to the principal axis of rotation. If no principal axis exists, σ_h is defined as the plane of the molecule. A σ_v plane (vertical mirror plane) and a σ_d plane (dihedral mirror plane) are defined so as to contain a principal axis of rotation and to be perpendicular to a σ_h plane, if it exists. When both σ_v and σ_d occur in the same system, the distinction between the types is made by defining σ_v to contain the greater number of atoms or to contain a principal axis of a reference Cartesian coordinate system (x or y axis). Any σ_d planes typically will contain bond angle bisectors. In cases with only one type of vertical plane, either σ_v or σ_d may be the conventional notation, depending on the total symmetry of the molecule. For example, in the eclipsed conformation of ethane the three vertical mirror planes, which intersect along the C_3 axis of the molecule and contain the two carbon atoms and two hydrogen atoms (one on each end), are conventionally designated σ_v. In the staggered conformation of ethane the three mirror planes, which also intersect along the C_3 axis of the molecule and contain the two carbon atoms and two hydrogen atoms, are conventionally designated σ_d. Fortunately, knowing whether a plane is to be called σ_v or σ_d in such cases is not crucial to applying symmetry arguments to physical problems. The conventions become apparent when using the standard tables that list (among other things) all the symmetry operations for a particular system (cf. Appendix A).

The operation of *inversion* is defined relative to the central point within the molecule, through which all symmetry elements must pass. This is usually taken as the origin of a Cartesian coordinate system. Relative to this system, if inversion exists for the molecule, for every point with coordinates (x, y, z) there will be an equivalent point at coordinates $(-x, -y, -z)$. The central point at $(0, 0, 0)$, which is the element associated with the inversion operation, is called an *inversion center* or *center of symmetry*. Viewed in a different way, if inversion symmetry exists, a line drawn from any atom through the center of the molecule will connect with an equivalent atom at an equal distance on the other side of the molecule. The point at which all such connecting lines intersect is the inversion center. Molecules that have inversion symmetry are said to be *centrosymmetric*.

Both the element and the operation of inversion are given the symbol i. Note that performing inversion twice in succession would bring every point (x, y, z) back into itself, equivalent to identity; that is, $ii = i^2 = E$. Thus, like a reflection plane, an inversion center has only one operation associated with it. Furthermore, since the inversion center is always located at the central point in the molecule, there can be only one inversion center in any system.

Planar MX_4 is centrosymmetric. Relative to the starting configurations of either Fig. 1.1 or Fig. 1.5, inversion interchanges the X atoms labeled X_A and X_C and also those labeled X_B and X_D. The central M atom, which is located at the inversion center, is not affected. This example is somewhat restricted, since all relations between equivalent points occur within the plane of the molecule itself (the xy plane). Perhaps a better example, illustrating the three-dimensional character of the inversion operation, is provided by an octahedral MX_6 molecule (Fig. 1.6). Here the equivalent atoms lie equidistant in positive and negative directions along each of the three axes of the coordinate system (i.e., $\pm x$, $\pm y$, $\pm z$).

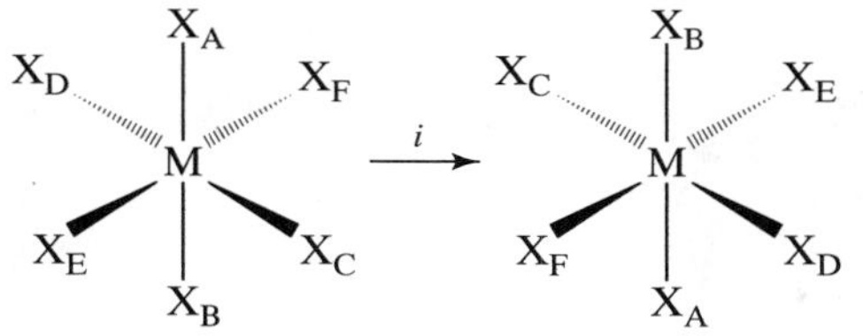

$$X_D \cdots \quad X_A \quad X_F$$
$$\text{M} \xrightarrow{\;i\;}$$
$$X_E \quad X_C$$
$$X_B$$

$$X_C \cdots \quad X_B \quad X_E$$
$$\text{M}$$
$$X_F \quad X_D$$
$$X_A$$

Figure 1.6 Effect of inversion (i) on an octahedral MX_6 molecule ($X_A = X_B = X_C = X_D = X_E = X_F$).

Another, even less restricted, example is ethane in the staggered configuration (Fig. 1.7). The inversion center is at the midpoint between the two carbon atoms. The operation of inversion relates the two carbon atoms to each other, and it relates pairs of hydrogen atoms on opposite ends of the molecule. Note that there is no inversion center for ethane in the eclipsed configuration. (Building models of the two conformations may be helpful to see the difference.)

Rotation–reflection, as its name implies, is a compound operation. It is also called *improper rotation,* to distinguish it from proper rotation, C_n. Rotation–reflection consists of a proper rotation followed by a reflection in a plane perpendicular to the axis of rotation. Actually, the order of performing rotation and reflection in a plane perpendicular to the rotation axis can be reversed, giving the same result. Most practitioners, however, take their cue from the name rotation–reflection and perform the two parts of the operation in that order. The axis of this operation is called an *improper axis.* Both the operation and the element are given the symbol S_n, where n refers to the initial rotation by $2\pi/n = 360°/n$.

The two parts of an improper rotation (C_n and σ_h) may be genuine operations of the molecule in their own right, but often they are not. If both a C_n rotation and a reflection perpendicular to the rotation (σ_h) do exist independently, then the improper rotation S_n must also exist, since both its parts are present. For example, in planar MX_4, both C_4 and σ_h exist; therefore, the operation S_4 exists. The S_4 improper axis in this case is collinear with the principal proper rotational axis (C_4). However, the presence of both C_4 and σ_h is not a *sine qua non* for the existence of S_4. A good example of this can be found in a tetrahedral MX_4 molecule, where each X–M–X angle bisector lies along an S_4 axis. Figure 1.8 shows the effects of the two steps of the S_4 operation about one such axis. As before, the four equivalent X atoms have been

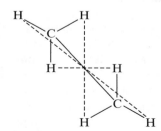

Figure 1.7 Ethane in the staggered configuration. The inversion center is at the midpoint along the C–C bond. Hydrogen atoms related by inversion are connected by dotted lines, which intersect at the inversion center. The two carbon atoms are also related by inversion.

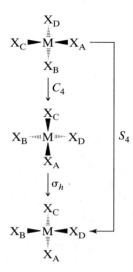

Figure 1.8 S_4 improper rotation of a tetrahedral MX_4 molecule ($X_A = X_B = X_C = X_D$). The improper axis is perpendicular to the page. Rotation is arbitrarily taken in a clockwise direction. Note that neither C_4 nor σ_h are genuine symmetry operations of tetrahedral MX_4.

labeled X_A, X_B, X_C, and X_D so that the effects of the operations can be followed. Notice that neither the C_4 step nor the σ_h step by itself results in a configuration that is indistinguishable (i.e., equivalent) to the configuration prior to executing each operation. However, the net result of these two steps in succession leads to a configuration which is, indeed, equivalent to the starting configuration, if we disregard the artificial A, B, C, D labels.

As these two MX_4 cases demonstrate, an S_4 axis must exist when C_4 and σ_h exist (e.g., planar MX_4), but it may also exist if neither C_4 nor σ_h exist (e.g., tetrahedral MX_4). As an exercise to see these kinds of possibilities in other cases, you might build models of ethane in both the eclipsed and staggered conformations and examine the effects of their improper rotations. Note that in the eclipsed conformation there is a C_3 axis collinear with the C–C bond and a σ_h mirror plane that bisects the C–C bond and lies perpendicular to it. Therefore, there must be an S_3 axis collinear with the C_3 axis. In the staggered configuration there is no σ_h plane, although the C_3 axis persists. However, there is an S_6 axis collinear with the C_3 axis, even though neither C_6 nor σ_h exist.

Like proper rotations, a series of improper rotations can be performed about the same axis. Figure 1.9 shows the results for successive S_4 rotations about an improper axis of tetrahedral MX_4. The direction of rotation (clockwise or counterclockwise) is unimportant, as long as successive rotation–reflection operations are carried out in the same direction. Note that successively carrying out two S_4 operations has the effect of a single C_2 operation about the same axis; that is, $S_4^2 = C_2$. Thus the S_4 axis is collinear with a C_2 axis. Carrying out a third S_4 operation, written S_4^3, results in an equivalent configuration, which also could have been reached by a single S_4 operation in the opposite direction (here, counterclockwise), shown as S_4^{-1} in Fig. 1.9. Keeping with the chosen direction of rotation and to avoid ambiguity, the notation S_4^3

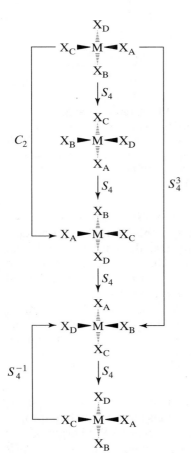

Figure 1.9 Successive S_4 operations on a tetrahedral MX_4 molecule ($X_A = X_B = X_C = X_D$). Rotations are clockwise, except S_4^{-1}, which is equivalent to the clockwise operation S_4^3.

is preferred. A fourth S_4 operation results in the original configuration; that is, $S_4^4 = E$. Consistent with the convention of designating operations by their simplest notation, only S_4 and S_4^3 are taken as improper rotations. The operations of S_4^2 and S_4^4 are equivalent to C_2 and E, respectively, and are so designated. Thus, there are two S_4 operations (S_4 and $S_4^3 = S_4^{-1}$) about this axis.

Figure 1.10 shows a tetrahedral molecule inscribed in a cube, a useful way of presenting such molecules, as we shall see throughout this text. From this we can see that the three equivalent S_4 axes, which lie along the bisectors of pairs of X–M–X angles, are collinear with the three C_2 axes. These axes define the x, y, and z directions of a reference Cartesian coordinate system for a tetrahedral MX_4 molecule. In this highly symmetric system all three directions are equivalent and indistinguishable, which implies that the three S_4 improper axes are geometrically indistinguishable from each other. Each improper axis has two operations associated with it, S_4 and $S_4^3 = S_4^{-1}$. Consequently, the three axes give rise to a total of six operations, which belong to a class designated $6S_4$ in tabular listings of the symmetry of a tetrahedral system.

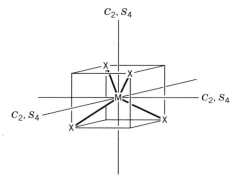

Figure 1.10   A tetrahedral MX_4 molecule inscribed in a cube. A C_2 axis, collinear with an S_4 axis, passes through the centers of each pair of opposite cube faces and through the center of the molecule.

The lowest-order improper rotation that is not a simpler operation is S_3. If we were to imagine a hypothetical S_1 operation, this would be a "onefold" rotation (i.e., E) followed by reflection. The first step does nothing, so the net effect is simply the second step, the reflection. Thus, $S_1 = \sigma$. If we were to imagine a hypothetical S_2 operation, this would be a C_2 operation followed by a σ_h operation. If we define the C_2 axis as z, the effect of the first step would be to convert every point (x, y, z) into an equivalent point at $(-x, -y, z)$. The second step, a reflection across the xy plane, would transpose every coordinate z into $-z$; that is, $(-x, -y, z) \rightarrow (-x, -y, -z)$. The net effect of the two steps is to convert every point (x, y, z) into an equivalent point at $(-x, -y, -z)$. This is the single result of an inversion operation, so $S_2 = i$.

For improper axes with $n \geq 3$, each S_n element generates a series of S_n operations. As we have seen with S_4, not all of these are uniquely improper rotations (e.g., $S_4^2 = C_2$, $S_4^4 = E$). For a series of successive S_n operations, the pattern of equivalences with simpler operations depends upon whether n is even or odd. The following general relationships for S_n^m operations, where $n \geq 3$ and $m = 1, 2, \ldots, n, \ldots, 2n$, may be useful at certain points:

1. If n is even, $S_n^n = E$.
2. If n is odd, $S_n^n = \sigma$ and $S_n^{2n} = E$.
3. If m is even, $S_n^m = C_n^m$ when $m < n$ and $S_n^m = C_n^{m-n}$ when $m > n$.
4. If S_n with even n exists, then $C_{n/2}$ exists.
5. If S_n with odd n exists, then both C_n and σ perpendicular to C_n exist.

1.2   Defining the Coordinate System

In many of our discussions it will be necessary to define a coordinate system for a molecular species. In all cases we will adopt a standard Cartesian coordinate system with axes x, y, and z defined by the so-called *"right-hand rule."* By this convention, the positive directions of the three cardinal axes are defined in the same sense as the thumb, index finger, and middle fingers of the right hand when extended so that they are mutually perpendicular to each other. Hold your right hand so that your thumb is pointing up, extend your index finger as if pointing a gun, and bend your middle finger so that it is perpendicular to the other two. Your thumb, index finger, and middle finger now correspond, respectively, to the z, x, and y directions of the Cartesian coordinate system (cf. Fig. 1.11). Alternately, take the thumb, index finger, and middle fingers as x, y, and z, respectively, in which case the palm must be rotated face up to present the z axis in its usual perpendicular orientation (cf. Fig. 1.11). Either mnemonic gives the same relative axis orientations. These orientations are retained regardless of whether the system is shown with the z axis pointing up or in some other orientation.

The orientation of a molecule's bonds and bond angles relative to the coordinate system is often defined on the basis of the symmetry of the molecule. Generally, the following conventions are observed:

1. The origin of the coordinate system is located at the central atom or the center of the molecule.

2. The z axis is collinear with the highest-order rotational axis (the principal axis). If there are several highest-order rotational axes, z is usually taken as the axis passing through the greatest number of atoms. However, for a tetrahedral molecule, the x, y, and z axes are defined as collinear with the three C_2 axes (collinear with the three S_4 axes).

3. For planar molecules, if the z axis as defined above is perpendicular to the molecular plane, the x axis lies in the plane of the molecule and passes through the greatest number of atoms (e.g., square planar XeF_4). If the z axis lies in the plane of the molecule, then the x axis stands perpendicular to the plane (e.g., bent H_2O).

4. For nonplanar molecules, once the z axis has been defined, the x axis is usually chosen so that the xz plane contains as many atoms as possible. If there are two or more such planes containing identical sets of atoms,

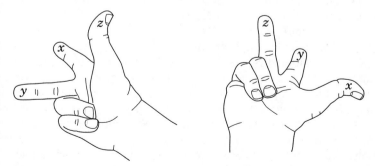

Figure 1.11   Two ways of assigning the thumb, index finger, and middle finger of the right hand as mnemonics for the axis orientations of a Cartesian coordinate system.

any one may be taken as the xz plane. Where a decision about the orientation of the x axis cannot be made on this basis, the distinction between x and y is usually not important or is not generally fixed by convention.

It is important to realize that these are conventions and not rules. When comparing texts, you may occasionally find different choices of axis orientation for the same molecular system. In such cases, the quantitative and qualitative results based on symmetry arguments will be the same, varying only in the labels used to describe them. Correlating the differences is generally straightforward. At times it can be useful to choose a nonconventional orientation of the coordinate system in order to emphasize particular relationships between different compounds or different geometries of the same compound.

1.3   Combining Symmetry Operations

We have seen that successive application of the same or different symmetry operations sometimes has the same effect as a single operation. For example, performing two S_4 rotations in succession about the same axis has the net effect of a single C_2 rotation. Likewise, performing a C_4 rotation followed by a σ_h reflection has by definition the same effect as the improper rotation S_4. This kind of combination is called *multiplication,* although in this context the term has a slightly different meaning than in the customary arithmetic sense.

The end result of carrying out different operations in succession may depend on the order in which they are performed. In other words, *combinations of symmetry operations do not always commute.* For this reason we need to adopt a standard way of writing such multiplications so as to imply the order of the operations. By convention, combinations of symmetry operations are written in a right-to-left order. For example, BA means "do A first, then B." If the result of BA is the same as could be achieved by a single operation X, then we write $BA = X$. Thus, $S_4 S_4 = C_2$, and $\sigma_h C_4 = S_4$.

As an example of noncommutative operations, consider a tetrahedron on which we perform S_4 followed by σ_v, and for comparison σ_v followed by S_4 (Fig. 1.12). The final configurations in both cases are indistinguishable, as they must be for genuine symmetry operations, but the actual positions of the four equivalent X atoms (labeled X_A, X_B, X_C, and X_D) are not the same. To get from one final configuration to the other we must perform a C_2 rotation about the same axis as the S_4 improper rotation. Thus, $S_4 \sigma_v \neq \sigma_v S_4$, but rather we see that $S_4 \sigma_v = C_2 \sigma_v S_4$. The fact that commutation is not generally observed does not mean that it is never observed. Indeed, many operations do commute with one another; for example, $C_4 \sigma_h = \sigma_h C_4 = S_4$.

Let us examine the complete set of symmetry operations for a particular molecule and determine all the binary combinations of the symmetry operations it possesses. Our example will be CBr_2Cl_2, shown with its symmetry el-

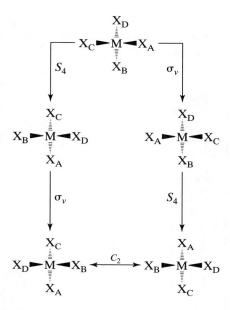

Figure 1.12   The order of performing S_4 and σ_v, shown here for a tetrahedral MX$_4$ molecule, affects the result. The final positions in each case are not the same, but they are related to each other by C_2.

ements in Fig. 1.13. The pairs of Br atoms and Cl atoms have been distinguished from one another by subscript a and b (Br$_a$, Br$_b$ and Cl$_a$, Cl$_b$) to facilitate following the effects of the operations. The complete set of symmetry operations for the molecule consists of identity (E), a twofold principal axis of rotation (C_2), and two reflections about different mirror planes (σ_v and σ_v'). As shown in Fig. 1.13, the two Cl atoms lie in the σ_v plane, and the two Br atoms lie in the σ_v' plane. Rather than depict the effect of each operation on the molecule, let us introduce a column matrix notation to indicate the positions of atoms before and after each operation. The carbon atom is unaffected by any operation, because it lies at the center point of the system, through which all symmetry elements pass. Only the Br and Cl atoms are moved in any operation, so our matrices need only describe the positions of

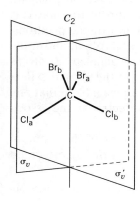

Figure 1.13   Symmetry elements of CBr$_2$Cl$_2$.

those atoms. Thus, a 1×4 matrix will suffice to describe the locations of these four atoms. Each position in the 1×4 column matrix should be read as a particular position in space, occupied by the designated atom. Using this notation, we obtain the following results:

$$[E] \times \begin{bmatrix} Br_a \\ Br_b \\ Cl_a \\ Cl_b \end{bmatrix} = \begin{bmatrix} Br_a \\ Br_b \\ Cl_a \\ Cl_b \end{bmatrix}$$

$$[C_2] \times \begin{bmatrix} Br_a \\ Br_b \\ Cl_a \\ Cl_b \end{bmatrix} = \begin{bmatrix} Br_b \\ Br_a \\ Cl_b \\ Cl_a \end{bmatrix}$$

$$[\sigma_v] \times \begin{bmatrix} Br_a \\ Br_b \\ Cl_a \\ Cl_b \end{bmatrix} = \begin{bmatrix} Br_b \\ Br_a \\ Cl_a \\ Cl_b \end{bmatrix}$$

$$[\sigma_v'] \times \begin{bmatrix} Br_a \\ Br_b \\ Cl_a \\ Cl_b \end{bmatrix} = \begin{bmatrix} Br_a \\ Br_b \\ Cl_b \\ Cl_a \end{bmatrix}$$

The symbols $[E]$, $[C_2]$, $[\sigma_v]$, and $[\sigma_v']$ represent operator matrices, which act on the original 1×4 matrices to convert them into the matrices that describe the orientations after the symmetry operations.* We will examine the explicit mathematical forms of these kinds of matrices in Chapter 2. At this point we need only be concerned with the results for CBr_2Cl_2. From the matrices above, we see that (1) identity, E, leaves all atoms in their original positions; (2) the twofold rotation, C_2, exchanges Br atoms with each other, and it exchanges Cl atoms with each other; (3) the reflection σ_v, exchanges Br atoms with each other, but leaves the Cl atoms in place; and (4) the second reflection, σ_v', exchanges Cl atoms, but leaves the Br atoms in place. Note that any atom that lies on a particular symmetry element is unaffected by the associated operation.

    Now let us consider the results for binary combinations of these operations, beginning with combinations in which each operation is performed

---

*Note that matrix equations are written in the usual mathematical way—that is, left-to-right notation.

twice. For each of these, performing the operation a second time brings the molecule back into its original configuration, essentially undoing the result of the first performance of the operation. Thus, we see $EE = E$, $C_2C_2 = E$, $\sigma_v\sigma_v = E$, and $\sigma_v'\sigma_v' = E$. Among the mixed binary combinations, notice that any combination with E, either as the first or second operation, gives the same result as the nonidentity operation alone. This must be, since E really does nothing to the molecule. Thus, we have $C_2E = EC_2 = C_2$, $\sigma_vE = E\sigma_v = \sigma_v$, $\sigma_v'E = E\sigma_v' = \sigma_v'$. The remaining mixed binary combinations are somewhat less obvious.

We can determine the net result of the product $C_2\sigma_v$ by the following procedure. First perform σ_v on the original configuration to obtain an intermediate configuration, and then perform C_2 on this configuration to obtain the final result. In matrix notation, the first step has the following effect:

$$[\sigma_v] \times \begin{bmatrix} \text{Br}_a \\ \text{Br}_b \\ \text{Cl}_a \\ \text{Cl}_b \end{bmatrix} = \begin{bmatrix} \text{Br}_b \\ \text{Br}_a \\ \text{Cl}_a \\ \text{Cl}_b \end{bmatrix}$$

From this configuration, we apply C_2 to obtain the following effect:

$$[C_2] \times \begin{bmatrix} \text{Br}_b \\ \text{Br}_a \\ \text{Cl}_a \\ \text{Cl}_b \end{bmatrix} = \begin{bmatrix} \text{Br}_a \\ \text{Br}_b \\ \text{Cl}_b \\ \text{Cl}_a \end{bmatrix}$$

This final configuration is the same as could be achieved by a σ_v' reflection on the original configuration of the molecule:

$$[\sigma_v'] \times \begin{bmatrix} \text{Br}_a \\ \text{Br}_b \\ \text{Cl}_a \\ \text{Cl}_b \end{bmatrix} = \begin{bmatrix} \text{Br}_a \\ \text{Br}_b \\ \text{Cl}_b \\ \text{Cl}_a \end{bmatrix}$$

From this we see that $C_2\sigma_v = \sigma_v'$.

In the case of these two operations, reversing the order produces the same final configuration. Applying the rotation first, we obtain

$$[C_2] \times \begin{bmatrix} \text{Br}_a \\ \text{Br}_b \\ \text{Cl}_a \\ \text{Cl}_b \end{bmatrix} = \begin{bmatrix} \text{Br}_b \\ \text{Br}_a \\ \text{Cl}_b \\ \text{Cl}_a \end{bmatrix}$$

Following this with a σ_v reflection yields

$$[\sigma_v] \times \begin{bmatrix} Br_b \\ Br_a \\ Cl_b \\ Cl_a \end{bmatrix} = \begin{bmatrix} Br_a \\ Br_b \\ Cl_b \\ Cl_a \end{bmatrix}$$

Once again, this is the same result as could have been achieved by a single σ_v' reflection. Since we reversed the order, this implies that combinations of C_2 and σ_v commute; that is, $C_2\sigma_v = \sigma_vC_2 = \sigma_v'$. In fact, all binary combinations of the four operations of CBr_2Cl_2 are commutative. You should verify that the following results are correct: $\sigma_v\sigma_v' = \sigma_v'\sigma_v = C_2$ and $C_2\sigma_v' = \sigma_v'C_2 = \sigma_v$.

The results we have just obtained can be summarized conveniently in a *multiplication table,* such as the following:

	E	C_2	σ_v	σ_v'
E	E	C_2	σ_v	σ_v'
C_2	C_2	E	σ_v'	σ_v
σ_v	σ_v	σ_v'	E	C_2
σ_v'	σ_v'	σ_v	C_2	E

Although all combinations of the present four operations commute, in general this is not so, and it will be important to observe the order of combination. For multiplication tables of this type, the assumed order of combination is row element (top) first, followed by column element (side). In our right-to-left notation, the product $BA = X$ would be read from a multiplication table as $B_{column}A_{row} \equiv B_{side}A_{top} = X$.

This multiplication table shows a number of features found in all such tables. For example, the first row of results duplicates the list of operations in the header row, and the first column of results likewise duplicates the list of operations in the label column. This must be so, because both involve combinations with E. Also, note that every row shows every operation once and only once, as does each column. Furthermore, the order of resultant operations in every row is different from any other row. The same is true of every column. Knowing these general features can greatly simplify constructing multiplication tables for other cases.

1.4   Symmetry Point Groups

The complete set of symmetry operations exhibited by any molecule defines a *symmetry point group.* As such, the set must satisfy the four requirements of a mathematical group: *closure, identity, associativity,* and *reciprocality.* Let us examine these requirements, continuing with the example of CBr_2Cl_2 and the multiplication table we developed in Section 1.3.

Closure. If A and B are elements of the group G, and if $AB = X$, then X is also in the group G. The term "element" is used in this context in its mathematical sense. The members of any mathematical group are the elements of the group. In the case of symmetry groups, with which we are concerned here, the group elements are the symmetry operations, not the symmetry elements. The symmetry elements associated with the operations in general do not fit the requirements for a mathematical group.

For symmetry groups, closure means that any combination of operations must be equivalent to an operation that is also a member of the group. Inspection of the multiplication table for the operations of CBr_2Cl_2 shows that all binary combinations equal either E, C_2, σ_v, or σ_v'. These four symmetry operations constitute the complete set of elements of a point group called C_{2v}. The multiplication table we developed in Section 1.3, then, is the multiplication table for the group C_{2v}. The number of operations (group elements) comprising the group defines the *order of the group,* designated h. Thus, the order of the group C_{2v}, is four (i.e., $h = 4$).

Identity. In any group G, there is an element E (the identity element), which commutes with any other element of the group, X, such that $EX = XE = X$. This requirement explains the need to define the symmetry operation of identity, which functions as the identity element for every symmetry group. As the C_{2v} multiplication table demonstrates, the identity operation does indeed meet the requirements for the identity element of a group.

Associativity. The associative law of combination is valid for all combinations of elements of the group. Thus, if A, B, C, and X are members of the group G and $C(BA) = X$, then $(CB)A = X$, too. When carrying out sequential multiplications of operations, we may group the combinations into any convenient pairs, so long as the order of combination is preserved. We must preserve the order, because commutation is not generally valid; for example, it may be that $CBA \neq BAC$.

Using the C_{2v}, multiplication table, we can demonstrate associativity for any combination of three or more operations. For example, we can show that $C_2(\sigma_v\sigma_v') = (C_2\sigma_v)\sigma_v'$. For the first combination we see

$$C_2(\sigma_v\sigma_v') = C_2C_2 = E$$

For the second combination we see

$$(C_2\sigma_v)\sigma_v' = \sigma_v'\sigma_v' = E$$

Since both methods of association lead to the same result, they must be equal to each other.

We should note that, contrary to generally observed behavior, all binary combinations of the group C_{2v} do commute. A limited number of other symmetry groups are similarly composed entirely of commuting operations. In the mathematics of groups, any group in which all combinations of elements commute is said to be *Abelian*. Therefore, the group C_{2v} is an Abelian group.

Reciprocality. Every element A of the group G has an *inverse, A^{-1}*, such that $AA^{-1} = A^{-1}A = E$. The meaning of the term "inverse" in this context is not the same as the familiar algebraic sense, where x and $1/x$ are the inverses of each other. In the theory of groups, two elements are the inverses of each other simply because their binary combination commutes and is equivalent to identity. Some elements may be their own inverses (i.e., $A = A^{-1}$). We have seen some examples of symmetry operations that are their own inverses; for example, $C_2C_2 = E$, $\sigma\sigma = E$, $ii = E$, $EE = E$. Note that for the point group C_{2v} every operation is its own inverse.

Among the operations that constitute a point group, there generally exist smaller sets that also obey the four requirements of a group. These smaller sets, which are groups in their own right, can be considered to be *subgroups* of the larger group from which the elements were culled. In general, if g is the order of a subgroup of a group whose order is h, then $h/g = n$, where n is an integer. In other words, the order of any subgroup must be an integer divisor of the order of the parent group. However, it is not a requirement that subgroups for all the allowed orders exist.

For C_{2v}, where $h = 4$, only subgroups with orders $g = 1$ and $g = 2$ are possible. In this simple case, subgroups of both exist. From the multiplication table for C_{2v}, we can identify the following subgroups, listed here with their standard group notations and the elements comprising them:

$$\begin{array}{ll} C_1 & \{E\} \\ C_2 & \{E, C_2\} \\ C_s & \{E, \sigma_v\} \\ C_s & \{E, \sigma_v'\} \end{array}$$

Since every value of h has 1 as an integer divisor and every group must contain the identity element, the group C_1, which consists of E only, is necessarily a subgroup of any other group. Indeed, it is the only possible group with order $h = 1$. A molecule that belongs to the group C_1 has no symmetry (other than identity) and is therefore *asymmetric*. A molecule with only a C_2 rotational axis belongs to the point group C_2, and one with only a mirror plane (σ) belongs to the point group C_s. In the list above, the group $C_s \{E, \sigma_v\}$ and the group $C_s \{E, \sigma_v'\}$ differ only in the choice of which reflection plane of C_{2v} is taken to define them. Outside of that context, the point group C_s is defined as the set $\{E, \sigma_h\}$, where the notation σ_h simply indicates the single reflection plane of the group.

Note that the sets $\{E, C_2, \sigma_v\}$, $\{E, C_2, \sigma_v'\}$, or $\{E, \sigma_v, \sigma_v'\}$ are neither subgroups of C_{2v} nor legitimate groups on their own. Aside from the fact that these sets would have orders of $g = 3$ (not an integer divisor of the group order $h = 4$), they do not show closure. In each case, combinations among the nonidentity elements give elements outside the set.

1.5 Point Groups of Molecules

The point group designations we have seen so far (C_{2v}, C_2, C_s, C_1) are examples of the *Schönflies notation*. This labeling system is used by most chemists and spectroscopists, and therefore we will use it throughout this text. By contrast, crystallographers prefer the *Hermann–Mauguin notation,* which is better suited for designating the 32 crystallographic point groups and the space groups used to describe crystal structures. Although we will not use Hermann–Mauguin notation, as examples of their form, here are the designations that correspond to the Schönflies labels we have seen: $C_{2v} = mm$, $C_s = m$, $C_2 = 2$, $C_1 = 1$.

Table 1.1 lists specific point groups and families of point groups that are important for classifying the symmetry of real molecules. As this shows, all the chemically important point groups fall within one of four general categories: *nonrotational, single-axis rotational, dihedral,* and *cubic.* In the descriptions of families of groups among the single-axis and dihedral categories, n is the order of the principal axis, which can have a value from 2 to infinity,

Table 1.1 Common Point Groups and Their Principal Operations

Symbol	Operations
Nonrotational Groups	
C_1	E (asymmetric)
C_s	E, σ_h
C_i	E, i
Single-Axis Groups ($n = 2, 3, \ldots, \infty$)	
C_n	E, C_n, $\ldots$, C_n^{n-1}
C_{nv}	E, C_n, $\ldots$, C_n^{n-1}, $n\sigma_v$ ($n/2$ σ_v and $n/2$ σ_d if n even)
C_{nh}	E, C_n, $\ldots$, C_n^{n-1}, σ_h
S_{2n}	E, S_{2n}, $\ldots$, S_{2n}^{2n-1}
$C_{\infty v}$	E, C_∞, $\infty\sigma_v$ (noncentrosymmetric linear)
Dihedral Groups ($n = 2, 3, \ldots, \infty$)	
D_n	E, C_n, $\ldots$, C_n^{n-1}, $nC_2(\perp C_n)$
D_{nd}	E, C_n, $\ldots$, C_n^{n-1}, S_{2n}, $\ldots$, S_{2n}^{2n-1}, $nC_2(\perp C_n)$, $n\sigma_d$
D_{nh}	E, C_n, $\ldots$, C_n^{n-1}, $nC_2(\perp C_n)$, σ_h, $n\sigma_v$
$D_{\infty h}$	E, C_∞, S_∞, $\infty C_2(\perp C_\infty)$, σ_h, $\infty\sigma_v$, i (centrosymmetric linear)
Cubic Groups	
T_d	E, $4C_3$, $4C_3^2$, $3C_2$, $3S_4$, $3S_4^3$, $6\sigma_d$ (tetrahedron)
O_h	E, $4C_3$, $4C_3^2$, $6C_2$, $3C_4$, $3C_4^3$, $3C_2(= C_4^2)$, i, $3S_4$, $3S_4^3$, $4S_6$, $4S_6^5$, $3\sigma_h$, $6\sigma_d$ (octahedron)
I_h	E, $6C_5$, $6C_5^2$, $6C_5^3$, $6C_5^4$, $10C_3$, $10C_3^2$, $15C_2$, i, $6S_{10}$, $6S_{10}^3$, $6S_{10}^7$, $6S_{10}^9$, $10S_6$, $10S_6^5$, 15σ (icosahedron, dodecahedron)

depending on the specific group. As a practical matter, the value of n usually is not greater than 8 for any real molecule one is likely to encounter, with the important exceptions of linear molecules for which $n = \infty$. Examples of molecules belonging to various specific point groups in each family are shown in Fig. 1.14.

The listings in Table 1.1 for the single-axis and dihedral groups show series of rotational axes as $C_n, \ldots, C_n^{n-1}$. In keeping with the practice of indicating operations as their simplest equivalents, some rotations in these series may be represented more conventionally as equivalent lower-order rotations. The same is true for series of improper rotations, as in the families of groups S_{2n} and D_{nd}, where certain members are equivalent to lower-order proper rotations (e.g., $S_8^2 = C_4$, $S_8^4 = C_2$, $S_8^6 = C_4^3$). Furthermore, in the families of groups C_{nh} and D_{nh}, which have both rotational axes and a horizontal mirror plane (σ_h) perpendicular to them, there are necessarily corresponding improper axes. Some of the improper rotations, however, are equivalent to other operations. As we have seen, $S_1 = \sigma_h$ and $S_2 = i$. Beyond these simple equivalences, some combinations of rotation with reflection give improper rotations and others give lower-order proper rotations. The equivalences for a series of C_n^m operations ($m = 1, 2, \ldots, n - 1$) combined with σ_h depends on whether n and m are odd or even numbers. Specifically, if n is even, $C_n^m \sigma_h = S_n^m$. If n is odd, $C_n^m \sigma_h = S_n^m$ when m is odd, and $C_n^m \sigma_h = S_n^{m+n}$ when m is even. The details of these series, however, are not generally important for determining the identity of a molecule's point group or for applying symmetry arguments to chemical problems. Complete listings of operations for all point groups appear at the top of the standard character tables found in virtually all books on chemical applications of group theory (see Appendix A). It is generally more useful to understand the characteristic components that distinguish one family of groups from the others.

Nonrotational Groups. With their low orders ($h = 1, 2$) and lack of an axis of symmetry, the nonrotational groups represent the lowest symmetry point groups. As previously noted, C_1 is the point group of asymmetric molecules. The group C_s describes the symmetry of bilateral objects that lack any symmetry other than E and σ_h. The external morphology of most mammals, for example, is nominally bilateral, and therefore described by the point group C_s. The trisubstituted borane shown in Fig. 1.14 is a molecular example of C_s symmetry. The group C_i, whose only nonidentity operation is inversion (i), is not a commonly encountered group, since most molecules that are centrosymmetric tend to have other symmetry, as well. The substituted cyclobutane species shown in Fig. 1.14 would have C_i symmetry.

Single-Axis Rotational Groups. The simplest family of these groups is C_n, which consists of the operations generated by a C_n rotation applied successively n times; that is, $C_n, C_n^2, \ldots, C_n^n = E$. These groups are examples of an

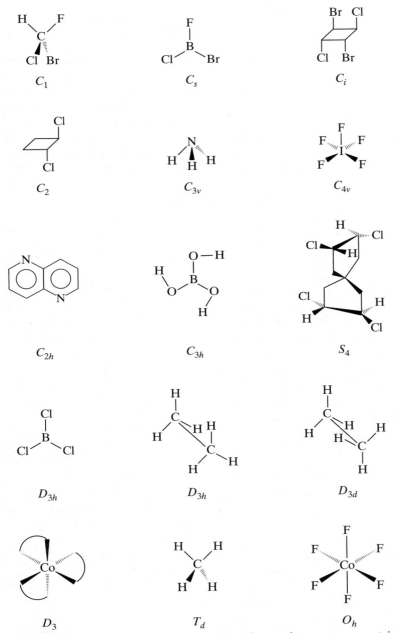

Figure 1.14 Examples of molecules with various point group symmetries.

important type of groups, called *cyclic groups*. In general, a cyclic group of order h is generated by taking a single element X through all its powers up to and including $X^h = E$. All cyclic groups are Abelian, since all their multiplications commute. Although some molecules belong to one of the C_n point groups (cf. Fig. 1.14), these groups are more frequently encountered because of their utility in simplifying problems of applied group theory. They often provide a more compact working group to solve problems for molecules that actually have significantly more symmetry than just the series of C_n rotations.

All cyclic groups, including the C_n groups, have multiplication tables with a characteristic form. As an illustration, consider the multiplication table for the point group C_4, whose elements are the operations E, C_4, $C_4^2 = C_2$, C_4^3.

C_4	E	C_4	C_2	C_4^3
E	E	C_4	C_2	C_4^3
C_4	C_4	C_2	C_4^3	E
C_2	C_2	C_4^3	E	C_4
C_4^3	C_4^3	E	C_4	C_2

The pattern of the products in each row, while unique, lists each element in the same sequential order as every other row. It is as if the elements were on a continuous scroll, which moves one position to the left with each successive row. This takes the first element of a row and makes it the last element of the next row. The columns scroll in a similar manner, moving elements up one and bumping the first element to the last position with each succeeding column. As a result, same elements appear along right-to-left diagonals of the table. Knowing this pattern makes it very easy to construct the multiplication table for any cyclic group.

To the rotations of the corresponding C_n groups the family of C_{nv} groups adds n vertical mirror planes, which intersect at the C_n axis. Molecular examples of these groups abound. The point group C_{2v}, to which CBr_2Cl_2 belongs, is an example of this family. Other examples include NH_3 (pyramidal, C_{3v}) and IF_5 (square pyramidal, C_{4v}). (See Fig. 1.14.) The point group $C_{\infty v}$, which has an infinite-fold C_∞ principal axis, is an important member of this family. It is the point group of all noncentrosymmetric linear molecules (e.g., HCl, ClBeF).

To generate any of the C_{nh} groups, we need only add a horizontal mirror plane to the series of C_n rotations of the appropriate cyclic C_n group. However, since $C_n\sigma_h = S_n$ and $C_2\sigma_h = S_2 = i$, these groups also have n-fold improper axes when $n > 2$, and they are centrosymmetric when n is even. A simple example of this family is the C_{3h} symmetry of trigonal planar boric acid, $B(OH)_3$, in which the B–O–H groups are presumed to be bent in the same direction (cf. Fig. 1.14).

The groups of the S_{2n} family are generated from $2n$ successive $2n$-fold improper rotations about a single axis. The last operation in the series is equivalent to identity, $S_{2n}^{2n} = E$. Since $S_2 = i$ and $S_{2n}^n = C_n$, inversion and proper rotations may be among the operations of the group. As the $2n$ notation implies, only groups of this type with even-order principal improper axes exist. The collection of operations generated by an odd-order S_n axis is the same as that generated by the combination C_n and σ_h, which defines groups of the type C_{nh}. Examples of molecules belonging to any of the S_{2n} groups are not common. An example of a molecule that would belong to the group S_4 is shown in Fig. 1.14.

Dihedral Groups. The dihedral groups have n twofold axes perpendicular to the principal n-fold axis. These C_2 axes are called the *dihedral axes*. Every symmetry operation of a point group also acts on the symmetry elements of the other operations. Therefore, the number and arrangement of the dihedral axes are dictated by the n-fold order of the principal axis. For example, suppose the principal axis of a dihedral group is C_3. Then, once we define one perpendicular C_2 axis, the operations C_3 and C_3^2 performed about the principal axis will generate the other two C_2 axes. Consistent with the C_3 angle of rotation, all three C_2 axes will be arranged at 120° from one another.

There are three families of dihedral groups: D_n, D_{nd}, and D_{nh}. The D_n groups may be thought of as C_n groups to which n dihedral C_2 operations have been added. Unlike the C_n groups, the D_n groups are not cyclic. The complex ion tris(ethylenediamine)cobalt (III), $[Co(en)_3]^{3+}$, in which bridging ligands occupy *cis*-related positions about the octahedrally coordinated cobalt ion, is an example of D_3 symmetry (cf. Fig. 1.14).

In the same manner, the D_{nd} groups are related by dihedral axes to the C_{nv} groups. In D_{nd} groups the combination of rotational operations and vertical mirror reflections (designated σ_d in these groups) generates a series of S_{2n} operations about an axis collinear with the principal axis. The staggered conformation of ethane is an example of D_{3d} symmetry (cf. Fig. 1.14).

The D_{nh} family of groups bears the same kind of dihedral relationship to the C_{nh} family. Like the corresponding single-axis groups, the D_{nh} groups include n-fold improper axes when $n > 2$ and are centrosymmetric when n is even. Planar BCl_3 and the eclipsed conformation of ethane are examples of D_{3h} symmetry. The group $D_{\infty h}$, in which the principal axis is an infinite-fold C_∞, is an important member of this family. It is the group of all linear, centrosymmetric molecules (e.g., H_2, CO_2).

The lowest members of the three dihedral families of groups, where $n = 2$, are in one respect unique. They all have three mutually perpendicular twofold axes, one of which is taken as the principal axis and hence the z axis of the reference Cartesian coordinate system. In older works, the D_2 designation was given the special symbol V. Therefore, the groups D_2, D_{2d}, and D_{2h} were formerly identified as V, V_d, and V_h, respectively. This notation is no longer used but may be encountered when consulting older references.

Cubic Groups. The cubic groups are associated with polyhedra that are geometrically related to the cube (cf. Fig. 1.15). All are characterized by the

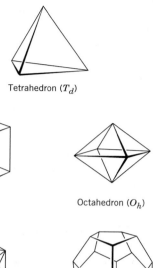

Tetrahedron (T_d)

Cube (O_h) Octahedron (O_h)

Icosahedron (I_h) Dodecahedron (I_h)

Figure 1.15 Polyhedra belonging to the groups T_d, O_h, and I_h.

presence of multiple, intersecting, high-order rotational axes. There are seven groups of this type, three of which are frequently encountered. These three (T_d, O_h, I_h) represent some of the most important geometries in chemistry.

The perfect tetrahedron defines the group T_d, comprised of the following 24 operations, listed by classes: E, $8C_3$ ($= 4C_3$, $4C_3^2$), $3C_2$, $6S_4$ ($= 3S_4$, $3S_4^3$), $6\sigma_d$. A threefold axis, generating the operations C_3 and C_3^2, emerges from each of the four triangular faces of a tetrahedron (cf. Fig. 1.15). When a tetrahedral molecule is inscribed in a cube, as in Fig. 1.10, a C_2 axis collinear with the bisector of opposing bond angles emerges from each pair of opposite cube faces. Three S_4 axes, each associated with S_4 and S_4^3 operations, are collinear with these C_2 axes. Any one of these can be taken as the z axis of a reference Cartesian coordinate system. Also, when the tetrahedron is shown inscribed in a cube, each of the four C_3 axes passes through a pair of opposite corners along a cube diagonal, collinear with an M–X bond (cf. Fig. 1.10). With $h = 24$, T_d represents one of the higher symmetries frequently encountered in structural chemistry.

Two related cubic groups, T and T_h, also have four intersecting threefold axes. The group T is a subgroup of T_d and consists of only the rotational operations of the higher symmetry group. Molecules with T and T_h are virtually unknown. As we shall see later, the group T is sometimes used as a simpler operating group to reduce the mathematical work of group theory applications for molecules that actually have T_d symmetry. The complete sets of operations for these groups may be found in Appendix A.

The octahedron and cube (Fig. 1.15) both belong to the point group O_h, which is composed of the following 48 operations, grouped by class: E, $8C_3(=4C_3, 4C_3^2)$, $6C_4(=3C_4, 3C_4^3)$, $6C_2$, $3C_2(=3C_4^2)$, i, $6S_4(=3S_4, 3S_4^3)$, $8S_6(=4S_6, 4S_6^5)$, $3\sigma_h(=\sigma_{xy}, \sigma_{yz}, \sigma_{xz})$, $6\sigma_d$. In the octahedron, a fourfold axis emerges from each pair of opposite apices, whereas a threefold axis emerges from each pair of opposite triangular faces. In the cube, a fourfold axis emerges from each pair of opposite faces, whereas a threefold axis emerges from each pair of opposite corners, extending the diagonals of the cube. Although molecules with cubic symmetry are rare, octahedral species are common among six-coordinated transition metal complexes. A rarely encountered subgroup of O_h is the rotational group O, whose operations are simply the proper rotations that comprise O_h (see Appendix A). Like the T_d rotational subgroup T, O is sometimes used as a simpler operating group for problems of molecules that actually have O_h symmetry.

Both the regular icosahedron and dodecahedron (Fig. 1.15) belong to the point group I_h, composed of the following operations, listed by classes: E, $12C_5$, $12C_5^2$, $20C_3$, $15C_2$, i, $12S_{10}$, $12S_{10}^3$, $20S_6$, 15σ. With $h = 120$, it is the highest symmetry that one is likely to encounter (aside from $C_{\infty v}$ and $D_{\infty h}$, for which $h = \infty$). In the icosahedron (Fig. 1.15) a fivefold axis emerges from each apex, while in the dodecahedron a fivefold axis emerges from the center of each face. The icosahedral ion $B_{12}H_{12}^{2-}$ is an example of an I_h species. There are currently no known examples of molecules with regular dodecahedral I_h symmetry.* However, buckminsterfullerene, C_{60}, is an example of an even higher-order polyhedron with I_h symmetry (Fig. 1.16). This soccer-ball-shaped carbon species consists of a spherical net composed of five-membered rings surrounded by six-membered rings. A fivefold rotational axis emerges from each pentagonal face, and a threefold axis emerges from each hexagonal face. Note that there are no sixfold axes, because each hexagon is surrounded alternately by three pentagons and three hexagons.

As with the other cubic groups, I_h has a purely rotational subgroup, designated I. The complete set of operations of this group is listed in Appendix A. This is not an important group for considerations of molecular symmetry.

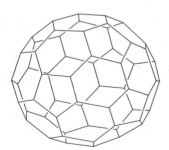

Figure 1.16 Buckminsterfullerene ("bucky ball"), C_{60}, an example of I_h symmetry.

*Dodecahedral eight-coordinate (CN8) complexes, such as $[Mo(CN)_8]^{4-}$ and $[ZrF_8]^{4-}$, actually have D_{2d} symmetry. [See J. L. Hoard and J. V. Silverton, *Inorg. Chem.* **1963**, *2*, 235; and D. R. Sears and J. H. Burns, *J. Chem. Phys.* **1964**, *41*, 3478.]

1.6   Systematic Point Group Classification

Identifying the point group of a molecule is a necessary first step for almost all applications of group theory in chemistry. Given that the number of operations can be as high as 120 or even infinity, identifying a molecule's point group by finding all the symmetry operations is not a practical approach. Fortunately, it is also not necessary. As we have seen, many operations are associated with a common symmetry element (e.g., C_3 requires C_3^2 about the same axis); the presence of certain operations in combination requires the presence of other operations (e.g., C_3 and σ_h require S_3); and each group can be seen as a member of a family of groups with a characteristic set of symmetry elements (e.g., the dihedral groups all have a single n-fold principal axis and n twofold axes perpendicular). The task, then, is reduced to finding the essential symmetry elements of the molecule that classify its point group unequivocally.

The best way to determine a molecule's point group is to look for key symmetry elements in a prescribed sequence. Such a sequence is illustrated by the "inverted tree" flow chart shown in Fig. 1.17. In using the flow chart, look sequentially for the symmetries indicated on the perpendicular lines, following the right or left branches according to whether a particular kind of symmetry is present ("Yes") or absent ("No").

The sequence begins by determining if the molecule has one of the readily identifiable structures associated with certain "special groups." These are noncentrosymmetric linear $C_{\infty v}$ (e.g., HCl, FBeCl), centrosymmetric linear $D_{\infty h}$ (e.g., CO_2, $BeCl_2$), tetrahedral T_d (e.g., CH_4, SO_4^{2-}), octahedral O_h (e.g., SF_6, $[Fe(CN)_6]^{4-}$), and icosahedral or other polyhedral I_h (e.g., $B_{12}H_{12}^{2-}$, C_{60}). Note that the geometry must be perfect for a molecule to belong to one of these groups. For example, while it is common in chemical discussions to refer to the shapes of CH_4, CH_3Cl, CH_2Cl_2, $CHCl_3$, and CCl_4 as "tetrahedral," only CH_4 and CCl_4 have the complete set of symmetry operations that define T_d. In this strict sense, the others are not truly tetrahedral. Both CH_3Cl and $CHCl_3$ belong to the group C_{3v}, and CH_2Cl_2 belongs to the group C_{2v}.

If it is apparent that the molecule is not one of these readily identifiable structures, look for a principal axis of rotation. If there is no axis of rotation whatsoever, the molecule must belong to one of the low-symmetry nonrotational groups—that is, C_s, C_i, or C_1. The specific group from among these can be determined by first looking for a mirror plane (often the plane of the molecule itself), which, if found, indicates C_s. If there is no mirror plane, look for inversion symmetry, which, if found, indicates C_i. The absence of both reflection and inversion indicates C_1, the point group of an asymmetric species.

Often the molecule will have one or more rotational axes. If so, it is necessary to identify the highest-order rotational axis, the principal axis, which will be taken as the z axis of the coordinate system (cf. Section 1.2). If a principal axis exists, it will interrelate the greatest number of atoms or groups of atoms, which will define one or more parallel planes (not necessarily mirror

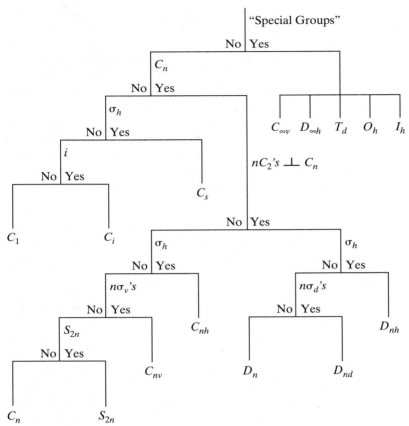

Figure 1.17 Flow chart for systematically determining the point group of a molecule.

planes) perpendicular to the axis of rotation. The number of equivalent atoms lying in any such plane will equal the order of the principal axis of rotation (i.e., n value of C_n). If the molecule happens to belong to D_2, D_{2d}, or D_{2h}, the highest-order axes will be three mutually perpendicular twofold axes. In these cases, choose as the principal axis the C_2 axis passing through the largest number of atoms.

Regardless of its order, once a principal axis has been identified, determine whether the molecule belongs either to one of the dihedral groups (D groups) or to one of the single-axis rotational groups (C groups). To make the distinction, look for n twofold axes (dihedral axes) perpendicular to the principal C_n axis (indicated as "nC_2's $\perp C_n$" on the flow chart, Fig. 1.17). These must lie in a common plane, which may or may not be a mirror plane of the molecule. For example, if the principal axis is C_3, look for three C_2 axes at 120° to each other in a plane perpendicular to the C_3 axis. The dihedral axes must intersect at the center of the molecule, but they are not required to pass through any of the atoms (although they often do).

If there are n twofold axes perpendicular to C_n, then the molecule belongs to one of the dihedral groups. To decide which one, begin by looking for a horizontal mirror plane (σ_h), which must contain the n twofold axes. If one is found, the molecule belongs to the D_{nh} group corresponding to the value of n for the principal axis. For example, if C_3 is the principal axis, the group is D_{3h}. If no horizontal mirror plane exists, look for n vertical mirror planes intersecting at the principal axis (indicated as "$n\sigma_d$'s" on the flow chart, Fig. 1.17). If found, these mirror planes indicate the D_{nd} group corresponding to the value of n for the principal axis. It is crucial to look for a horizontal mirror plane before looking for these σ_d planes, because both D_{nh} and D_{nd} groups have vertical mirror planes, which therefore provide no basis for distinction. If neither horizontal nor vertical mirror planes are found, then by default the appropriate dihedral group is D_n.

If the molecule has a principal axis but lacks n dihedral axes, it belongs to one of the single-axis groups. To distinguish between the possible groups, begin as with the dihedral groups by looking for a horizontal mirror plane (σ_h). If one is found, the molecule belongs to a C_{nh} group, where the value of n corresponds to the order of the principal axis. Thus, if the principal axis is C_3, the group is C_{3h}. If there is no horizontal mirror plane, look for n vertical mirror planes intersecting at the principal axis (indicated "$n\sigma_v$'s" on the flow chart, Fig. 1.17). If found, these mirror planes indicate the C_{nv} group appropriate to the order of the C_n principal axis. If there are no mirror planes, the molecule belongs to a C_n or S_{2n} group. As a practical matter, it is more likely to be the former than the latter. To decide, look for an improper axis S_{2n} collinear with the principal C_n axis. With the possible exception of an inversion center (e.g., if the group is S_6), the C_n and S_{2n} axes should be the only symmetry elements (other than identity) if the molecule belongs to one of the S_{2n} groups. If the molecule lacks both mirror planes and an improper rotation axis, it belongs to one of the cyclic, purely rotational groups, C_n.

Among the single-axis groups, the C_{nh} groups are not frequently encountered. As previously noted, boric acid is a simple example of C_{3h} symmetry (cf. Fig. 1.14). By contrast, the C_{nv} groups are probably the most frequently encountered single-axis groups, and with practice they are among the easiest to identify quickly. Both C_n and S_{2n} groups are so infrequently encountered among simple molecules that properly assigning a molecule that genuinely does belong to one of these groups can be tricky. As the S_4 example in Fig. 1.14 suggests, molecules with S_{2n} symmetry tend to have somewhat exotic structures. As an operating principle, one is well advised to double check that certain elements have not been missed when the procedure seems to suggest one of these rare groups. A well-known medical aphorism, offering sage advice to physicians attempting to reach a diagnosis from a set of symptoms, is equally relevant to the process of determining a molecule's point group: "When you hear hoof beats, think horses, not zebras."

As an illustration of the procedure, let us consider the systematic classification of the point group of PF_5 (Fig. 1.18). You may find it helpful to refer to a physical model as we search for the key symmetry elements. At the start, we recognize that PF_5 is a trigonal bipyramid, and therefore not a shape belonging to one of the "special groups" (e.g., linear, tetrahedral, octahedral, icosahedral). Therefore, we begin our systematic search for key symmetry elements by looking for a principal axis of rotation. Either from Fig. 1.18 or a model we may note that there are two symmetrically distinct kinds of bonds in PF_5. Two bonds (the axial positions) lie $180°$ from each other and $90°$ from the three other kinds of bonds (the equatorial positions). The three equatorial bonds lie $120°$ from each other within a plane. We can define a rotational axis, collinear with the axial bonds, that relates the three equatorial bonds by the series of rotations C_3, C_3^2, and $C_3^3 = E$. By contrast, taking any one of the equatorial bonds as an axis relates only the two axial bonds by the series of rotations C_2 and $C_2^2 = E$. Thus, the axial bonds lie along a C_3 axis, and the equatorial bonds lie along C_2 axes. There are no other axes of rotation, so the C_3 is the highest-order rotational axis of the molecule, which identifies it as the principal axis. We now know that PF_5 belongs to one of the rotational groups.

We next determine if the group is single-axis (a C group) or dihedral (a D group). To decide, we look for the three C_2 axes perpendicular to the C_3 principal axis ($nC_2 \perp C_n$, where $n = 3$), the dihedral axes required of a D group. We have already noted that the three equatorial bonds lie along C_2 axes, so PF_5 must belong to one of the D groups (i.e., D_{3h}, D_{3d}, or D_3).

To decide among the possible D groups, we look for a horizontal mirror plane (σ_h), perpendicular to the C_3 principal axis. The plane of the three equatorial bonds is a σ_h plane. Its reflection operation interchanges the two axial positions and reflects the three equatorial positions into themselves. At this point we can conclude that the point group is D_{3h}.

In determining that PF_5 belongs to the point group D_{3h}, we have ignored some of the other symmetry elements of the group. The classification procedure only concentrates on finding the characteristic elements that uniquely define a group. For example, we have not sought out the three vertical mirror planes (σ_v) that are among the elements whose operations comprise the D_{3h} group. Each of the σ_v planes passes through the two axial positions and one of the equatorial positions of the trigonal bipyramid. The operation reflects these three positions into themselves, and it interchanges the other two equatorial positions, which lie on opposite sides of the plane. These mirror planes are not unique to D_{3h}. If we had found these planes before finding the σ_h plane, we might have incorrectly assigned the point group as D_{3d}, which also has vertical mirror planes. In other words, the order of carrying out the classification procedure is critical to a correct determination. We might also

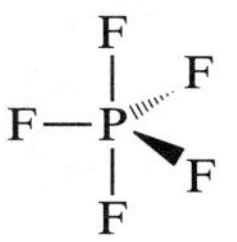

Figure 1.18 Trigonal bipyramidal structure of PF_5.

note that D_{3h} has an S_3 axis collinear with the C_3 axis. This must be so, because both C_3 and σ_h exist, and $\sigma_h C_3 = S_3$ by definition. Our classification procedure does not require finding the S_3 axis, since the point group in this case is determined by the presences of other necessary elements.

Grouped by classes, the complete set of symmetry operations that comprise the point group D_{3h} is E, $2C_3$ ($= C_3, C_3^2$), $3C_2$, σ_h, $2S_3$ ($= S_3, S_3^2$), $3\sigma_v$. There are 12 operations in this list, so the group order is $h = 12$. The 12 operations are associated with 10 symmetry elements: the object itself for E; the C_3 proper axis for the C_3 and C_3^2 rotations; the three dihedral C_2 axes, one for each C_2 rotation; the horizontal mirror plane for the σ_h reflection; the S_3 improper axis (collinear with the C_3 axis) for the S_3 and S_3^2 operations; and the three vertical mirror planes for the three σ_v reflections.

The flow chart of Fig. 1.17 is a convenient mnemonic for learning systematic point group classification. However, with practice, reference to the chart should become unnecessary.* Indeed, chemists experienced in symmetry classification often can determine the point group of a molecule almost instantly, even without needing to carry out a sequential analysis. In part, this is possible because certain structure patterns recur frequently and become associated in the mind with their appropriate point groups.

The three conformations of ethane exemplify one type of recurring geometrical pattern. To see the pattern, imagine the two sets of three hydrogen atoms on each end of ethane as two triangles separated by a distance (cf. Fig. 1.19). Looking down the principal axis (C_3), we find that the point group is D_{3h} when the triangles are aligned (eclipsed) and D_{3d} when they are perfectly staggered. We note further that if the triangles are slightly misaligned (skewed configuration), the vertical mirror plane symmetry is destroyed and the point

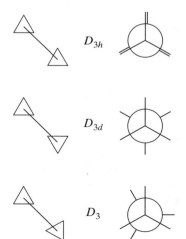

Figure 1.19   Representations of the three conformations of ethane as two triangles separated along the C_3 axis. The corresponding Newman projections are shown on the right.

*To help develop your facility with systematic classification, you might try verifying the point group assignments given for the molecules shown in Fig. 1.14.

group is D_3 (cf. Fig. 1.19).* We can generalize these relationships for any molecule that can be represented as two regular polygons separated along a principal axis of rotation. When the polygons are eclipsed, the group is D_{nh}; when they are perfectly staggered, it is D_{nd}; and when they are skewed, it is D_n. As a test of this, you might want to make a model of ferrocene, $Fe(C_5H_5)_2$, and systematically determine the point groups of the configurations resulting from the three types of ring alignment.

1.7 Optical Activity and Symmetry

Many molecules can exist in either of two optically active isomers, called *enantiomers*. When plane-polarized light is passed through separate samples of enantiomers, one (the dextrorotatory isomer) will cause clockwise rotation of the polarity, and the other (the levorotatory isomer) will cause counterclockwise rotation of the polarity. Compounds that can exist as enantiomeric pairs are called *chiral;* and from the standpoint of symmetry considerations, they are said to be *dissymmetric*. Dissymmetric is not the same as asymmetric (despite what some dictionaries may suggest). *Asymmetric* means the absence of any symmetry other than identity. In contrast, dissymmetric molecules may have significant, albeit limited, symmetry. Asymmetric molecules are simply the least symmetric of dissymmetric molecules.

The hallmark of enantiomers is that they are nonsuperimposable mirror images of each other. Figure 1.20 shows two examples of such dissymmetric enantiomer pairs. As expected, asymmetric CHBrClF has two nonsuperimposable isomers and therefore is chiral. However, the complex ion $[Co(en)_3]^{3+}$ also exists as two enantiomers but has significant symmetry, belonging to

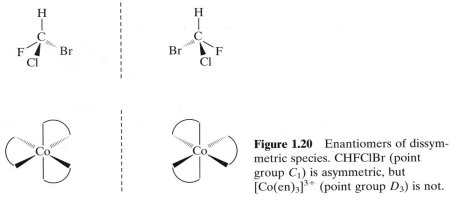

Figure 1.20 Enantiomers of dissymmetric species. CHFClBr (point group C_1) is asymmetric, but $[Co(en)_3]^{3+}$ (point group D_3) is not.

*You may want to build models of all three conformers to verify these point group assignments. The dihedral axes can be difficult to see for the staggered and skewed configurations. Looking at the Newman projections (cf. Fig. 1.19), you can find the C_2 axes as bisectors of the angles between pairs of C–H bonds on the "front" and "back" of the projection.

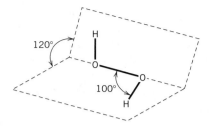

Figure 1.21 The structure of hydrogen peroxide (point group C_2).

point group D_3. As this illustrates, chiral molecules may have some symmetry. In general, the restriction is that *a molecule is dissymmetric and may be chiral either if it is asymmetric or if it has no other symmetry than proper rotation.* The point group D_3 satisfies the latter criterion, since the group consists of the operations E, $2C_3$, $3C_2$. In light of these restrictions, only molecules belonging to certain point groups are candidates for chirality. The possible chiral groups are C_1 (asymmetric), C_n, and D_n.*

The traditional test of chirality is to draw or build models of the suspect species to see if nonsuperimposable, mirror-image isomers exist. As a practical matter, this exercise need only be carried out if the molecule belongs to one of the possibly chiral point groups. In most cases, molecules belonging to any of these groups will be chiral, but in rare cases enantiomeric pairs cannot actually exist. For example, hydrogen peroxide, H_2O_2, whose structure is shown in Fig. 1.21, belongs to the point group C_2 but is not optically active. While it is possible to build two models that are not superimposable, the distinction does not exist for the actual molecular structure, because relatively free rotation about the O–O bond continuously interconverts the two hypothetical enantiomers. Here, stereochemical nonrigidity precludes chirality.

Problems

1.1 For the following molecules, which are shown in Fig. 1.14, sketch the locations of all the symmetry elements, and list all operations associated with each symmetry element: (a) NH_3 (C_{3v}), (b) IF_5 (C_{4v}), (c) $B(OH)_3$ (C_{3h}), (d) BCl_3 (D_{3h}), (e) C_2H_6 in the staggered conformation (D_{3d}).

1.2 Given the set of operations $\{E, C_4, \sigma_h\}$, determine the other operations that must be present to form a complete point group. [*Hint:* Consider all the products of the given elements with themselves and with each other.] Identify the point group for the complete set of operations. What is the order of the group?

1.3 Aside from the trivial group C_1, the point group formed from the complete set of operations from Problem 1.2 has six subgroups. Identify the subgroups and give the order of each.

---

*The rotational cubic groups T, O, and I can also be added to this list, but no molecules with these symmetries are likely to be encountered.

1.4 Cyclic groups are formed by taking the series of powers on a single element up to the order of the group, such that $G = [X, X^2, \ldots, X^h = E]$. Taking each of the following operations as the base element of a cyclic group, determine the series of all operations that constitutes the group, identify the group, and develop its multiplication table: (a) C_3, (b) C_6, (c) S_4. Identify all the subgroups of these cyclic groups.

1.5 Develop the multiplication table for the group C_{2h}, which consists of the operations E, C_2, i, and σ_h. [*Hint:* Determine the effects of the operations on an arbitrary point whose initial coordinates are x, y, z.] Is this group Abelian?

1.6 Determine the point group of each of the following shapes:

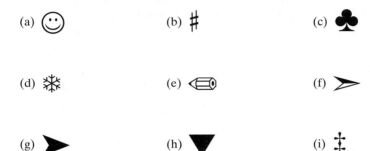

(j) a Styrofoam coffee cup (no decoration)
(k) a ceramic coffee mug with handle (no decoration)
(l) a dumbbell (no markings)
(m) a tennis ball, including the seams (one color, no markings)
(n) an airplane propeller with four blades
(o) a soccer ball, including the seams (one color, no markings)

1.7 Determine the point group of each of the following molecules or ions, whose shapes can be determined by use of valence-shell electron-pair repulsion (VSEPR) theory: (a) SeF_5^-, (b) ClF_2^+, (c) AsF_4^-, (d) XeF_2, (e) XeF_4, (f) BeF_3^-, (g) SiF_6^{2-}, (h) OCN^-, (i) $AsCl_4^+$, (j) OSF_4, (k) *trans*-FNNF, (l) *cis*-FNNF, (m) ClSSCl (nonplanar), (n) $S_2O_3^{2-}$, (o) *trans*-$(OH)_4XeO_2$.

1.8 Consider the following ideal geometries for MX_n molecules ($n = 3$–6) and the distortions described for each. What are the point groups of the ideal geometry and the distorted geometry?

(a) MX_3 trigonal planar distorted by lifting the M atom out of the plane.
(b) MX_4 tetrahedral distorted by slightly flattening the molecule along one of the C_2 axes.
(c) MX_4 square planar distorted by equally elongating two *trans*-related M–X bonds.
(d) MX_5 trigonal bipyramidal distorted by equally elongating or shortening the two axial bonds.
(e) MX_5 trigonal bipyramidal distorted by elongating one of the equatorial bonds.
(f) MX_5 trigonal bipyramidal distorted by elongating one of the equatorial bonds and shortening one of the axial bonds (or vice versa).

(g) MX_6 octahedral distorted by elongating two *trans*-related bonds.

(h) MX_6 octahedral distorted by slightly closing the 90° angles between the three M–X bonds in both sets of *cis*-related positions.

1.9 Identify the point group for each of the following Fe^{3+} complexes with the bidentate oxalate ligand (ox = $C_2O_4^{2-}$). Where allowed by the point group, determine which complexes are chiral. [*Hint:* Use models to help identify the point group and to verify the existence of enantiomers.]

(a) tris(oxalato)ferrate(III), $[Fe(ox)_3]^{3-}$

(b) *trans*-dichlorobis(oxalato)ferrate(III), $[FeCl_2(ox)_2]^{3-}$

(c) *cis*-dichlorobis(oxalato)ferrate(III), $[FeCl_2(ox)_2]^{3-}$

(d) *cis*-dibromo-*trans*-dichlorooxalatoferrate(III), $[FeBr_2Cl_2ox]^{3-}$

(e) *cis*-dibromodichlorooxalatoferrate(III), $[FeBr_2Cl_2ox]^{3-}$.

1.10 Identify the point group of each of the following structures.

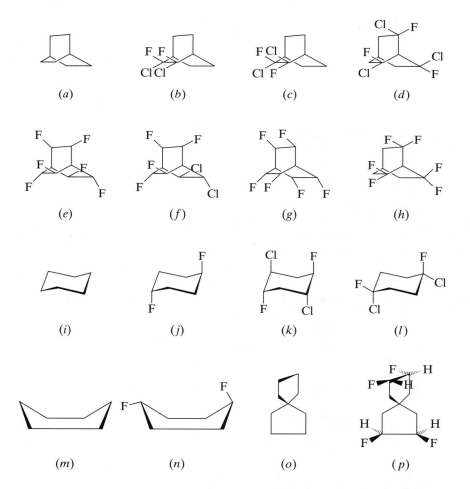

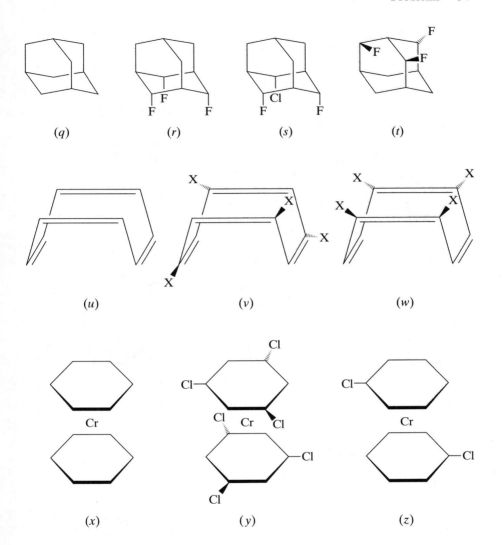

(q)   (r)   (s)   (t)

(u)   (v)   (w)

(x)   (y)   (z)

1.11 Identify the point group of each of the following cyclopentane derivatives in the configurations shown.

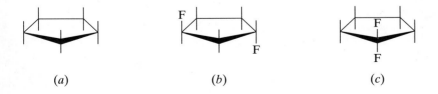

(a)   (b)   (c)

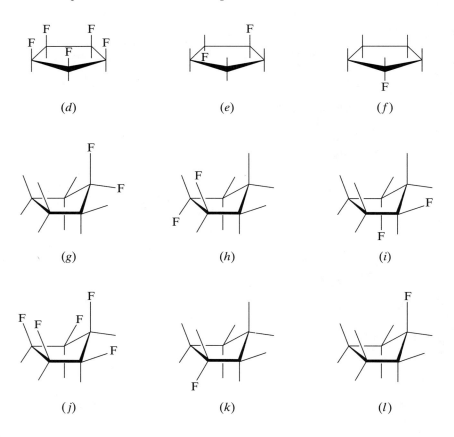

1.12   Identify the point group of each of the following structures.

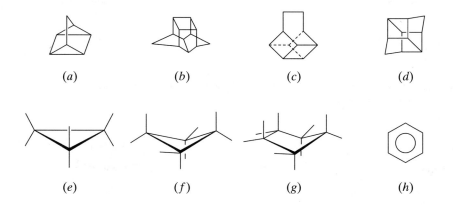

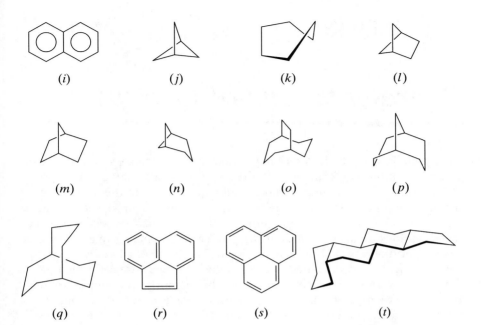

(i)   (j)   (k)   (l)

(m)   (n)   (o)   (p)

(q)   (r)   (s)   (t)

CHAPTER 2

Representations of Groups

As we saw in Chapter 1, describing a molecule's symmetry in terms of symmetry operations conforms with the requirements of a mathematical group. This allows us to apply mathematical techniques of group theory to describe and analyze some of the molecule's chemically interesting properties. The approach we take, in general, is to define a set of imagined vectors on the molecule's various atoms to represent the properties of interest (e.g., atomic orbitals, hybrid orbitals, vibrational motions). These vectors, like the properties they represent, may be related to one another in specific ways through the effects of the various symmetry operations that comprise the molecule's point group. The ways in which the vectors behave as a result of the operations yield a symmetry description of the property of interest, which in turn facilitates defining the property in quantum mechanical terms.

A set of vectors, such as we have just described, can be used to elucidate a molecule's physical properties because it forms the basis for a mathematical representation of the point group. Consequently, in order to apply symmetry arguments to the solution of molecular problems, we need an understanding of mathematical representations of groups—their construction, meaning, and manipulation. Our focus in this text is on applying group theory, and not on rigorous mathematical development. Therefore, the approach we will take will be decidedly nonrigorous. We will develop the concept of representations and the techniques for manipulating them through specific examples that illustrate fundamental principles and general results. Proofs of the theorems and their consequential general results can be found in more advanced texts.

2.1 Irreducible Representations

For our purposes, we may define a *representation* of a group as a set of symbols that will satisfy the multiplication table for the group. The symbols themselves are called the *characters* of the representation.* In the case of symmetry point groups, with which we are concerned, the characters may be positive or negative integers, numeric values of certain trigonometric functions, imaginary numbers involving the integer $i = \sqrt{-1}$, or even square matrices. By

---

*The term "character" can have a double meaning. As we shall see, most often the characters of a representation are characters of certain matrices.

Table 2.1 Multiplication Table for C_{2v}

C_{2v}	E	C_2	σ_v	σ_v'
E	E	C_2	σ_v	σ_v'
C_2	C_2	E	σ_v'	σ_v
σ_v	σ_v	σ_v'	E	C_2
σ_v'	σ_v'	σ_v	C_2	E

way of introduction, we will first consider representations whose characters are simply positive or negative integers. Recall the multiplication table we developed for C_{2v}, in Section 1.3, shown here as Table 2.1. If we make a set of substitutions for the four operations of the group such that the substituted characters also obey the general relationships of this multiplication table, then the set of characters will be a representation of C_{2v}. We can find one such representation by making the seemingly trivial substitution of the integer 1 for each of the four operations; that is, $E = 1$, $C_2 = 1$, $\sigma_v = 1$, $\sigma_v' = 1$. With these substitutions, the multiplication table becomes

C_{2v}	$E = 1$	$C_2 = 1$	$\sigma_v = 1$	$\sigma_v' = 1$
$E = 1$	1	1	1	1
$C_2 = 1$	1	1	1	1
$\sigma_v = 1$	1	1	1	1
$\sigma_v' = 1$	1	1	1	1

As uninteresting and obvious as these multiplication results are, they nonetheless do obey the same combinational relationships of the group elements themselves, albeit with extreme redundancy. Thus, the set of substitutions of 1 for every operation is a genuine representation of the group. If this is true for C_{2v}, we can see that it would be valid for any other point group, regardless of its order or the operations of which it is composed. Indeed, the set of all 1 characters makes the most fundamental representation for any point group. This is called the *totally symmetric representation*. In the case of the group C_{2v}, the totally symmetric representation is designated by the symbol A_1.

Somewhat less trivial substitutions, composed of both positive and negative values of the integer 1, can be used to construct other representations of C_{2v}. One set that obeys the relations of the multiplication table is

$$E = 1, \qquad C_2 = 1, \qquad \sigma_v = -1, \qquad \sigma_v' = -1$$

For this set the multiplication table takes on the following form:

C_{2v}	$E = 1$	$C_2 = 1$	$\sigma_v = -1$	$\sigma_v' = -1$
$E = 1$	1	1	-1	-1
$C_2 = 1$	1	1	-1	-1
$\sigma_v = -1$	-1	-1	1	1
$\sigma_v' = -1$	-1	-1	1	1

Although the connection between the products and the original operations is sometimes ambiguous (e.g., an answer of $+1$ could mean either E or C_2), the results are nonetheless consistent with the original multiplication table (Table 2.1). Thus, the set of characters 1, 1, -1, -1 is a representation of C_{2v}. This representation is given the symbol A_2. Other sets of characters that satisfy the requirements of the multiplication table (which you should verify) and therefore form valid representations are

$$E = 1, \qquad C_2 = -1, \qquad \sigma_v = 1, \qquad \sigma_v' = -1$$

$$E = 1, \qquad C_2 = -1, \qquad \sigma_v = -1, \qquad \sigma_v' = 1$$

These representations of C_{2v} are designated B_1 and B_2, respectively.

These four sets of ± 1 characters are the only sets that meet the criterion of conformability with the multiplication table of the group. For example, suppose we try the set

$$E = -1, \qquad C_2 = 1, \qquad \sigma_v = 1, \qquad \sigma_v' = -1$$

With these substitutions the multiplication table takes on the form

C_{2v}	$E = -1$	$C_2 = 1$	$\sigma_v = 1$	$\sigma_v' = -1$
$E = -1$	1	-1	-1	1
$C_2 = 1$	-1	1	1	-1
$\sigma_v = 1$	-1	1	1	-1
$\sigma_v' = -1$	1	-1	-1	1

This is not the same as the results of the multiplication table for the original operations of C_{2v}. For example, we know that $EE = E$, but with $E = -1$ the table above suggests that E combines with itself to give either C_2 or σ_v, both of which have been substituted by $+1$. Other discrepancies of this sort occur throughout this table and show that the set -1, 1, 1, -1 is not a genuine representation of C_{2v}.

If we attempt to substitute other integers, say 0 or ± 2, the results also will not satisfy the requirements of the multiplication table. However, for other point groups, representations with these characters may be possible. For C_{2v} the representations we have found (viz., A_1, A_2, B_1, B_2) are the simplest and most fundamental representations of C_{2v}. For this reason they are called *irreducible representations* of the group.

We can list the characters and related properties of the irreducible representations of C_{2v} in a tabular form, called a *character table*. A simplified character table, based on the representations we have just developed, would look like Table 2.2. Each row lists the individual character of each operation for the representation named in the left-hand column. The labels for the irreducible representations (A_1, A_2, B_1, and B_2) are the standard *Mulliken symbols*. Their form and meaning are explained in Section 2.4.

Table 2.2 Partial Character Table for C_{2v}

C_{2v}	E	C_2	σ_v	σ'_v
A_1	1	1	1	1
A_2	1	1	−1	−1
B_1	1	−1	1	−1
B_2	1	−1	−1	1

2.2 *Unit Vector Transformations*

In applying group theory to chemical problems we will need to use vector representations for molecular properties. Therefore, we now turn our attention to the behavior of vectors—particularly unit vectors—when subjected to the operations of the molecule's point group. For illustration, we will continue to use C_{2v}. For our purposes, a unit vector is designated by an arrow of an arbitrary fundamental unit length (representing a unit of the vector property), with its base at the origin of a Cartesian coordinate system [i.e., the point (0, 0, 0)] and its tip pointing in some particular direction.

Figure 2.1 shows a unit vector **z** relative to a Cartesian coordinate system. In a molecule with C_{2v} symmetry, such a vector would lie collinear with the C_2 axis. Let us examine the effects of applying the four operations of C_{2v} on this vector. Clearly, the identity operation leaves **z** in place, since E does nothing. The C_2 operation is performed about the same axis on which **z** lies, so it too leaves the vector unaffected; that is, its orientation before and after the operation is identical. The σ_v and σ'_v planes intersect along **z**, so the vector lies in both of them. As a result, the reflections also leave **z** nonshifted. Mathematically, we could express these results, the new positions of **z** after each operation, by multiplying the unit vector **z** by +1 for each operation. Later we will want to express such transformations for more general vectors and sets of vectors, and these situations will require the use of $n \times n$ matrices (where n is an integer). To be consistent with the methodology of more general cases, let us express the present four multiplications by +1 for the vector **z** as multiplications by trivial 1×1 matrices, symbolized by the use of

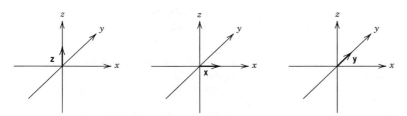

Figure 2.1 Unit vectors **z**, **x**, and **y** (left to right).

square brackets ([ ]). In this case, each matrix will be [+1]. We can summarize the effects of C_{2v} on **z** as follows:

Operation	z becomes	In matrix notation
E	z	[+1]z
C_2	z	[+1]z
σ_v	z	[+1]z
σ_v'	z	[+1]z

The four 1×1 operator matrices, which express the effects of the four operations of C_{2v} on the unit vector **z**, are called *transformation matrices*. In this case, each transformation matrix is identical to the character of the operation listed for the irreducible representation A_1 in the C_{2v} character table (Table 2.2). In other words, the characters of the A_1 representation express the transformation properties of a unit vector **z** under the operations of C_{2v}. We say, then, that **z** transforms as A_1 in C_{2v}. Irreducible representations are sometimes referred to as *species* (or, more explicitly, *symmetry species*), so another way of saying this would be that **z** belongs to the A_1 species of C_{2v}.*

A less trivial set of transformations occurs when we apply the operations of C_{2v} to a unit vector **x**, as shown in Fig. 2.1. This vector lies in the σ_v plane (the xz plane), which intersects with the σ_v' plane (the yz plane) along the C_2 axis (cf. Fig. 1.13). Considering the operations in order, we note once again that identity does nothing to the position of **x**. However, this time the operation C_2 reverses the direction of **x**, making it point in the $-x$ direction; that is, **x** becomes $-$**x**. The vector lies in the σ_v plane, so reflection in that plane leaves it unaffected. Reflection in the σ_v' plane, to which the vector is orthogonal, reverses its direction, so that again **x** is transformed into $-$**x**. We can summarize these results as follows:

Operation	x becomes	In matrix notation
E	x	[+1]x
C_2	$-$x	[$-$1]x
σ_v	x	[+1]x
σ_v'	$-$x	[$-$1]x

This time we see that the four transformation matrices are identical to the characters of the B_1 irreducible representation of C_{2v} (cf. Table 2.2).

---

*Use of the term "species" as a synonym for "irreducible representation" dates to the earliest applications of group theory to problems of quantum mechanics. At 10 syllables, "irreducible representation" is unwieldy (especially when spoken) but unambiguous. "Species," with only two syllables, is succinct but potentially ambiguous. The question can arise, "Are we referring to the molecule or the symmetry of one of its properties?" In this text, when there might be potential for confusion, we will use the more explicit terms "symmetry species" or "irreducible representation."

Therefore, we can say that $\mathbf{x}$ transforms as B_1 in C_{2v}, or, alternately, that $\mathbf{x}$ belongs to the B_1 species of C_{2v}.

Finally, consider the effects of the C_{2v} operations on a unit vector $\mathbf{y}$, as shown in Fig. 2.1. This vector lies in the σ'_v plane (the yz plane). The effects of the four operations can be summarized as follows:

Operation	$\mathbf{y}$ becomes	In matrix notation
E	$\mathbf{y}$	$[+1]\mathbf{y}$
C_2	$-\mathbf{y}$	$[-1]\mathbf{y}$
σ_v	$-\mathbf{y}$	$[-1]\mathbf{y}$
σ'_v	$\mathbf{y}$	$[+1]\mathbf{y}$

Here the transformation matrices are the same as the characters of the B_2 irreducible representation of C_{2v} (cf. Table 2.2). Thus we see that $\mathbf{y}$ transforms as B_2 in C_{2v}, or equivalently that $\mathbf{y}$ belongs to the B_2 species of C_{2v}.

In addition to linear unit vectors, we will occasionally need to consider vectors that suggest rotations about the three Cartesian axes. For example, free gaseous molecules are constantly tumbling, and these motions can be resolved as rotations about the x, y, and z axes. In certain applications, identifying these components of rotational motion with their related symmetry species is important. Figure 2.2 shows a curved vector, $\mathbf{R}_z$, representing rotation about the z axis. To analyze the effects of the operations of C_{2v} on this vector, we will assume that if an operation reverses the sense of rotation, the vector has been transformed into the negative of itself. The operations E and C_2 do not affect the sense of rotation of $\mathbf{R}_z$, so the operator matrices in both cases are $[+1]$. To envision the effects of reflection, imagine the $\mathbf{R}_z$ vector encountering an actual reflecting mirror interrupting its circular path. As the vector approaches the mirror, it will "see" its reflected image, coming toward it with the opposite sense of rotation. Thus, the operator matrices for both σ_v and σ'_v are $[-1]$. These results can be summarized as follows:

Operation	$\mathbf{R}_z$ becomes	In matrix notation
E	$\mathbf{R}_z$	$[+1]\mathbf{R}_z$
C_2	$\mathbf{R}_z$	$[+1]\mathbf{R}_z$
σ_v	$-\mathbf{R}_z$	$[-1]\mathbf{R}_z$
σ'_v	$-\mathbf{R}_z$	$[-1]\mathbf{R}_z$

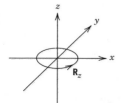

Figure 2.2   A rotational vector, $\mathbf{R}_z$.

Table 2.3 Character Table for C_{2v} with Linear and Rotational Vector Transformations

C_{2v}	E	C_2	σ_v	σ_v'	
A_1	1	1	1	1	z
A_2	1	1	-1	-1	$\mathbf{R}_z$
B_1	1	-1	1	-1	$x, \mathbf{R}_y$
B_2	1	-1	-1	1	$y, \mathbf{R}_x$

As with the cases of linear unit vectors, we can compare the set of 1×1 matrices with the characters of the various irreducible representations in Table 2.2. Here we see that they are identical to the characters of A_2. Thus, we can say that $\mathbf{R}_z$ transforms as A_2 in C_{2v}, or that $\mathbf{R}_z$ belongs to the A_2 species in C_{2v}. In similar fashion, we could show that the rotational vectors $\mathbf{R}_x$ and $\mathbf{R}_y$ transform as B_2 and B_1, respectively, in C_{2v}.

The transformation properties of both linear and rotational vectors are matters of recurring interest in applying point group theory to chemical problems. Therefore, it is customary to list these properties in the character table in a column to the immediate right of the character listings. For C_{2v}, the addition of this information to the basic character table (Table 2.2) gives the expanded character table, shown as Table 2.3.

2.3 Reducible Representations

Molecular properties are not always conveniently located along the axes of a Cartesian coordinate system. More often they are oriented in general directions that can be resolved into vector components along the three cardinal directions of the coordinate system. To see how the transformation properties of general vectors relate to the irreducible representations of a group, consider the vector $\mathbf{v}$, whose base is at the origin of the system [the point (0, 0, 0)] and whose tip is at a general point with coordinates (x, y, z) as shown in Figure 2.3. The orientation of the vector is arbitrarily chosen, and therefore the coordinates (x, y, z) have no special values. As before, we will examine the effects of the operations of C_{2v}, on this vector.

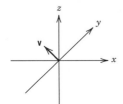

Figure 2.3 A general vector $\mathbf{v}$ with arbitrary orientation.

The base of the vector **v** lies at the origin of the system and will be unaffected by any operation. Therefore, we need only concern ourselves with the position of the tip of the vector. We will represent the three coordinates of the tip, both before and after each operation, in a column matrix form. For example, identity changes nothing, so we may write

$$[E] \times \begin{bmatrix} x \\ y \\ z \end{bmatrix} = \begin{bmatrix} x \\ y \\ z \end{bmatrix}$$

The symbol $[E]$ represents a transformation matrix, whose specific form we need to define. In the cases of the unit vectors we considered in Section 2.2 the transformation matrices had a trivial 1×1 format, since we only had to account for the position of one coordinate in each case. For the vector **v**, we must account for all three coordinates simultaneously. This requires that the transformation matrices for each operation have 3×3 dimensions.

Before proceeding further with construction of the necessary transformation matrices, let us review some of the general properties of matrices and the rules for multiplying them. A matrix is a rectangular array of numbers that combines with other such arrays according to specific rules. The individual numbers are the *elements* of the matrix. Each element may be designated by its row and column positions in the matrix. The following general matrix illustrates the customary indexing system:

$$\begin{bmatrix} a_{11} & a_{12} & \cdots & a_{1n} \\ a_{21} & a_{22} & \cdots & a_{2n} \\ \cdot & \cdot & \cdots & \cdot \\ \cdot & \cdot & \cdots & \cdot \\ \cdot & \cdot & \cdots & \cdot \\ a_{m1} & a_{m2} & \cdots & a_{mn} \end{bmatrix}$$

The dimensions of a matrix are specified as the number of rows by the number of columns. Thus, the general matrix above is an $m \times n$ matrix. The column matrix for the tip of the general vector **v**, for example, is a 3×1 matrix.

We will be concerned primarily with multiplication of matrices, for which the order of combination is left to right. If two matrices are to be multiplied together they must be *conformable*, which means that the number of columns in the first (left) matrix must be the same as the number of rows in the second (right) matrix. The product matrix has as many rows as the first matrix and as many columns as the second matrix. For example,

$$\begin{bmatrix} a_{11} & a_{12} \\ a_{21} & a_{22} \\ a_{31} & a_{32} \end{bmatrix} \begin{bmatrix} b_{11} & b_{12} & b_{13} \\ b_{21} & b_{22} & b_{23} \end{bmatrix} = \begin{bmatrix} c_{11} & c_{12} & c_{13} \\ c_{21} & c_{22} & c_{23} \\ c_{31} & c_{32} & c_{33} \end{bmatrix} \qquad (2.1)$$

In general, if we multiply an $n \times m$ matrix with an $m \times l$ matrix, the result will be an $n \times l$ matrix. The elements of the product matrix, c_{ij}, are the sums of the products $a_{ik}b_{kj}$ for all values of k from 1 to m; that is,

$$c_{ij} = \sum_{k=1}^{m} a_{ik}b_{kj} \tag{2.2}$$

From Eq. (2.2), the elements in the product matrix of Eq. (2.1) are

$$\begin{bmatrix} c_{11} & c_{12} & c_{13} \\ c_{21} & c_{22} & c_{23} \\ c_{31} & c_{32} & c_{33} \end{bmatrix} = \begin{bmatrix} (a_{11}b_{11} + a_{12}b_{21}) & (a_{11}b_{12} + a_{12}b_{22}) & (a_{11}b_{13} + a_{12}b_{23}) \\ (a_{21}b_{11} + a_{22}b_{21}) & (a_{21}b_{12} + a_{22}b_{22}) & (a_{21}b_{13} + a_{22}b_{23}) \\ (a_{31}b_{11} + a_{32}b_{21}) & (a_{31}b_{12} + a_{32}b_{22}) & (a_{31}b_{13} + a_{32}b_{23}) \end{bmatrix} \tag{2.3}$$

Given the requirement of conformability, it follows that matrix multiplication, like combination of symmetry operations, is not in general commutative. However, matrix multiplication is associative.

Returning now to the transformations of the vector $\mathbf{v}$ under the operations of C_{2v}, we can see that the effect of the identity operation is expressed by the equation

$$\begin{bmatrix} 1 & 0 & 0 \\ 0 & 1 & 0 \\ 0 & 0 & 1 \end{bmatrix} \begin{bmatrix} x \\ y \\ z \end{bmatrix} = \begin{bmatrix} \{(1)x + (0)y + (0)z\} \\ \{(0)x + (1)y + (0)z\} \\ \{(0)x + (0)y + (1)z\} \end{bmatrix} = \begin{bmatrix} x \\ y \\ z \end{bmatrix} \tag{2.4}$$

Thus, the matrix that describes the effect of E on the coordinates (x, y, z), which we shall call the transformation matrix for the identity operation, is

$$\begin{bmatrix} 1 & 0 & 0 \\ 0 & 1 & 0 \\ 0 & 0 & 1 \end{bmatrix}$$

The operation C_2 does not affect the z coordinate of the vector, because that component is collinear with the axis of rotation. However, C_2 reverses the sense of the vector's x and y coordinates, transforming them into $-x$ and $-y$, respectively. In matrix notation this is

$$\begin{bmatrix} -1 & 0 & 0 \\ 0 & -1 & 0 \\ 0 & 0 & 1 \end{bmatrix} \begin{bmatrix} x \\ y \\ z \end{bmatrix} = \begin{bmatrix} \{(-1)x + (0)y + (0)z\} \\ \{(0)x + (-1)y + (0)z\} \\ \{(0)x + (0)y + (1)z\} \end{bmatrix} = \begin{bmatrix} -x \\ -y \\ z \end{bmatrix} \tag{2.5}$$

The 3×3 matrix on the left of Eq. (2.5) is the transformation matrix for the operation C_2.

For σ_v, which lies in the xz plane, the reflection will transform the y coordinate into the negative of itself. The x and z coordinates, which lie in the plane, will not be affected. In matrix notation this is

$$\begin{bmatrix} 1 & 0 & 0 \\ 0 & -1 & 0 \\ 0 & 0 & 1 \end{bmatrix}\begin{bmatrix} x \\ y \\ z \end{bmatrix} = \begin{bmatrix} \{(1)x + (0)y + (0)z\} \\ \{(0)x + (-1)y + (0)z\} \\ \{(0)x + (0)y + (1)z\} \end{bmatrix} = \begin{bmatrix} x \\ -y \\ z \end{bmatrix} \qquad (2.6)$$

Again, the 3×3 matrix on the left in Eq. (2.6) is the transformation matrix for σ_v.

Finally, for σ_v', which lies in the yz plane, the reflection will transform the x coordinate into the negative of itself and leave the y and z coordinates unaffected. In matrix notation this is

$$\begin{bmatrix} -1 & 0 & 0 \\ 0 & 1 & 0 \\ 0 & 0 & 1 \end{bmatrix}\begin{bmatrix} x \\ y \\ z \end{bmatrix} = \begin{bmatrix} \{(-1)x + (0)y + (0)z\} \\ \{(0)x + (1)y + (0)z\} \\ \{(0)x + (0)y + (1)z\} \end{bmatrix} = \begin{bmatrix} -x \\ y \\ z \end{bmatrix} \qquad (2.7)$$

Once again, the 3×3 matrix on the left of Eq. (2.7) is the transformation matrix for the operation of σ_v'.

In each of the cases of the three unit vectors **x**, **y**, and **z**, we saw that the set of 1×1 transformation matrices that described the effects of the four operations of C_{2v} constituted the characters of a representation of the group (cf. Section 2.2). In those cases the representations were identical to irreducible representations of C_{2v}. Likewise, the four 3×3 transformation matrices we obtain from Eq. (2.4) through (2.7), which describe the effects of the operations of C_{2v} on the coordinates of the general vector **v**, constitute the characters of a representation of the group. The test of this assertion is to show that these matrices combine with one another in the same ways as their corresponding operations, as dictated by the multiplication table of the group (Table 2.1). For example, we know from Table 2.1 that $C_2E = C_2$. Using the corresponding transformation matrices for E [Eq. (2.4)] and C_2 [Eq. (2.5)] we obtain the following product*:

$$\begin{bmatrix} 1 & 0 & 0 \\ 0 & 1 & 0 \\ 0 & 0 & 1 \end{bmatrix}\begin{bmatrix} -1 & 0 & 0 \\ 0 & -1 & 0 \\ 0 & 0 & 1 \end{bmatrix} = \begin{bmatrix} -1 & 0 & 0 \\ 0 & -1 & 0 \\ 0 & 0 & 1 \end{bmatrix} \qquad (2.8)$$

The product is the transformation matrix for C_2, consistent with the product for the operations themselves.

---

*Remember that the order of writing combinations of operations is right to left, but the order of multiplying matrices is left to right. Thus, in Eq. (2.8), the E transformation matrix is written to the left of the C_2 transformation matrix for the product C_2E.

Similarly, we know that $\sigma_v C_2 = \sigma'_v$. Using the transformation matrices from Eq. (2.5) and (2.6), we obtain

$$\begin{bmatrix} -1 & 0 & 0 \\ 0 & -1 & 0 \\ 0 & 0 & 1 \end{bmatrix}\begin{bmatrix} 1 & 0 & 0 \\ 0 & -1 & 0 \\ 0 & 0 & 1 \end{bmatrix} = \begin{bmatrix} -1 & 0 & 0 \\ 0 & 1 & 0 \\ 0 & 0 & 1 \end{bmatrix} \tag{2.9}$$

Again, the product matrix is the transformation matrix for the σ'_v operation, consistent with the result for the combination of operations. All other combinations give results consistent with those of the C_{2v} character table, a fact which you should be able to verify. Thus, the set of four transformation matrices satisfies the criterion for a representation of C_{2v}.

We can list the four transformation matrices as characters of a representation in a tabular form, as we did for the irreducible representations. As such, our representation for the transformation of the vector **v**, which in full matrix form we shall designate Γ_m, would appear as follows:

C_{2v}	E	C_2	σ_v	σ'_v
Γ_m	$\begin{bmatrix} 1 & 0 & 0 \\ 0 & 1 & 0 \\ 0 & 0 & 1 \end{bmatrix}$	$\begin{bmatrix} -1 & 0 & 0 \\ 0 & -1 & 0 \\ 0 & 0 & 1 \end{bmatrix}$	$\begin{bmatrix} 1 & 0 & 0 \\ 0 & -1 & 0 \\ 0 & 0 & 1 \end{bmatrix}$	$\begin{bmatrix} -1 & 0 & 0 \\ 0 & 1 & 0 \\ 0 & 0 & 1 \end{bmatrix}$

It is apparent that this representation is not one of the four irreducible representations of the group (cf. Table 2.2). Rather, as we shall see, Γ_m is a *reducible representation* that can be broken up as the sum of certain irreducible representations.

In its present form, Γ_m is extremely unwieldy. We can recast it in a more convenient form by noting a common property of all the transformation matrices of which it is composed. If we examine the diagonal of each matrix running from upper left to lower right, we see that the elements in succession express how x, y, and z, respectively, are transformed into themselves or the negative of themselves. We can combine this information into one character for each operation by taking the sum along the diagonal of each transformation matrix, known as the *trace* or *character of the matrix*. Doing this, we obtain a different form of the representation Γ_m, which we shall call Γ_v, the representation of characters:

C_{2v}	E	C_2	σ_v	σ'_v
Γ_v	3	-1	1	1

Being composed of single digits, this representation looks more like the irreducible representations we have associated with the unit vectors **x**, **y**, and **z**. However, the character 3 under the E operation makes it apparent that Γ_v is not an irreducible representation of C_{2v}. Rather, Γ_v is a reducible representation, which is the sum of three of the irreducible representations of C_{2v}. With a little trial and error we could deduce that the irreducible representations A_1, B_1, and B_2 uniquely add to give the representation Γ_v; that is, $\Gamma_v = A_1 + B_1 + B_2$. We can see this by adding up the characters for each operation of the three irreducible representations:

C_{2v}	E	C_2	σ_v	σ_v'
A_1	1	1	1	1
B_1	1	−1	1	−1
B_2	1	−1	−1	1
Γ_v	3	−1	1	1

Relating this result to the vector **v**, which forms the basis for the reducible representation, we see that the symmetry of the vector is a composite of three irreducible representations (symmetry species), which represent the symmetries of more fundamental vectors. Note that the three irreducible representations of which Γ_v is composed are the three symmetry species by which the unit vectors **z** (A_1), **x** (B_1), and **y** (B_2) transform. This result makes sense when we consider that any vector in three-dimensional space can be resolved into components along the three axes of a Cartesian coordinate system.

Although it is more convenient to deal with a representation of characters, such as Γ_v, the same reduction into irreducible representations can be seen from a full matrix representation, such as Γ_m. As we have noted, each diagonal element, c_{ii}, expresses how one of the coordinates x, y, or z is transformed by the operation. If we mark off a series of 1×1 matrices along the diagonals of all the 3×3 matrices, so as to exclude all zero elements (a process called *block diagonalization*), we can isolate all the c_{ii} elements. Each c_{11} element expresses the transformation of the x coordinate, each c_{22} element expresses the transformation of the y coordinate, and each c_{33} element expresses the transformation of the z coordinate for each operation. Our matrix representation Γ_m after block diagonalization would look like the following:

C_{2v}	E	C_2	σ_v	σ_v'
Γ_m	$\begin{bmatrix} 1 & 0 & 0 \\ 0 & 1 & 0 \\ 0 & 0 & 1 \end{bmatrix}$	$\begin{bmatrix} -1 & 0 & 0 \\ 0 & -1 & 0 \\ 0 & 0 & 1 \end{bmatrix}$	$\begin{bmatrix} 1 & 0 & 0 \\ 0 & -1 & 0 \\ 0 & 0 & 1 \end{bmatrix}$	$\begin{bmatrix} -1 & 0 & 0 \\ 0 & 1 & 0 \\ 0 & 0 & 1 \end{bmatrix}$

Reading across the first row (c_{11} components), we see that the set of 1×1 matrices is the same as the set of characters of the B_1 irreducible representation, by which **x** transforms. Likewise, the second and third rows correspond, respectively, to B_2, the symmetry species of **y**, and A_1, the symmetry species of **z**. Thus we see by block diagonalization that Γ_m reduces into the same three irreducible representations as does Γ_v.

In a representation of matrices, such as Γ_m, the *dimension of the representation* is the order of the square matrices of which it is composed. For a representation of characters, such as Γ_v, the dimension is the value of the character for the identity operation. This occurs because all c_{ii} elements are $+1$ for the identity operation in all representations. By either definition, we see that the dimension of the reducible representation for the transformation of the vector $\mathbf{v}$ in C_{2v} is 3. The dimension of the reducible representation must equal the sum of the dimensions of all the irreducible representations of which it is composed. In general, for a reducible representation with a dimension d_r we obtain

$$d_r = \sum_i n_i d_i \qquad (2.10)$$

where d_i is the dimension of each irreducible representation, and n_i is the number of times each irreducible representation contributes to the reducible representation. For our reducible representation for the vector $\mathbf{v}$, the three contributing irreducible representations each have a dimension of 1 ($d_i = 1$), and each contributes only once ($n_i = 1$) to the total reducible representation.

When we apply group theory to molecular problems, we customarily generate a reducible representation that reflects the symmetry characteristics of the property of interest. The subsequent reduction of this reducible representation into its component irreducible representations will be an important step in developing solutions to the problem. In simple cases, such as the example we have just seen, the reduction can be accomplished by inspection or trial-and-error techniques. This is most often the case when the dimension of the reducible representation is small. For representations with larger dimensions (generally, $d_r \geq 4$), a systematic method of determining the component irreducible representations is more efficient. Before discussing this technique (cf. Section 3.1), we need to become familiar with some additional features of representations and character tables, especially those for more complicated groups than C_{2v}.

2.4 More Complex Groups and Standard Character Tables

Appendix A shows character tables for all point groups that one is likely to encounter in dealing with problems of real molecules. Some of these tables have features we have not encountered in our examination of C_{2v}. A case in point is the character table for C_{3v}, Table 2.4. The group C_{3v} describes, for example, the symmetry of ammonia and other such pyramidal MX_3 species. It consists of six operations ($h = 6$) grouped into three classes. There can be only one identity in any group, so, as always, E stands in a class of its own. The notation $2C_3$ in the character table stands for C_3 and C_3^2, which together comprise a class. Similarly, the notation $3\sigma_v$ indicates that the three vertical

Table 2.4 Character Table for the Point Group C_{3v}

C_{3v}	E	$2C_3$	$3\sigma_v$		
A_1	1	1	1	z	$x^2 + y^2, z^2$
A_2	1	1	-1	R_z	
E	2	-1	0	$(x, y)\,(R_x, R_y)$	$(x^2 - y^2, xy)(xz, yz)$

mirror planes, which intersect at the C_3 axis with a 120° angle between them, form a class of their own. From a geometrical point of view, operations in the same class can be converted into one another by changing the axis system through application of some symmetry operation of the group. In the case of the group C_{3v}, the operations of C_3 and C_3^2 are interconverted by one of the vertical mirror planes. Likewise, the three mirror planes are seen to form a class because they are interconverted by C_3 and C_3^2.

A mathematically more general definition of class requires defining an equality called the *similarity transform*. The elements A and B belong to the same class if there is an element X within the group such that $X^{-1}AX = B$, where X^{-1} is the inverse of X (i.e., $XX^{-1} = X^{-1}X = E$). If $X^{-1}AX = B$, we say that B is the similarity transform of A by X, or that A and B are *conjugate* to one another. We should note that the element X may in some cases be the same as either A or B.

In C_{3v}, and indeed in any group, E forms a class by itself, because all similarity transforms result in E. With itself, the similarity transform $EEE = E$. With all other elements of the group the similarity transforms are $X^{-1}EX = X^{-1}X = E$. To verify the other class groupings ($2C_3$ and $3\sigma_v$) we will need a multiplication table for the group C_{3v}. To avoid ambiguity, we define C_3 and C_3^2 in a clockwise direction, and we arbitrarily label the three σ_v planes as shown in Fig. 2.4. Following the convention that an operation at the top is performed before an operation at the side, we obtain Table 2.5, the multiplication table for C_{3v}. We note that C_3 and C_3^2 are the inverses of each other,

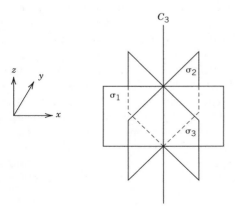

Figure 2.4 Defining the orientations of the mirror planes of C_{3v}.

Table 2.5 Multiplication Table for the Group C_{3v}

C_{3v}	E	C_3	C_3^2	σ_1	σ_2	σ_3
E	E	C_3	C_3^2	σ_1	σ_2	σ_3
C_3	C_3	C_3^2	E	σ_3	σ_1	σ_2
C_3^2	C_3^2	E	C_3	σ_2	σ_3	σ_1
σ_1	σ_1	σ_2	σ_3	E	C_3	C_3^2
σ_2	σ_2	σ_3	σ_1	C_3^2	E	C_3
σ_3	σ_3	σ_1	σ_2	C_3	C_3^2	E

since $C_3C_3^2 = C_3^2C_3 = E$. All other elements are their own inverses. Furthermore, we can see that the order of performing operations is important, because certain pairs of operations do not commute. For example, in right-to-left notation, $\sigma_1\sigma_3 = C_3^2$, but $\sigma_3\sigma_1 = C_3$. Thus, C_{3v} is not Abelian.

Let us now take all the similarity transforms on C_3 to discover what elements are in the same class with it.

$$EC_3E = C_3$$

$$C_3^2C_3C_3 = C_3^2C_3^2 = C_3$$

$$C_3C_3C_3^2 = C_3E = C_3$$

$$\sigma_1C_3\sigma_1 = \sigma_1\sigma_3 = C_3^2$$

$$\sigma_2C_3\sigma_2 = \sigma_2\sigma_1 = C_3^2$$

$$\sigma_3C_3\sigma_3 = \sigma_3\sigma_2 = C_3^2$$

We see that all similarity transforms generate either C_3 or C_3^2, which means that these two operations are members of the same class. The same general result would have been obtained had we taken all the similarity transforms on C_3^2 instead of C_3. You should be able to verify this and also verify in a similar manner that the three mirror planes belong to a class of their own.

It is a general relationship of group theory (cf. Section 2.5) that the number of classes equals the number of irreducible representations of the group. In the case of C_{3v}, we see from the character table (Table 2.4) that there are three irreducible representations, consistent with the three classes of the group. One of these, labeled with the Mulliken symbol E (which should not be confused with the identity element), has a dimension of 2 ($d_i = 2$), as evidenced by the character of 2 for the identity operation. The irreducible representation E is a *doubly degenerate representation*. Doubly degenerate and even *triply degenerate representations* are found in many chemically important point groups with principal axes greater than twofold.* The group D_{2d} is a notable exception, being a group with a twofold principal axis and yet having a doubly degenerate irreducible representation (see the character table for D_{2d} in Appendix A). As an example of triply degenerate irreducible representations ($d_i = 3$), note the representations with the Mulliken symbols T_1 and T_2 of the point group T_d, shown in Appendix A. In all these cases, it is important to understand that these are irreducible representations and not reducible representations.

We can show the need to have a doubly degenerate irreducible representation for the point group C_{3v} by examining the transformation properties of a general vector $\mathbf{v}$ under the operations of the group, much in the same way we did for the group C_{2v} (Section 2.3). We can simplify the process in this case by recognizing that the projection of $\mathbf{v}$ on the z axis is unaffected by any of the symmetry operations of the group. Consequently, in the 3×3 transformation matrices for each operation, the matrix elements c_{31} and c_{32} will always be 0, and the element c_{33} will always be 1. Furthermore, all operations affect x and y independently of z, so in each transformation the matrix elements c_{13} and c_{23} will always be 0. Thus each operation involves a matrix multiplication of the general form

$$
\begin{bmatrix} ? & ? & 0 \\ ? & ? & 0 \\ 0 & 0 & 1 \end{bmatrix} \begin{bmatrix} x \\ y \\ z \end{bmatrix} = \begin{bmatrix} ? \\ ? \\ z \end{bmatrix} \tag{2.11}
$$

With the z component's transformation properties identified, we can turn our attention to the x and y components by looking at the projection of $\mathbf{v}$ in the xy plane. This will give us the remaining elements in each 3×3 transformation matrix.

In general, the transformation matrix for any operation in a class has a form that is unique from the matrices of the other members of the class. However, for any representation, whether irreducible or reducible, the character of the transformation matrix for a given operation is the same as that for any other operation in the same class. In the case of C_{3v}, the characters for C_3 and C_3^2 are the same, and the characters for σ_1, σ_2, and σ_3 are the same for any representation of the group. This permits, for example, listing one character for each class of operations for each irreducible representation in the character table. For our consideration of the transformations of the general vector $\mathbf{v}$ in C_{3v}, this means we need only construct the matrix for one rep-

---

*The groups I and I_h have irreducible representations with dimensions up to $d_i = 5$. Also, the full rotational group R_3, which describes the symmetry of a sphere, has no limits on the dimensions of its irreducible representations. These are not important groups for considerations of molecular structures one is likely to encounter. However, the group R_3 has considerable theoretical importance, as we shall see in Chapter 7.

resentative operation in each class in order to obtain a set of characters for the reducible representation. Furthermore, we are at liberty to choose any convenient operation among those comprising the class. In the class composed of C_3 and C_3^2, we will choose the simpler of the two, the C_3 rotation. Consistent with normal trigonometric practice, we will in this instance define the C_3 rotation in a counterclockwise direction. The effect of the C_3 rotation on the projection of the vector **v** in the xy plane is shown in Fig. 2.5. In the class of three vertical mirror planes ($3\sigma_v$), it will simplify our considerations to assume that our chosen mirror plane lies in the xz plane of the coordinate system (σ_1 in Fig. 2.4). The other two mirror planes (which we are ignoring) lie at 120° and 240° from the chosen plane.

For the identity operation, we obtain the same transformation matrix that would be found for a vector **v** in any point group [cf. Eq. (2.4)]:

$$
\begin{bmatrix} 1 & 0 & 0 \\ 0 & 1 & 0 \\ 0 & 0 & 1 \end{bmatrix}
\begin{bmatrix} x \\ y \\ z \end{bmatrix}
=
\begin{bmatrix} x \\ y \\ z \end{bmatrix}
\tag{2.12}
$$

For σ_v, since we have defined our chosen plane as σ_{xz}, the x coordinate will be unaffected, and the y coordinate will be changed into the negative of itself. This gives the following matrix equation:

$$
\begin{bmatrix} 1 & 0 & 0 \\ 0 & -1 & 0 \\ 0 & 0 & 1 \end{bmatrix}
\begin{bmatrix} x \\ y \\ z \end{bmatrix}
=
\begin{bmatrix} x \\ -y \\ z \end{bmatrix}
\tag{2.13}
$$

The transformation by C_3 is less straightforward. As can be seen from Fig. 2.5, the new coordinates of the projection vector's tip, the point (x', y'), cannot be described solely in terms of x or y, independently of each other. Rather, each new coordinate requires an expression in both x and y. From trigonometry we can write

$$
x' = \cos \frac{2\pi}{3} x - \sin \frac{2\pi}{3} y = -\frac{1}{2} x - \frac{\sqrt{3}}{2} y
\tag{2.14}
$$

$$
y' = \sin \frac{2\pi}{3} x + \cos \frac{2\pi}{3} y = \frac{\sqrt{3}}{2} x - \frac{1}{2} y
\tag{2.15}
$$

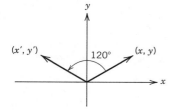

Figure 2.5 Projection of **v** in the xy plane and the effect of C_3 in a counterclockwise direction.

The need to have both x and y terms to define either x' or y' causes the transformation matrix to have off-diagonal elements that are nonzero:

$$
\begin{bmatrix} -\dfrac{1}{2} & -\dfrac{\sqrt{3}}{2} & 0 \\ \dfrac{\sqrt{3}}{2} & -\dfrac{1}{2} & 0 \\ 0 & 0 & 1 \end{bmatrix} \begin{bmatrix} x \\ y \\ z \end{bmatrix} = \begin{bmatrix} \left(-\dfrac{x}{2} - \dfrac{\sqrt{3}y}{2}\right) \\ \left(\dfrac{\sqrt{3}x}{2} - \dfrac{y}{2}\right) \\ z \end{bmatrix} = \begin{bmatrix} x' \\ y' \\ z' \end{bmatrix} \tag{2.16}
$$

Equation (2.16) is a specific case of the transformation of a point (x, y, z) through a counterclockwise rotation about the z axis through an angle $\theta = 120°$. In general, for a counterclockwise rotation about z through any angle θ we may write

$$
\begin{bmatrix} \cos \theta & -\sin \theta & 0 \\ \sin \theta & \cos \theta & 0 \\ 0 & 0 & 1 \end{bmatrix} \begin{bmatrix} x \\ y \\ z \end{bmatrix} = \begin{bmatrix} x' \\ y' \\ z' \end{bmatrix} \tag{2.17}
$$

If the rotation is taken in the clockwise sense, the signs on the $\sin \theta$ terms will be reversed in Eq. (2.17).

Gathering together our three 3×3 transformation matrices from Eq. (2.12), (2.13), and (2.16), we obtain the following*:

C_{3v}	E	C_3	σ_v
Γ_m	$\begin{bmatrix} 1 & 0 & 0 \\ 0 & 1 & 0 \\ 0 & 0 & 1 \end{bmatrix}$	$\begin{bmatrix} -1/2 & -\sqrt{3}/2 & 0 \\ \sqrt{3}/2 & -1/2 & 0 \\ 0 & 0 & 1 \end{bmatrix}$	$\begin{bmatrix} 1 & 0 & 0 \\ 0 & -1 & 0 \\ 0 & 0 & 1 \end{bmatrix}$

We can reduce this into its component irreducible representations by taking block diagonals of each matrix. The blocks we take must be the same size across all three matrices. The presence of nonzero, off-diagonal elements in the transformation matrix for C_3 restricts us to diagonalization into a 2×2 block and a 1×1 block. As a result, for all three matrices we must adopt a scheme of block diagonalization that yields one set of 2×2 matrices and another set of 1×1 matrices, as follows:

C_{3v}	E	C_3	σ_v
Γ_m	$\begin{array}{cc|c} 1 & 0 & 0 \\ 0 & 1 & 0 \\ \hline 0 & 0 & 1 \end{array}$	$\begin{array}{cc|c} -1/2 & -\sqrt{3}/2 & 0 \\ \sqrt{3}/2 & -1/2 & 0 \\ \hline 0 & 0 & 1 \end{array}$	$\begin{array}{cc|c} 1 & 0 & 0 \\ 0 & -1 & 0 \\ \hline 0 & 0 & 1 \end{array}$

*The full representation would also show a matrix for C_3^2 and two matrices for the other σ_v planes. Our goal is to obtain representations of characters. Therefore, we can ignore these missing matrices, since their characters will be the same as those of the representative operations we have chosen from their classes.

Our inability to reduce this matrix representation into three irreducible representations of 1×1 matrices, equivalent to characters, is a consequence of the mixing of x and y by the operation C_3.

We can still obtain irreducible representations of characters by summing along the traces (upper left to lower right diagonals) of the block matrices in Γ_m. This gives us the following two representations:

C_{3v}	E	$2C_3$	$3\sigma_v$
$\Gamma_{x,y}$	2	-1	0
Γ_z	1	1	1

The first of these, $\Gamma_{x,y}$, is the doubly degenerate irreducible representation E of C_{3v} (cf. Table 2.4). In the third column of the character table, where the transformations of linear and rotational vectors are listed, we see a notation (x, y) along the row for the E representation. This indicates that the unit vectors **x** and **y** transform as a degenerate pair. This means that in C_{3v} there is no difference in symmetry between the x and y directions, and they may be treated as equivalent and indistinguishable. Thus, if a molecule with C_{3v} symmetry possesses a property along x there will be an equivalent and indistinguishable property along y, which means they are *degenerate*. For example, in a molecule with C_{3v} symmetry, p_x and p_y orbitals on a central atom are required by symmetry to have the same energy and be indistinguishable; that is, they are a degenerate pair.

The second representation, Γ_z, is the totally symmetric representation of C_{3v}, given the standard Mulliken symbol A_1 (cf. Table 2.4). The notation z in the third column of the character table indicates that a unit vector **z** would transform by the totally symmetric representation in C_{3v}. The fact that z transforms by a different species than the degenerate pair (x, y) means that any property along z will be unique from properties along x and y. For example, we have noted that p_x and p_y orbitals on a central atom in a molecule with C_{3v} symmetry would be degenerate. Since properties along z transform by a different species, a p_z orbital in such a molecule would be expected to have a different energy and be distinguishable from the degenerate pair of p_x and p_y orbitals. Unlike a free atom, the three p orbitals would not be fundamentally degenerate in such a system and would have the same energies only by a fortuitous combination of circumstances, a condition known as *accidental degeneracy*. In other words, placing an atom in a C_{3v} environment lifts the threefold degeneracy of the p orbitals.

The third column of the character table for C_{3v} also indicates the species by which the rotational vectors transform. As the character table indicates, $\mathbf{R}_z$ transforms as A_2, and $\mathbf{R}_x$ and $\mathbf{R}_y$ transform together as a degenerate pair by E, the doubly degenerate representation.

Throughout the character tables (cf. Appendix A), the use of parentheses around two or three vectors means that they transform as a degenerate set. When two vectors are so grouped, the indicated pair transforms as a doubly degenerate representation (e.g., x and y in C_{3v}). When three vectors are included, they transform as a triply degenerate representation. For example, x, y, and z transform degenerately as T_2 in the group T_d (see the character table in Appendix A). Sometimes two or more vectors transform by the same species but are not degenerate. In such cases, the vectors are not grouped within parentheses. For example, in the group C_{2h} both x and y transform as the nondegenerate species B_u (see the character table in Appendix A). The fact that B_u has a dimension of 1 ($d_i = 1$) precludes two properties from being degenerately transformed by that symmetry species. In C_{2h}, the transformation of x is independent of y and vice versa, but both have the same symmetry properties as the B_u irreducible representation.

The last column of a character table lists the transformation properties of the binary direct products of linear vectors. These are important because, among other things, they correspond to the transformation properties of d orbitals. Thus, a notation such as z^2, $x^2 - y^2$, xy, xz, or yz can be taken as indicating the species by which the d orbital of the same designation transforms. In this connection, a notation $2z^2 - x^2 - y^2$, such as listed for the species E_g in the point group O_h, can be taken as indicating the transformation property of a d_{z^2} orbital (cf. O_h character table in Appendix A). Conversely, notations such as $x^2 + y^2 + z^2$ (e.g., A_{1g} in O_h) or $x^2 + y^2$ (e.g., A_1 in C_{3v}) do not correspond to d orbital transformations in the usual formulation of the orbitals. These notations occur because in some point groups the direct products for certain pairs of vectors are spread across two representations. We will examine direct products in more detail in Section 3.5.

You may have noticed in perusing the character tables in Appendix A that some groups have representations that use the imaginary integer $i = \sqrt{-1}$, or the symbols $\epsilon = \exp(2\pi i/n)$ and $\epsilon^* = \exp(-2\pi i/n)$, where n is the order of the principal axis. For example, the C_n groups with $n \geq 3$ show these kinds of characters (cf. Appendix A). Other groups with such characters include the C_{nh} groups with $n \geq 3$; the improper axis groups, S_{2n}; and two of the cubic groups, T and T_h. All these groups contain one or more pairs of irreducible representations that are complex conjugates of one another. The paired representations appear on successive lines in the character tables, joined by braces ({ }). Each pair is given the single Mulliken symbol of a doubly degenerate representation (e.g., E, E_1, E_2, E', E'', E_g, E_u). Despite this symbolism, each of the paired complex-conjugate representations is an irreducible representation in its own right. They arise in these groups because of the fundamental theorem of group theory (discussed in Section 2.5) that requires the number of representations in any group to be equal to the number of classes in the group.

When dealing with applications in one of the groups with imaginary characters, it is sometimes convenient to add the two complex-conjugate representations to obtain a representation of real characters. When the paired representations have i and $-i$ characters, the addition is straightforward; that is, $i + (-i) = 0$. When they have ϵ and ϵ^* characters, where $\epsilon = \exp(2\pi i/n)$, the following identities are used in taking the sum:

$$\epsilon^p = \exp(2\pi p i/n) = \cos 2\pi p/n + i \sin 2\pi p/n \tag{2.18}$$

$$\epsilon^{*p} = \exp(-2\pi p i/n) = \cos 2\pi p/n - i \sin 2\pi p/n \tag{2.19}$$

Combining Eqs. (2.18) and (2.19), we have

$$\epsilon^p + \epsilon^{*p} = 2 \cos 2\pi p/n \tag{2.20}$$

Thus all complex-conjugate characters in the two irreducible representations add to give real-number characters.

For example, the E representation of C_3 appears in the character table as follows:

C_3	E	C_3	C_3^2
E	$\begin{Bmatrix} 1 \\ 1 \end{Bmatrix}$	$\begin{matrix} \epsilon \\ \epsilon^* \end{matrix}$	$\begin{matrix} \epsilon^* \\ \epsilon \end{matrix}$

where $\epsilon = \exp(2\pi i/3)$. Using Eq. (2.20), the sum of the complex-conjugate imaginary characters is $\epsilon + \epsilon^* = 2 \cos 2\pi/3$. Using this result when adding the complex-conjugate irreducible representations gives

C_3	E	C_3	C_3^2
$\{E\}$	2	$2 \cos 2\pi/3$	$2 \cos 2\pi/3$

This combined representation, composed of all real-number characters, can be used instead of the two imaginary-character representations in various applications. However, as a sum of two genuine irreducible representations, the real-number representation is not an irreducible representation. Rather, it is a reducible representation and must be handled as such. To emphasize the distinction, a real-character reducible representation formed by summing complex-conjugate irreducible representations will be designated in this text by surrounding the Mulliken symbol of the pair in braces; e.g., $\{E\}$.

Having examined irreducible representations of a variety of groups, we now can describe the meaning of the Mulliken symbols. For groups of finite order ($h \neq \infty$), nondegenerate representations ($d_i = 1$) are labeled A or B. As previously noted, doubly degenerate irreducible representations ($d_i = 2$) are designated by E, and triply degenerate irreducible representations ($d_i = 3$) are designated by T. In older texts you may find triply degenerate irreducible representations designated by F, a notation that is no longer used.

An A representation is symmetric with respect to the principal rotation (C_n of highest order in the system). As a result, in the character table of the group any A representation always has a $+1$ character for the principal axis

C_n operation. This means that a property that transforms as an A representation will retain its sense of sign after the C_n operation. For example, in C_{2v} the unit vector **z** transforms as A_1, because its sense remains unchanged by the C_2 operation. A B representation is antisymmetric with respect to the principal rotation, which is indicated by the character -1 under the principal axis C_n operation in the character table of the group. In this case, a property that transforms as a B representation will have a change of sign as a result of the C_n operation. For example, in C_{2v} the unit vectors **x** and **y** transform as B_1 and B_2, respectively, because their directional senses are reversed by C_2.

The infinite-order point groups $C_{\infty v}$ and $D_{\infty h}$, to which linear molecules belong, use a Greek letter notation for the Mulliken symbols of the irreducible representations. Nondegenerate representations are designated by Σ. The doubly degenerate representations are given the symbols Π, Δ, and Φ.

Any of these primary symbols may be modified by subscript or superscript notations. These modifying notations indicate symmetry or antisymmetry with respect to some symmetry operation other than the principal rotation. In the case of nondegenerate representations, this symmetry or antisymmetry results in a character of $+1$ or -1, respectively, for the referenced operation. In centrosymmetric groups, a subscript g (for German *gerade* = even) indicates symmetry and a subscript u (for German *ungerade* = uneven) indicates antisymmetry with respect to inversion (i). For finite-order groups, subscripts 1 and 2 respectively indicate symmetry and antisymmetry with respect to a nonprincipal rotation (lower order than the principal axis) or to a vertical mirror plane (σ_v). For nondegenerate representations of the infinite-order groups $C_{\infty v}$ and $D_{\infty h}$, superscript $+$ and $-$ have the same meanings as do 1 and 2 in the finite-order groups. Addition of prime ($'$) or double prime ($''$) to a primary symbol indicates symmetry or antisymmetry with respect to a horizontal mirror plane (σ_h).

Regardless of the group, the first-listed irreducible representation in the character table is the totally symmetric representation of the group. The totally symmetric representation is always composed of $+1$ characters for all operations of the group. Depending on the group, the Mulliken symbol for the totally symmetric representation will be A, A_1, A', A_1', A_g, A_{1g}, Σ^+, or Σ_g^+.

2.5   *General Relationships of Irreducible Representations*

As you will appreciate, our development of the concept of representations in the foregoing sections of this chapter has been by design mathematically nonrigorous. In a more mathematically respectable development we would invoke the strictures of the Great Orthogonality Theorem to generate the limited set of irreducible representations that are allowed for each group. A presentation of the theorem itself is unnecessary for our purposes, but is readily available in more advanced treatments.* Nonetheless, the Great Orthogonality Theorem results in a number of important relationships among the characters of representations and among the representations themselves. Some of these have practical consequences. Therefore, we will simply state without proof some of the general relationships arising from the Great Orthogonality Theorem and seek to illustrate them with specific examples. You will note that we have already used some of these.

1. The sum of the squares of the dimensions of all the irreducible representations is equal to the order of the group; that is,

$$\sum_i d_i^2 = h \tag{2.21}$$

where d_i is the dimension of the ith irreducible representation and h is the order of the group.

The character for the operation E in the ith irreducible representation, $\chi_i(E)$, is equal to the order of the representation. Therefore, in any group we can verify this rule simply by squaring the characters for E and summing over all representations; that is,

$$\sum_i [\chi_i(E)]^2 = h \tag{2.22}$$

For example, consider the partial character table for T_d, shown below.

T_d	E	$8C_3$	$3C_2$	$6S_4$	$6\sigma_d$	$\Rightarrow h = 24$
A_1	1	1	1	1	1	$\Rightarrow d_i = 1$
A_2	1	1	1	-1	-1	$\Rightarrow d_i = 1$
E	2	-1	2	0	0	$\Rightarrow d_i = 2$
T_1	3	0	-1	1	-1	$\Rightarrow d_i = 3$
T_2	3	0	-1	-1	1	$\Rightarrow d_i = 3$

Adding the operations in each class indicates that the group order is 24. Using the characters under the E operation for the irreducible representations as values of d_i, we obtain by Eq. (2.22)

$$\sum_i d_i^2 = \sum_i [\chi_i(E)]^2 = 1^2 + 1^2 + 2^2 + 3^2 + 3^2$$

$$= 1 + 1 + 4 + 9 + 9$$

$$= 24 = h$$

2. The number of irreducible representations of a group is equal to the number of classes.

We have seen this in our examination of character tables (cf. Section 2.4). In the partial character table for T_d above, we see that the 24 oper-

---

*A presentation of the theorem without proof may be found in F. A. Cotton, *Chemical Applications of Group Theory,* 3rd ed., John Wiley & Sons, New York, 1990, p. 81. Proof of the theorem may be found in texts such as H. Weyl, *The Theory of Groups and Quantum Mechanics,* Dover Publications, New York, 1950, 157; and H. Eyring, J. Walter, and G. E. Kimbal, *Quantum Chemistry,* John Wiley & Sons, New York, 1944, p. 371.

ations are grouped into five classes. Consequently, there are five irreducible representations: A_1, A_2, E, T_1, T_2.

3. In a given representation (irreducible or reducible) the characters for all operations belonging to the same class are the same.

This fact, which we have seen demonstrated for the group C_{3v} (cf. Section 2.4), permits great economy in writing the character tables and also simplifies generating reducible representations in various applications, as we shall see.

4. The sum of the squares of the characters in any irreducible representation equals the order of the group; that is,

$$\sum_R [\chi_i(R)]^2 = h \tag{2.23}$$

where R is any operation of the group and the summation is taken over all operations.

We can simplify Eq. (2.23) by using the fact that all operations in a class have the same character (point 3 above). A class of g_c operations R will contribute the same value of $\chi_i(R)$ to Eq. (2.23), a total of g_c times. Therefore we may write

$$\sum_{R_c} g_c[\chi_i(R_c)]^2 = h \tag{2.24}$$

where R_c is a class of operations R, and g_c is the order of the class R_c. The summation in Eq. (2.24) is taken over all classes of operations.

As an example, consider the T_2 representation of T_d, as listed in the character table above (point 1). Applying Eq. (2.24), we have

$$\sum_{R_c} g_c[\chi_{T_2}(R_c)]^2 = (3)^2 + 8(0)^2 + 3(-1)^2 + 6(-1)^2 + 6(-1)^2$$

$$= 9 + 0 + 3 + 6 + 6$$

$$= 24 = h$$

5. Any two different irreducible representations are orthogonal, which means

$$\sum_{R_c} g_c\chi_i(R_c)\chi_j(R_c) = 0 \tag{2.25}$$

where χ_i and χ_j are characters for the class of operations R_c of two different irreducible representations of the group (i.e., $i \neq j$).

Physically, this means two vectors that transform by different irreducible representations are orthogonal. In general, this can be stated mathematically for any two vectors, **a** and **b**, in p-dimensional space by the expression

$$\sum_{l=1}^{p} a_l b_l = 0 \tag{2.26}$$

where a_l and b_l are the projections of **a** and **b** in the lth dimension. Our considerations are limited to three-dimensional space, so the summation in Eq. (2.26) extends to $p = 3$.

As an example, consider the irreducible representations A_2 and E in the group T_d, as shown in the character table above (point 1). By Eq. (2.25) we obtain

$$\sum_{R_c} g_c\chi_i(R_c)\chi_j(R_c) = 1(1 \times 2) + 8(1 \times -1) + 3(1 \times 2) + 6(-1 \times 0) + 6(-1 \times 0)$$

$$= 2 - 8 + 6 + 0 + 0$$

$$= 0$$

Points 4 and 5, as represented by Eq. (2.24) and (2.25), can be combined into one equation if we introduce a function notation that is used frequently in quantum mechanics. The *Kronecker delta* function, δ_{ij}, is defined such that $\delta_{ij} = 0$ if $i \neq j$, and $\delta_{ij} = 1$ if $i = j$. Using the Kronecker delta, we can combine Eqs. (2.24) and (2.25) into the following expression:

$$\sum_{R_c} g_c\chi_i(R_c)\chi_j(R_c) = h\delta_{ij} \tag{2.27}$$

Problems

2.1   The operations of the group C_{2h} are E, C_2, i, and σ_h.

    (a) Without consulting the C_{2h} character table, determine the sets of characters comprising the irreducible representations by which the unit vectors **x**, **y**, and **z** transform in C_{2h}.

    (b) Do the same for the rotational vectors $\mathbf{R}_x$, $\mathbf{R}_y$, and $\mathbf{R}_z$.

2.2   Consider a general vector **v**, whose base is at $(0, 0, 0)$ and whose tip is at (x, y, z), in the point group C_{2h}.

    (a) Derive the set of four 3×3 transformation matrices that constitute the reducible representation, Γ_m, by which **v** transforms.

    (b) Reduce Γ_m into its component irreducible representations by block diagonalization.

    (c) Write the reducible representation of characters, Γ_v, that corresponds to the matrix representation, Γ_m.

    (d) Show that Γ_v reduces to the same irreducible representations as Γ_m.

    (e) Show that the four transformation matrices comprising Γ_m obey the same combinational relationships as the operations of C_{2h}. [*Hint:* You will need to work out the multiplication table for C_{2h}.]

2.3   Consider the three p orbitals p_x, p_y, and p_z, which are degenerate for an isolated atom M. If M is surrounded by several X atoms, the electrostatic field they create may lift the degeneracy among the p orbitals. By consulting the appropriate character table, describe the degree of degeneracy among p orbitals allowed by symmetry for each of the following structures: (a) MX_2, linear; (b) MX_2, bent;

(c) MX_3, trigonal planar; (d) MX_3, pyramidal; (e) MX_3, T-shaped (as in ClF_3); (f) MX_4, tetrahedral; (g) MX_4, square planar; (h) MX_4, irregular tetrahedral (as in SbF_4^-); (i) MX_5, square pyramidal; (j) MX_5, trigonal bipyramidal; (k) MX_6, octahedral.

2.4 Using the C_{3v} multiplication table (Table 2.5), verify that the three σ_v planes belong to the same class.

2.5 In C_{3v} both C_3 and C_3^2 belong to the same class, listed as $2C_3$ in the character table. As members of the same class their characters for any representation are the same.

(a) Demonstrate that both C_3 and C_3^2 have a character of 1 for the A_1 representation, by which z transforms.

(b) Demonstrate that both C_3 and C_3^2 have a character of -1 for the E representation, by which x and y transform degenerately. [*Hint:* Write the 2×2 transformation matrices describing the actions of the operations on a point (x, y) and determine their characters.]

2.6 Describe the implied symmetry of the following irreducible representations on the basis of their Mulliken symbols: (a) A_g in C_{2h}, (b) B_2 in C_{4v}, (c) E in D_3, (d) A_1'' in D_{3h}, (e) E' in D_{3h}, (f) B_{1g} in D_{4h}, (g) E_u in D_{4h}, (h) T_g in T_h.

2.7 Construct real-number representations by combining the complex-conjugate paired irreducible representations in the following point groups: (a) C_4, (b) C_6, (c) C_5, (d) C_7.

2.8 Fill in the missing characters in the character table below, which is presented in standard format. The symbols A, B, C, and D represent certain symmetry operations, and E is identity.

	E	$2A$	B	$2C$	$2D$
Γ_1					
Γ_2	1		1	-1	-1
Γ_3	1	-1	1	1	-1
Γ_4	1	-1	1		1
Γ_5		0		0	0

2.9 For the point group O (cf. Appendix A), show that the group conforms to the five generalizations from the Great Orthogonality Theorem presented in Section 2.5.

2.10 Consider the point group C_3, consisting of the operations E, C_3, and C_3^2.

(a) Write the three transformation matrices for a general vector $\mathbf{v}$ under the operations of the group, thereby forming a reducible representation Γ_v.

(b) By block diagonalization, reduce Γ_v into two representations, Γ_z and $\Gamma_{x,y}$, by which the unit vector $\mathbf{z}$ and the pair of unit vectors $\mathbf{x}$ and $\mathbf{y}$ transform, respectively.

(c) Rewrite $\Gamma_{x,y}$ as a representation of characters, and show that it is equivalent to the complex-conjugate pair of irreducible representations designated E in the C_3 character table.

(d) Explain why the representation E in the group C_3 must be a pair of complex-conjugate irreducible representations.

CHAPTER 3

Techniques and Relationships for Chemical Applications

In this chapter we will consider some important techniques and relationships that are frequently employed in chemical applications of group theory. Chief among these is the systematic reduction of reducible representations, which is necessary for nearly all applications. Other topics considered in this chapter are important in certain circumstances, which one will encounter with less regularity. Those wishing to progress immediately to applications can profitably study Section 3.1 and then skip to the pertinent material in Chapters 4 through 7, returning to sections of Chapter 3 as needed. Those preferring a more complete grounding before considering chemical applications may wish to study all of this chapter first.

3.1 Systematic Reduction of Reducible Representations

We noted in Chapter 2 that applying group theory to chemical problems generally involves constructing a set of vectors on a molecule's atoms to represent a particular property. The set of vectors is said to form a basis for a representation in the point group of the molecule. The representation itself is generated by subjecting the vector basis to all the operations of the group, much in the manner that we carried out for the single general vector $\mathbf{v}$ in C_{2v} in Section 2.3. The dimension of the reducible representation, d_r, is proportional to the number of vectors in the basis set, which may be quite large for a complex molecule. As Eq. (2.10) shows, the sum of dimensions of the component reducible representations, d_i, equals d_r. Therefore, for a large dimension representation, the component irreducible representations and the number of times each contributes may be difficult to ascertain by inspection. Fortunately, reducing a representation into its component irreducible representations can be accomplished systematically.

For all groups of finite order, we can accomplish the reduction of a representation by applying the equation

$$n_i = \frac{1}{h} \sum_c g_c \chi_i \chi_r \tag{3.1}$$

where n_i is the number of times the irreducible representation i occurs in the reducible representation; h is the order of the group; c is the class of opera-

66

tions; g_c is the number of operations in the class; χ_i is the character of the irreducible representation for the operations of the class; and χ_r is the character for the reducible representation for the operations of the class. Consistent with Eq. (2.10), the sum of the products of the dimensions of the irreducible representations multiplied by the number of times they contribute, $n_i d_i$, must equal the dimension of the reducible representation, d_r.

In applying Eq. (3.1) to decompose a reducible representation, Γ_r, we are in effect asking the following question for each irreducible representation of the group: "How many times does this species contribute to the collection that adds up to Γ_r?" The answer in each case, n_i, may be zero (in which case the irreducible representation is not a part of Γ_r) or any integer for which $n_i \leq d_r/d_i$, consistent with Eq. (2.10). In principle, we successively apply Eq. (3.1) for each and every irreducible representation of the group, starting with the totally symmetric representation and working through all the others one by one. Actually, we can stop at any point where we have found a sufficient number of irreducible representations to account for the dimension d_r of Γ_r. To find n_i for a particular irreducible representation, we proceed class by class to multiply together the character of the irreducible representation for the class, the character for the reducible representation for the class, and the number of operations in the class, to give the product $g_c \chi_i \chi_r$. Then we sum all these products for all the classes and divide by the order of the group, h.

We could write out all the terms of Equation (3.1) for each irreducible representation in the conventional mathematical manner. However, this process can become quite cumbersome, particularly if the group has many classes. For example, the group O_h, the point group of octahedral species, consists of 10 classes and consequently 10 irreducible representations. Equation (3.1) for each irreducible representation will consist of 10 terms of the product $g_c \chi_i \chi_r$. Since there are 10 irreducible representations, the whole process involves writing 100 such terms. Needless to say, all this writing of mathematical terms invites error. Fortunately, there is a better way of organizing the work, called the *tabular method*.*

To illustrate the technique, consider the following reducible representation of the point group T_d:

T_d	E	$8C_3$	$3C_2$	$6S_4$	$6\sigma_d$
Γ_r	8	-1	4	-2	0

This trial representation has no particular meaning and has been constructed simply for illustrative purposes. In actual applications, a reducible representation would be developed by subjecting a set of vectors for the property of interest (a basis set) to the operations of the group. With the reducible representation in hand, one would then proceed to decompose it into its component species by Eq. (3.1).

*R. L. Carter, *J. Chem. Educ.* **1991,** *68,* 373–374.

To reduce our trial representation, we will need a character table for the group, which is shown in Table 3.1. The last two columns of the character table have been omitted from Table 3.1 (cf. the character table for T_d in Appendix A), since we do not need information on vector transformation properties to reduce Γ_r. Adding up the numbers of operations in all the classes, we note that the order of the group is $h = 24$.

The arithmetic of applying Eq. (3.1) will be recorded on a tabular work sheet with the form shown in Table 3.2. We will fill in the large central portion of Table 3.2 with the various products $g_c\chi_i\chi_r$ along the rows for the irreducible representations. Then we will sum across each row to obtain Σ values, and finally divide each sum by the group order, h, which for T_d is 24. By this procedure, we can generate the complete work sheet shown in Table 3.3.

Table 3.1   Partial Character Table of T_d

T_d	E	$8C_3$	$3C_2$	$6S_4$	$6\sigma_d$
A_1	1	1	1	1	1
A_2	1	1	1	-1	-1
E	2	-1	2	0	0
T_1	3	0	-1	1	-1
T_2	3	0	-1	-1	1

Table 3.2   Work Sheet for Reducing Γ_r

T_d	E	$8C_3$	$3C_2$	$6S_4$	$6\sigma_d$		
Γ_r	8	-1	4	-2	0	Σ	$\Sigma/24$
A_1							
A_2							
E							
T_1							
T_2							

Table 3.3   Completed Work Sheet Showing the Reduction of Γ_r

T_d	E	$8C_3$	$3C_2$	$6S_4$	$6\sigma_d$		
Γ_r	8	-1	4	-2	0	Σ	$\Sigma/24$
A_1	8	-8	12	-12	0	0	0
A_2	8	-8	12	12	0	24	1
E	16	8	24	0	0	48	2
T_1	24	0	-12	-12	0	0	0
T_2	24	0	-12	12	0	24	1

From this we see that $\Gamma_r = A_2 + 2E + T_2$. The proof that this is correct is that the characters of the irreducible representations of this combination add to give the characters of Γ_r, as shown below.

T_d	E	$8C_3$	$3C_2$	$6S_4$	$6\sigma_d$
A_2	1	1	1	-1	-1
E	2	-1	2	0	0
E	2	-1	2	0	0
T_2	3	0	-1	-1	1
Γ_r	8	-1	4	-2	0

When actually working out a work sheet such as Table 3.3, one moves back and forth from the work sheet to the character table. To find n_i for any species, take the characters of the irreducible representation from the character table (for our example, Table 3.1), multiply them by the corresponding characters for Γ_r, as shown on the work sheet (Table 3.2 or 3.3), and multiply each of those products by the number of operations in its class. Once all the products in a line have been obtained, sum across the row (Σ) and divide by the group order to obtain n_i for that irreducible representation. Note, however, that the first line of products on the work sheet, which is the line for the totally symmetric representation, always involves χ_i values that are $+1$. Thus, the first line of a work sheet really only involves multiplying the numbers of operations in each class by the characters of Γ_r for each class, equivalent to $g_c \chi_r$. Proceeding to the next and following lines, notice from the character table that successive irreducible representations differ from one another by simple changes in sign or multiplications of characters from the preceding line (cf. Table 3.1). Recognizing this makes it easy to generate the successive lines of the work sheet. We simply make the appropriate sign changes or multiplications, as dictated by the character table, from either the first or previous rows in the work sheet. In the case of T_d, for example, the difference between A_1 and A_2 is a sign change for the characters of $6S_4$ and $6\sigma_d$. Thus our second line of products is the same as the first except that the last two entries have a negative sign (-1). Similar simple changes are made in successive rows as we work through the n_i values for all the irreducible representations of the group.

In many cases, the reducible representation will have one or more 0 characters, as is the case in our example. As Table 3.3 shows, any column of products under a 0 character in Γ_r will likewise be 0 and can be ignored in working out the n_i values. Recognizing shortcuts such as this can greatly speed the work, compared to writing out every explicit term of Eq. (3.1).

Laying out the work in tabular form aids in checking for arithmetic errors. For example, we know that the sum across any row must be divisible by the group order. If it is not, probably an error has been made in one or more of the terms in the row. It is also possible that the original reducible representation was constructed incorrectly (e.g., incomplete basis set of vectors, or incorrect interpretation of the effects of the operations on the basis set). The error usually can be detected by checking that the correct sign changes or multiplications were made from one row to the next. For example, in Table 3.3, along the row for E, if we had inadvertently failed to change the sign of the product for $8C_3$ from the row for A_2, the sum would have been $16 - 8 + 24 + 0 + 0 = 32$, which is not divisible by 24. The error could be caught by referring back to the character table and seeing that the character of $+1$ for A_2 becomes -1 for E. While the same error could be detected in the standard mathematical expression for n_E, in the tabular form there is no hunting for which term corresponds to which class of operations. If no arithmetic error has been made in moving from one line to another, then the original reducible representation may be faulty. Alternately, an error may have been made in generating the first row, where the products are $g_c\chi_r$ (e.g., failing to multiply by the number of operations in the class, g_c).

With all the n_i results lined up in a column in the work sheet, it is easy to verify that the sum of dimensions of the component irreducible representations is the same as the dimension, d_r, of the reducible representation, as required by Eq. (2.10). In our example, Eq. (2.10) works out as $d_r = (1)(1) + (2)(2) + (1)(3) = 8$, which is the dimension of Γ_r, as shown by its character for the identity operation (cf. Table 3.2 or 3.3).

As one proceeds through the work sheet, it is a good idea to keep an eye on the running total of the dimensions of the found species. If at any point it exceeds the dimension of the reducible representation, an error probably has been made in one or more of the n_i calculations. The error would be of a type that would give a sum fortuitously divisible by the order of the group. Beyond this kind of error detection, noting the running sum of the dimensions may in some cases prevent carrying out unnecessary work. The reducible representation may consist of species that are listed in the upper lines of the character table. If so, the dimension of the reducible representation will be satisfied before reaching the bottom row of the work sheet. Therefore, continue row by row in the work sheet only until a sufficient number of irreducible representations has been found to equal the dimension of the reducible representation, consistent with Eq. (2.10). Unfortunately in our example the last-listed irreducible representation contributed to Γ_r, so we had to continue through the entire range of species. Regardless of whether one can stop the work early or not, it is always prudent to verify, as we did, that the resulting collection of irreducible representations does, indeed, add to give the reducible representation.

3.2   Handling Representations with Imaginary Characters

We noted in Section 2.4 that certain groups have irreducible representations with characters that involve the imaginary integer $\sqrt{-1}$. These representations are grouped by braces as complex-conjugate pairs in the character table of the group. By long-standing convention, the pair is given the Mulliken symbol of a doubly degenerate representation (e.g., E), even though each imaginary-character representation is an irreducible representation in its own right.

We noted further in Section 2.4 that for many applications it may be convenient to combine the complex-conjugate pair into one real-number representation by adding the two characters for each class of operations in the group. Where necessary, the trigonometric identities of Eq. (2.18) and (2.19), or their sum as Eq. (2.20), may be used for this purpose. This procedure is justifiable in applications of group theory to real molecules, since the reducible representation in such cases must contain both complex-conjugate irreducible representations in equal number, if it contains either.* However, one must not lose sight of the fact that the combined representation—although a genuine representation—is not an irreducible representation of the group. To emphasize this in this text, we will surround the Mulliken symbol for a combined pair with braces (e.g., $\{E\}$).

Using combined, real-number representations can avoid the complications of dealing with imaginary characters and minimize the number of times n_i needs to be calculated when reducing a representation of one of the imaginary-character groups by Eq. (3.1). However, the fact that the combined pair is a reducible representation, and not an irreducible representation, causes the answer for n_i to be twice what it should be for any combined representation.[†] To illustrate this, we will construct a reducible representation in C_{4h} by adding together several irreducible representations of the group, including a complex-conjugate pair. Then, we will take the resulting reducible representation, whose components we already know, and decompose it by Eq. (3.1) to see if the result is the same. For this purpose we will use the character table for C_{4h} shown as Table 3.4, which has been modified by combining the two complex-conjugate pairs of representations, E_g and E_u, into real-number representations, designated $\{E_g\}$ and $\{E_u\}$, respectively (cf. the standard character table for C_{4h} in Appendix A).

Table 3.4 Modified Character Table for C_{4h}

C_{4h}	E	C_4	C_2	C_4^3	i	S_4^3	σ_h	S_4
A_g	1	1	1	1	1	1	1	1
B_g	1	-1	1	-1	1	-1	1	-1
$\{E_g\}$	2	0	-2	0	2	0	-2	0
A_u	1	1	1	1	-1	-1	-1	-1
B_u	1	-1	1	-1	-1	1	-1	1
$\{E_u\}$	2	0	-2	0	-2	0	2	0

---

*A fuller exposition of this point can be found in M. Tinkham, *Group Theory and Quantum Mechanics,* McGraw-Hill, New York, 1964, p. 147; or M. Hammermesh, *Group Theory and Its Application to Physical Problems,* Addison-Wesley, Reading, MA, 1962, p. 118.
†A mathematical justification of this can be found in R. L. Carter, *J. Chem. Educ.* **1993**, *70,* 17–19.

Let the test representation be $\Gamma_r = 2B_g + \{E_g\} + A_u$. By adding the characters from these chosen, component representations we obtain the characters of Γ_r:

C_{4h}	E	C_4	C_2	C_4^3	i	S_4^3	σ_h	S_4
B_g	1	-1	1	-1	1	-1	1	-1
B_g	1	-1	1	-1	1	-1	1	-1
$\{E_g\}$	2	0	-2	0	2	0	-2	0
A_u	1	1	1	1	-1	-1	-1	-1
Γ_r	5	-1	1	-1	3	-3	-1	-3

Now, using the tabular method (Section 3.1), let us systematically reduce Γ_r. This yields the following work sheet.

C_{4h}	E	C_4	C_2	C_4^3	i	S_4^3	σ_h	S_4	Σ	$\Sigma/8$
Γ_r	5	-1	1	-1	3	-3	-1	-3	Σ	$\Sigma/8$
A_g	5	-1	1	-1	3	-3	-1	-3	0	0
B_g	5	1	1	1	3	3	-1	3	16	2
$\{E_g\}$	10	0	-2	0	6	0	2	0	16	2
A_u	5	-1	1	-1	-3	3	1	3	8	1
B_u	5	1	1	1	-3	-3	1	-3	0	0
$\{E_u\}$	10	0	-2	0	-6	0	-2	0	0	0

This result suggests (erroneously) that $\Gamma_r = 2B_g + 2\{E_g\} + A_u$. The n_i results, appearing in the $\Sigma/8$ column of the work sheet, clearly are correct for the non-degenerate species and for the absent combined representation $\{E_u\}$. However, the answer for $\{E_g\}$ is twice what it should be. Even if we had not known the composition of Γ_r beforehand, we could have recognized that the overall result for all n_i is incorrect, because the sum of the dimensions of the found species exceeds that of the reducible representation, in violation of Eq. (2.10).

This doubling problem will always occur for combined, real-character representations if they are used with Eq. (3.1) as if they were irreducible representations. There is no problem if the individual imaginary-character representations are used to calculate n_i. For example, for our test representation Γ_r, the values of n_i for E_g and E_u can be calculated by Eq. (3.1) with the individual complex-conjugate representations, here designated $E_g(1)$, $E_g(2)$, $E_u(1)$, and $E_u(2)$. The work sheet for these representations in separated form is as follows:

C_{4h}	E	C_4	C_2	C_4^3	i	S_4^3	σ_h	S_4	Σ	$\Sigma/8$
$E_g(1)$	5	$-i$	-1	i	3	$-3i$	1	$3i$	8	1
$E_g(2)$	5	i	-1	$-i$	3	$3i$	1	$-3i$	8	1
$E_u(1)$	5	$-i$	-1	i	-3	$3i$	-1	$-3i$	0	0
$E_u(2)$	5	i	-1	$-i$	-3	$-3i$	-1	$3i$	0	0

This correctly shows that Γ_r contains $E_g(1) + E_g(2) \equiv \{E_g\}$.

From this it might seem that one should always use the individual, imaginary-character representations with Eq. (3.1). Although that approach will always avoid the doubling problem, in most point groups with imaginary characters (except groups like C_4, C_{4h}, and S_4) the complex-conjugate representations involve the functions $\epsilon = \exp(2\pi i/n)$ and $\epsilon^* = \exp(-2\pi i/n)$. These are very cumbersome to manipulate with Eq. (3.1), either in their exponential form or as their trigonometric equivalents [Eqs. (2.18) and (2.19)]. Therefore, it is usually more practical to use the combined, real-character representations, as long as one is aware of the doubling problem. If this practice is followed, the true number of times a complex-conjugate pair contributes to the reducible representation can be found by dividing the n_i result from Eq. (3.1) by 2. For all other, genuinely irreducible representations the n_i results will be correct as calculated.

3.3 Group–Subgroup Relationships: Descent and Ascent in Symmetry

One motivation for studying chemical applications of group theory is to understand how molecular properties change as structure changes. In many cases, atomic substitution or molecular deformation leads to a new structure that belongs to a higher-order or lower-order group. If the new point group is higher order, *ascent in symmetry* has occurred; if the new point group is lower order, *descent in symmetry* has occurred. In either event, there is often a group–subgroup relationship between the old and new point groups. If ascent in symmetry occurs, certain properties that were distinguishable may become degenerate in the new higher-order point group. Conversely, if descent in symmetry occurs, where the new point group is a subgroup of the old, degeneracies that existed in the old structure may be lifted, and formerly equivalent properties may become distinguishable in the new configuration. Knowing how the irreducible representations of a group correlate to the irreducible representations of its subgroups can aid in predicting how certain properties may change with changes in symmetry. This in turn can enable chemists to infer structural changes on the basis of observed data, or to interpret data more accurately when a known structural change has occurred.

Consider the resulting changes in symmetry to an initially octahedral molecule, MA_6, when certain of the A atoms are substituted by B and C atoms, as shown in Fig. 3.1. The operations of the point groups of the five structures of Fig. 3.1 are summarized in Table 3.5. Examining the operations in Table 3.5, we can see that structures II through V all belong to point groups composed of operations that are found in the point group O_h; that is, they belong to subgroups of O_h. Furthermore, structures II through IV are related to each other as successive subgroups of one another. Only structure V is not related to structures II through IV, because it retains the C_3 axis from O_h, which the others do not. However, like structures II through IV, structure V also belongs to a subgroup of O_h.

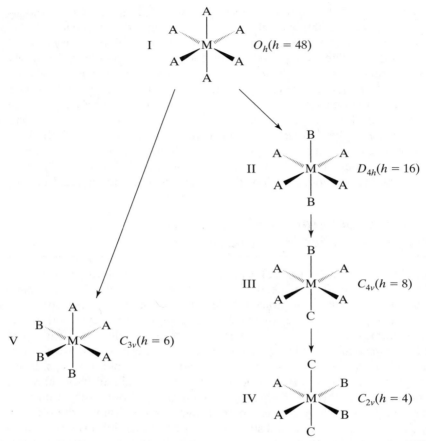

Figure. 3.1 Group–subgroup relationships among some substituted octahedral molecules.

Table 3.5 Symmetries of the Structures of Fig. 3.1

No.	Group	Operations	h
I	O_h	E, $8C_3$, $6C_2$, $6C_4$, $3C_2(=C_4^2)$, i, $6S_4$, $8S_6$, $3\sigma_h$, $6\sigma_d$	48
II	D_{4h}	E, $2C_4$, C_2, $2C_2'$, $2C_2''$, i, $2S_4$, σ_h, $2\sigma_v$, $2\sigma_d$	16
III	C_{4v}	E, $2C_4$, C_2, $2\sigma_v$, $2\sigma_d$	8
IV	C_{2v}	E, C_2, $\sigma_v(xz)$, $\sigma_v'(yz)$	4
V	C_{3v}	E, $2C_3$, $3\sigma_v$	6

What these progressive structural changes illustrate is a descent in symmetry through a hierarchy of related subgroups. The structures have been arranged in Fig. 3.1 to show the group–subgroup family relations in descending group order. Progressing from top (structure I) down either branch, we see that the successive subgroups are reached by loss of certain key elements (cf. Table 3.5). For example, in going from O_h of structure I to C_{3v} of structure V, the C_4 and C_2 axes, the σ_h plane, and the inversion center (i) are lost, along with other related elements.

Structural changes of this sort, in which the basic geometry of the molecule is essentially preserved, are fairly common. The continuity of group–subgroup relationships between the structures represents ascent or descent in symmetry. When the basic geometry changes abruptly, there may be no continuous relationship between the groups before and after the change. For example, when an MA_5 molecule changes from trigonal bipyramidal geometry (D_{3h}) to square pyramidal geometry (C_{4v}), or vice versa, there is no group–subgroup relationship between the two (cf. Fig. 3.2). The operations of C_{4v} ($h = 8$) are not a subset of the operations of D_{3h} ($h = 12$), nor the converse.

That D_{3h} and C_{4v} do not have a group–subgroup relationship should be apparent even without examining the operations of the two groups. Recall that the order of a subgroup must be an integer divisor of the order of the group to which it is related (cf. Section 1.4). With an order of 8, C_{4v} cannot be a subgroup of D_{3h}, whose order is 12. By the same reasoning, structure V of Fig. 3.1 cannot belong to a subgroup of any of the groups of structures II through IV.

When point groups are related as group and subgroup their irreducible representations are related, too. For example, consider D_{4h} and its subgroup C_{4v}. Table 3.6 shows a character table for D_{4h} in which the operations shared with C_{4v} have been blocked off, and only the characters for the shared operations are shown. For each irreducible representation of D_{4h}, the characters for the shared operations define an irreducible representation of C_{4v}, the subgroup. Thus, there is a correlation between the representations of D_{4h} and C_{4v}. There are only five irreducible representations in C_{4v}, compared to 10 for D_{4h}, so each representation of C_{4v} correlates to two of D_{4h}. This is shown by the correlation diagram in Fig. 3.3.

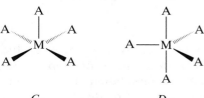

C_{4v}
($h = 8$)
D_{3h}
($h = 12$)

Figure. 3.2 Square pyramidal and trigonal bipyramidal structures of MA_5.

Table 3.6 Character Table of D_{4h} Showing Characters for Operations Shared with C_{4v}

D_{4h}	E	$2C_4$	C_2	$2C_2'$	$2C_2''$	i	$2S_4$	σ_h	$2\sigma_v$	$2\sigma_d$	C_{4v}
A_{1g}	1	1	1						1	1	A_1
A_{2g}	1	1	1						−1	−1	A_2
B_{1g}	1	−1	1						1	−1	B_1
B_{2g}	1	−1	1						−1	1	B_2
E_g	2	0	−2						0	0	E
A_{1u}	1	1	1						−1	−1	A_2
A_{2u}	1	1	1						1	1	A_1
B_{1u}	1	−1	1						−1	1	B_2
B_{2u}	1	−1	1						1	−1	B_1
E_u	2	0	−2						0	0	E

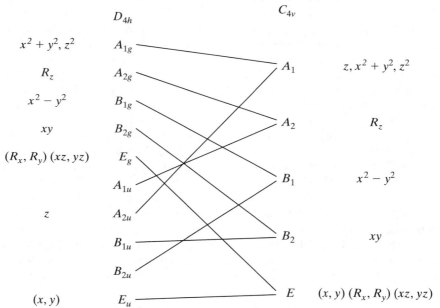

Figure. 3.3 Correlation diagram showing relationships between the species of D_{4h} and C_{4v}.

The physical significance of correlations, such as those between the species of D_{4h} and C_{4v}, is that a property that transforms as one representation in a group will transform as its correlated representation in a subgroup. The character, $\chi(R)$, resulting from a transformation matrix that describes the behavior of a vector or set of vectors that constitute the basis for a representation, depends only on R and its orientation relative to the coordinates of the system. Therefore, between a group and any one of its subgroups, representations arising from the same vector basis will have the same $\chi(R)$ values for all operations that occur in both groups.

Often, two or more bases of separate representations of a group yield the same set of $\chi(R)$ values for those operations that are carried over into the subgroup. In such cases, the separate representations of the larger group will correlate to a single representation of the subgroup. In other words, the separate bases in the group form a set of redundant bases for a single representation in the subgroup. We can see this by comparing the vector transformation properties of D_{4h} and C_{4v}, which are indicated in Fig. 3.3. For example, note that in D_{4h} the direct products $x^2 + y^2$ and z^2 transform as A_{1g}, and the linear vector z transforms as A_{2u}. When we examine these representations in Table 3.6, we see for the five classes of operations of D_{4h} that are shared with C_{4v} the characters are the same for both A_{1g} and A_{2u}. Thus, $x^2 + y^2$, z^2, and z all transform in C_{4v} by the same irreducible representation, which in this case is the totally symmetric representation, A. Likewise, all other pairs of representations of D_{4h} that correlate with a single representation in C_{4v} share the same set of bases.*

In many cases, degenerate representations of a group (Mulliken symbols of the type E or T) may become two or three distinguishable bases in a subgroup. When this happens the degenerate representation of the higher-order group will correlate to two or three irreducible representations of the subgroup. A doubly degenerate representation would split into two nondegenerate representations. A triply degenerate representation would split into either three nondegenerate representations or a combination of one nondegenerate representation and one doubly degenerate representation, depending on the group and subgroup.

In the case of descent from D_{4h} to C_{4v}, the bases for the doubly degenerate representations E_g and E_u retain their degeneracies and become the E representation in the subgroup. The fourfold rotational symmetry of both groups allows degenerate representations. However, if we descend further to C_{2v}, as when changing from structure III to IV in Fig. 3.1, the retention of only a twofold axis prohibits degenerate representations. Thus, the degenerate representation E of C_{4v} must be split into two nondegenerate representations in C_{2v}.

To determine the correlation, consider the C_{4v} character table, shown as Table 3.7, in which only the characters for operations shared with C_{2v} are listed. One immediately evident consequence of the descent in symmetry is that the class of reflections $2\sigma_v$ of C_{4v} becomes two separate classes, σ_v and σ_v', in C_{2v}. Therefore, the operations of C_{2v} are listed on a separate line in

---

*The species A_{1u}, B_{1u}, and B_{2u} of D_{4h} show no vector transformations in either Fig. 3.3 or the character table (cf. Appendix A). The basis for any of these species is simply not one of the usually tabulated vectors or direct products. Note, however, that the "missing" basis in each case is also "missing" for the correlated species in C_{4v}.

Table 3.7   Character Table for C_{4v} Showing Characters for Operations Shared with C_{2v}

C_{4v}	E	$2C_4$	C_2	$2\sigma_v$	$2\sigma_d$	
	E		C_2	σ_v, σ_v'		C_{2v}
A_1	1		1	1		A_1
A_2	1		1	-1		A_2
B_1	1		1	1		A_1
B_2	1		1	-1		A_2
E	2		-2	0		?

Table 3.7. Notice that there is no representation in C_{2v} with characters corresponding to the E representation of C_{4v}. We can determine the two representations of C_{2v} that correlate to this set of characters by treating them as comprising a reducible representation, Γ_E, in the smaller group. By inspection, we see that $\Gamma_E = B_1 + B_2$:

C_{2v}	E	C_2	σ_v	σ_v'
B_1	1	-1	1	-1
B_2	1	-1	-1	1
Γ_E	2	-2	0	0

Combining this information with the correlations that are evident in Table 3.7 leads to the correlation diagram shown in Fig. 3.4.

The important point of this illustration is that the degeneracy that exists in C_{4v} is lifted on descending to C_{2v}. This means that two properties that may be indistinguishable by reason of symmetry (i.e., degenerate) in a molecule of C_{4v} symmetry will be distinguishable (i.e., nondegenerate) if the symmetry is reduced to C_{2v}.

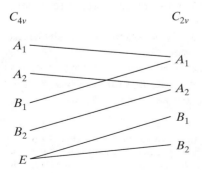

Figure. 3.4   Correlation diagram for C_{4v} and C_{2v} assuming that the two σ_v planes of C_{4v} become σ_v and σ_v' of C_{2v}.

It is worth noting that the correlation between C_{4v} and C_{2v} would be different if we had assumed that the two σ_d planes of C_{4v} became σ_v and σ'_v in C_{2v}. In this case, the correlations between B_1 and B_2 of C_{4v} with A_1 and A_2 of C_{2v} would have been reversed, becoming $B_1 \to A_2$ and $B_2 \to A_1$. Thus, we must take care to define which operations are retained when making correlations between groups and subgroups where more than one choice is possible. The choice, however, does not affect the fate of any properties of a molecule, merely the Mulliken labels attached to them.

Correlation tables relating most important groups with their subgroups are shown in Appendix B. Where correlations differ by choice of retained operations, the preserved elements are noted in the heading of the column for the subgroup.

3.4 Reducing Representations of Groups with Infinite Order

In Section 3.1 we noted that Eq. (3.1) can be used to reduce any reducible representation of a group with finite order. Unfortunately, the division by h, the group order, makes this equation unsuitable for representations of the infinite-order groups, namely $C_{\infty v}$ and $D_{\infty h}$. In many cases involving linear molecules, the reduction can be accomplished by inspection, but for representations with higher dimensions a work-around technique may be useful.

One of the most practical alternative techniques for reducing representations of infinite-order groups, originally suggested by Strommen and Lippincott,* takes advantage of group–subgroup relationships. Realizing that $C_{\infty v}$ and $D_{\infty h}$ are merely special cases of the family of groups C_{nv} and D_{nh}, respectively, it follows that all members of these families are subgroups of their respective infinite-order groups. Therefore, to avoid the problem of needing to divide by infinity in Eq. (3.1), we can set up the reducible representation in any convenient subgroup and correlate the component irreducible representations with the species for the infinite-order group. When applied to physical problems, this technique amounts to pretending that the molecule has a lower-order, finite group symmetry. Once the results are obtained, they are correlated with the appropriate species of the true, infinite-order group.

Realizing that the infinite-order groups have an infinite number of irreducible representations, we must concede that it is impossible to construct a complete correlation between any subgroup and its parent infinite-order group. However, for applications to physical problems such as we will consider in this text, most of our concern will be with species of the group and subgroup that are associated with unit vector or direct product bases, the kinds that are tabulated in the last two columns of the character tables. Within

---

*D. P. Strommen and E. R. Lippincott, *J. Chem. Educ.* **1972,** *49,* 341–342.

this limitation, it is possible to work out the meaningful correlations by matching vector bases in the two related groups. Some correlations beyond these are also easy to deduce.

The remaining question for implementing this strategy is which subgroup to choose. In both $C_{\infty v}$ and $D_{\infty h}$, the C_∞ axis, which would correspond to the axis of a linear molecule in either case, is taken as z of a Cartesian coordinate system. Hence, the x and y axes lie perpendicular to the molecular axis. In choosing the subgroup in which to work, it is usually most convenient to pick a group that does not intermix x and y coordinates (e.g., as would a group with a C_3 principal axis). A group with a C_2 or C_4 principal axis is geometrically compatible with the orientations of x and y to each other, and therefore either is generally a good choice for the working subgroup. The groups with the lower-order principal axis (C_2) have fewer operations and irreducible representations, which minimize some of the work. Therefore, for most applications, it will be most convenient to use C_{2v} as a working subgroup for a problem in $C_{\infty v}$, and to use D_{2h} for a problem in $D_{\infty h}$. Correlation tables for these group–subgroup relations are shown as Tables 3.8 and 3.9.

To illustrate the technique, consider the following representation of the group D_{2h}, assuming that it is constructed as a working-subgroup substitute for a representation in $D_{\infty h}$.

D_{2h}	E	$C_2(z)$	$C_2(y)$	$C_2(x)$	i	$\sigma(xy)$	$\sigma(xz)$	$\sigma(yz)$
Γ_r	15	-5	-1	-1	-3	1	5	5

With Eq. (3.1) and the tabular method (Section 3.1), we find

$$\Gamma_r = 2A_g + 2B_{2g} + 2B_{3g} + 3B_{1u} + 3B_{2u} + 3B_{3u}$$

To find which irreducible representations would comprise Γ_r in the parent group $D_{\infty h}$, we examine the correlations listed in Table 3.9. From Table 3.9

Table 3.8 Partial Correlation Between $C_{\infty v}$ and C_{2v}

$C_{\infty v}$	C_{2v}
$A_1 = \Sigma^+$	A_1
$A_2 = \Sigma^-$	A_2
$E_1 = \Pi$	$B_1 + B_2$
$E_2 = \Delta$	$A_1 + A_2$

Table 3.9 Partial Correlation Between $D_{\infty h}$ and D_{2h}

$D_{\infty h}$	D_{2h}
Σ_g^+	A_g
Σ_g^-	B_{1g}
Π_g	$B_{2g} + B_{3g}$
Δ_g	$A_g + B_{1g}$
Σ_u^+	B_{1u}
Σ_u^-	A_u
Π_u	$B_{2u} + B_{3u}$
Δ_u	$A_u + B_{1u}$

we see that A_g becomes Σ_g^+ and B_{1u} becomes Σ_u^+ in $D_{\infty h}$. Each pair $B_{2g} + B_{3g}$ becomes the doubly degenerate species Π_g, and each pair $B_{2u} + B_{3u}$ becomes the doubly degenerate species Π_u in $D_{\infty h}$. Thus, in $D_{\infty h}$ we have

$$\Gamma_r = 2\Sigma_g^+ + 2\Pi_g + 3\Sigma_u^+ + 3\Pi_u$$

Note that either in D_{2h} or $D_{\infty h}$ the sum of the dimensions of the component irreducible representations is the same as the dimension of Γ_r; that is, $d_r = 15$.

3.5   More About Direct Products

The direct product listings that appear in the last column of a character table show the symmetries of direct products between linear vectors. Actually, direct product combinations can be taken between any number of irreducible representations, whether or not they are associated with linear vectors. Indeed, it is not even necessary in many cases to know the specific basis for the representation. The relationships between individual irreducible representations and their direct products are relevant to certain applications of quantum mechanics, such as spectroscopic selection rules. Therefore, it is useful to understand how direct products are generated in group theory and to know some of the general results that have consequences for dealing with molecular problems.

For any direct product between two or more irreducible representations of a group, the result will also be a representation of the group. Thus,

$$\Gamma_a\Gamma_b\Gamma_c \ldots = \Gamma_{abc\ldots} \qquad (3.2)$$

where the characters of the direct product representation $\Gamma_{abc\ldots}$ for each operation R of the group are given by

$$\chi_a(R)\chi_b(R)\chi_c(R)\ldots = \chi_{abc\ldots}(R) \qquad (3.3)$$

To generate the representation for any direct product, we simply multiply together the characters of the component representations, operation by operation. The resulting representation may be either a reducible or irreducible representation. In either case the dimension of the product, d_p, is the product of the dimensions of all the component representations:

$$d_p = \prod_i d_i \qquad (3.4)$$

We will illustrate the procedure and some general results using the group C_{4v}, as shown in Table 3.10. We do not need the vector direct product listings, so they have been omitted from the table.

1. *If all the combined irreducible representations are nondegenerate, then the product will be a nondegenerate representation, too.*

Table 3.10 Partial Character Table for C_{4v}

C_{4v}	E	$2C_4$	C_2	$2\sigma_v$	$2\sigma_d$
A_1	1	1	1	1	1
A_2	1	1	1	−1	−1
B_1	1	−1	1	1	−1
B_2	1	−1	1	−1	1
E	2	0	−2	0	0

This follows directly from Eq. (3.4). For example, in C_{4v} the direct product B_1B_2, formed by multiplying the characters of the two representations, is the irreducible representation A_2, as shown below.

C_{4v}	E	$2C_4$	C_2	$2\sigma_v$	$2\sigma_d$
B_1	1	−1	1	1	−1
B_2	1	−1	1	−1	1
A_2	1	1	1	−1	−1

2. *The product of a nondegenerate representation and a degenerate representation is a degenerate representation.* The product may be either the original degenerate representation or another of the same order (if such exists).

This, too, is a consequence of Eq. (3.4). For example, in C_{4v} we obtain $B_2E = E$, as shown below.

C_{4v}	E	$2C_4$	C_2	$2\sigma_v$	$2\sigma_d$
B_2	1	−1	1	−1	1
E	2	0	−2	0	0
E	2	0	−2	0	0

In this case the product must be the original degenerate representation, because there is no other in the group.

3. *The direct product of any representation with the totally symmetric representation is the representation itself.*

This follows because the totally symmetric representation in any group consists of all +1 characters. For example, in C_{4v} we obtain $A_1E = E$, as shown below.

C_{4v}	E	$2C_4$	C_2	$2\sigma_v$	$2\sigma_d$
A_1	1	1	1	1	1
E	2	0	−2	0	0
E	2	0	−2	0	0

4. *The direct product of degenerate representations is a reducible representation.*

By definition, a degenerate representation is one for which $d_i \geq 2$, so the product of Eq. (3.2) will be $d_p \geq 4$ for any direct product of degenerate representations. Since no irreducible representation has an order greater than 3 (at least among the common finite groups), the product must be a reducible representation.* For example, in C_{4v} the product EE yields the following characters:

C_{4v}	E	$2C_4$	C_2	$2\sigma_v$	$2\sigma_d$
E	2	0	−2	0	0
E	2	0	−2	0	0
Γ_p	4	0	4	0	0

By applying Eq. (3.1), it can be shown that the product representation, Γ_p, decomposes as

$$\Gamma_p = A_1 + A_2 + B_1 + B_2$$

As always, the proof of this decomposition is that the sums of the characters for the irreducible representations are the characters of the reducible representation, Γ_p.

5. *The direct product of an irreducible representation with itself is or contains the totally symmetric representation.*

If the representation is nondegenerate, the product with itself must be the totally symmetric representation, because the product characters are all $(\pm 1)^2 = 1$. If the representation is degenerate, its product with itself will be a degenerate representation (point 4, above) whose component irreducible representations include the totally symmetric representation. (The proof of this is given under point 6, below.) The example of EE in C_{4v} shown above illustrates the point for the self-product of a degenerate representation.

6. *Only the direct product of a representation with itself is or contains the totally symmetric representation.*

This is most readily apparent for the product of two different nondegenerate representations, for which at least one of the $\chi_a(R)\chi_b(R)$ products must be negative. This is inconsistent with the totally symmetric representation. More generally, we can determine how many times the totally symmetric representation occurs in any Γ_p, regardless of dimen-

---

*The rarely encountered groups I and I_h have degenerate representations with $d_i = 3, 4, 5$, but none with $d_i = 2$. Hence, $d_p \geq 9$ for all direct products of degenerate representations in these groups. Here, too, any product of degenerate representations is a reducible representation.

sion, by solving Eq. (3.1). In the case of the totally symmetric representation, $\chi_i = \chi_A = 1$ for all classes of any group. Therefore Eq. (3.1) yields

$$hn_A = \sum_c g_c \chi_i \chi_r = \sum_c g_c \chi_A \chi_p = \sum_c g_c \chi_p \qquad (3.5)$$

All the characters of Γ_p are products of the characters of two irreducible representations Γ_a and Γ_b; that is, $\chi_p = \chi_a \chi_b$ for all classes of the group. Furthermore, Γ_a and Γ_b must be orthogonal, as required by Eq. (2.27). Therefore, we may rewrite Eq. (3.5) as

$$hn_A = \sum_c g_c \chi_a \chi_b = h \delta_{ab} \qquad (3.6)$$

from which it follows that $n_A = \delta_{ab}$. Thus, n_A will be zero for all $\Gamma_p = \Gamma_a \Gamma_b$, except when $\Gamma_a = \Gamma_b$. This means that only the direct product of an irreducible representation with itself can contain the totally symmetric representation. Moreover, Eq. (3.6) requires that the self-product contain the symmetric representation once, and only once.

In addition to the points above, we can make some generalizations about the Mulliken symbols of direct products between representations. Recall from Section 2.4 that g and u subscripts indicate symmetry and antisymmetry with respect to inversion. Likewise, prime ($'$) and double prime ($''$) indicate symmetry and antisymmetry with respect to a horizontal mirror plane. When representations with either of these symbols are combined as direct products, the product representation or representations will be symmetric to the referenced symmetry operation (g or $'$) when both irreducible representations are symmetric (g or $'$) or both are antisymmetric (u or$''$). If one is symmetric and the other is antisymmetric, the product will be antisymmetric (u or$''$). In terms of the Mulliken symbols, these results can be summarized as follows:

$$(g)(g) = (g) \qquad (u)(u) = (g) \qquad (g)(u) = (u)$$
$$(')(') = (') \qquad ('')('') = (') \qquad (')('') = ('')$$

These results have consequences for selection rules in spectroscopy, as well as other applications in quantum mechanics.

Problems

3.1 Reduce the following representations into their component species:

(a)
D_{2d}	E	$2S_4$	C_2	$2C_2'$	$2\sigma_d$
Γ_a	4	0	0	0	2

(b)
C_{3v}	E	$2C_3$	$3\sigma_v$
Γ_b	12	0	2

(c)

D_{3h}	E	$2C_3$	$3C_2$	σ_h	$2S_3$	$3\sigma_v$
Γ_c	5	2	1	3	0	3

(d)

C_{4v}	E	$2C_4$	C_2	$2\sigma_v$	$2\sigma_d$
Γ_d	18	2	−2	4	2

(e)

D_{6h}	E	$2C_6$	$2C_3$	C_2	$3C_2'$	$3C_2''$	i	$2S_3$	$2S_6$	σ_h	$3\sigma_d$	$3\sigma_v$
Γ_e	18	0	0	0	−2	0	0	0	0	6	0	2

(f)

T_d	E	$8C_3$	$3C_2$	$6S_4$	$6\sigma_d$
Γ_f	15	0	−1	−1	3

(g)

O_h	E	$8C_3$	$6C_2$	$6C_4$	$3C_2$ $(=C_4^2)$	i	$6S_4$	$8S_6$	$3\sigma_h$	$6\sigma_d$
Γ_g	21	0	−1	3	−3	−3	−1	0	5	3

(h)

D_{5h}	E	$2C_5$	$2C_5^2$	$5C_2$	σ_h	$2S_5$	$2S_5^3$	$5\sigma_v$
Γ_h	7	2	2	1	5	0	0	3

3.2 Reduce the following representations, from groups whose irreducible representations contain imaginary characters, into their component species:

(a)

C_4	E	C_4	C_2	C_4^3
Γ_a	15	1	−1	1

(b)

C_{3h}	E	C_3	C_3^2	σ_h	S_3	S_3^5
Γ_b	21	0	0	7	−2	−2

(c)

C_5	E	C_5	C_5^2	C_5^3	C_5^4
Γ_c	7	2	2	2	2

3.3 Using the transformation properties listed in the character tables, determine the correlations between species of the following groups and their indicated subgroups: (a) $C_{4v} \rightarrow C_4$, (b) $D_{3h} \rightarrow D_3$, (c) $D_{5d} \rightarrow C_{5v}$.

3.4 Determine the correlations between species of D_{4h} and D_{2d} (a) when $2C_2'$ and $2\sigma_d$ of D_{4h} become $2C_2'$ and $2\sigma_d$, respectively, of D_{2d}; and (b) when $2C_2''$ and $2\sigma_v$ of D_{4h} become $2C_2'$ and $2\sigma_d$, respectively, of D_{2d}. [*Hint:* For each species of D_{4h}, compare characters of the operations retained in D_{2d} with the characters listed for the various species of D_{2d}.]

3.5 The following representations were generated in C_{2v} and D_{2h} to avoid working in the true molecular point groups $C_{\infty v}$ and $D_{\infty h}$, respectively. Determine the species comprising the reducible representation of the infinite-order group in each case.

(a)

C_{2v}	E	C_2	σ_v	σ_v'
Γ_a	15	−5	5	5

(b)

D_{2h}	E	$C_2(z)$	$C_2(y)$	$C_2(x)$	i	$\sigma(xy)$	$\sigma(xz)$	$\sigma(yz)$
Γ_b	9	−3	−1	−1	−3	1	3	3

3.6 Determine the species of the following direct products: (a) in D_{4d}, $B_2 \times B_2$; (b) in T_d, $T_2 \times T_2$; (c) in D_{6d}, $A_1 \times E_5$; (d) in D_{2d}, $B_1 \times B_2$; (e) in C_{4h}, $B_g \times A_u$; (f) in D_{3h}, $A_1'' \times A_2''$; (g) in C_{4h}, $A_u \times E_u$; (h) in D_{3d}, $E_g \times E_u$.

3.7 Consider the following sequential structural changes (I → II → III). For each series, indicate (i) the point group of each structure, (ii) the specific symmetry elements that are lost or gained in the transitions I → II and II → III, (iii) whether the transition II → III represents descent or ascent in symmetry, (iv) whether the point groups of II and III bear a group–subgroup relationship to each other, and (v) whether the point groups of I and III bear a group–subgroup relationship to each other.

3.8 For each sequence of structure changes in Problem 3.7, construct a correlation diagram (similar to those shown in Figs. 3.3 and 3.4) that links the symmetry species of the point groups of structures I, II, and (if possible) III, in succession. Do not attempt to construct a correlation between point groups that do not have a group–subgroup relationship. Note any degenerate symmetry species whose degeneracy is lifted or any nondegenerate symmetry species that become degenerate as a result of the changes in point group with each transition.

3.9 Construct the correlation table for the group D_4, showing correlations of its species to those of its subgroups C_4 and C_2. Note that there are three possible ways of defining C_2, depending on whether $C_2(=C_4^2)$, C_2', or C_2'' from D_4 is retained in the subgroup. Show correlations for all possibilities.

3.10 Verify the correlations between species of the group I and its subgroups T, D_5, C_5, D_3, C_3, D_2, and C_2, as shown in Appendix B.

CHAPTER 4

Symmetry and Chemical Bonding

Atomic orbitals possess symmetry properties that can be associated with the irreducible representations of the point group of the molecular system. When two or more atoms form chemical bonds, these symmetries assume great importance in determining what kinds of interactions can and cannot occur. Moreover, symmetry and group theory can be used effectively as a means of simplifying the complex quantum mechanical problem of constructing appropriate wave function descriptions of the bonding in molecules.

At the simplest level, consideration of the symmetries of orbitals on atoms can help us understand the nature of the bonding interactions between atoms in molecules. At a more detailed level, group theory arguments can guide us toward construction of appropriate hybrid orbitals that are helpful in reconciling the observed shapes of molecules with the atomic orbitals that are available for bonding. At its most powerful, group theory can be used to guide construction of general molecular orbitals for the entire molecule. With quantitative data, this approach enables construction of a molecular orbital scheme that can elucidate the electronic structure of a molecule. Use of a special function in group theory called the projection operator (cf. Chapter 5) can even aid in deducing the mathematical forms of the molecular orbital wave functions.

4.1 Orbital Symmetries and Overlap

In 1939 Feyman and Hellman showed that the forces that hold positive nuclei together in bonds are essentially the same as would exist with a static distribution of negative charge. From this comes the association of a buildup of electron density (i.e., increased electron probability) with the formation of a chemical bond. In principle, the bonded state can be represented by a *Schrödinger wave equation* of the general form

$$\mathcal{H}\Psi = E\Psi \qquad (4.1)$$

where $\mathcal{H}$ is the *Hamiltonian operator*, which describes the kinetic and potential energies of the system as functions of the masses and positions of all particles; Ψ is a wave function solution to the equation, also called an *eigen-*

88

function of $\mathcal{H}$; and E is the total energy of the system associated with Ψ, also called an *eigenvalue*. For a system consisting of i particles with masses m_i, subject to a potential energy V, which is a function of the positions of the particles, the Hamiltonian is

$$\mathcal{H} = - \sum_i \frac{h^2}{8\pi^2 m_i} \left(\frac{\partial^2}{\partial x_i^2} + \frac{\partial^2}{\partial y_i^2} + \frac{\partial^2}{\partial z_i^2} \right) + V(x_i, y_i, z_i) \qquad (4.2)$$

As may be evident from Eq. (4.2), setting up and solving the wave equation for all but the simplest molecular systems is not a feasible approach. Instead, it is customary to construct approximate wave functions for the molecule from the atomic orbitals of the interacting atoms. By this approach, when two atomic orbitals overlap in such a way that their individual wave functions add constructively, the result is a buildup of electron density in the region around the two nuclei. This follows from the association between the probability, P, of finding the electron at a point in space and the product of its wave function and its complex conjugate*:

$$P \propto \Psi\Psi^* \qquad (4.3)$$

If the particle exists, the probability of finding it over all points throughout space is unity, which leads to the normalization condition

$$\int \Psi\Psi^* \, d\tau = 1 \qquad (4.4)$$

where $d\tau$ indicates that the integration is performed over all coordinates of space; that is, $d\tau = dx \, dy \, dz$. In order to ensure that Eq. (4.4) is satisfied for a trial wave function, ψ, we may find it necessary to multiply by a constant, N, called the *normalization constant*, such that

$$\int (N\psi)(N\psi^*) \, d\tau = N^2 \int \psi\psi^* \, d\tau = 1 \qquad (4.5)$$

If ψ is a solution to Eq. (4.1), then so too is $N\psi$, and the value of E in Eq. (4.1) is unaffected.

We will be interested in constructing wave functions for the molecular system using the wave functions of the individual atoms. The extent of overlap between two orbitals on atoms A and B, and hence the nature and effectiveness of their interaction, is given by the *Slater overlap integral*,

$$S = \int \Psi_A \Psi_B \, d\tau \qquad (4.6)$$

The value of S can be calculated for H_2, but otherwise it must be approximated. For our purposes, it is only necessary to have a qualitative appreciation for Slater overlap integrals and a sense of whether a particular integral is positive, negative, or zero. If $S > 0$, the overlap results in a *bonding* inter-

---

*This connection was first suggested by the German physicist Max Born in 1926, and it resulted in part in his subsequent award of a Nobel Prize in 1954. Born's interpretation was based on analogy with the wave theory of light, by which the intensity of a light wave is proportional to its amplitude; that is, $I \propto A^2$.

action, characterized by a reenforcement of the total wave function and a buildup of electron density around the two nuclei. If $S < 0$, overlap results in an *antibonding* interaction with decrease of electron density in the region around the two nuclei. If $S = 0$, overlap results in a *nonbonding* interaction, where the distribution of electron density is essentially the same as that described by atomic orbitals on two adjacent atoms that do not have bonding or antibonding interactions.

These kinds of interactions are relatively easy to visualize by bringing together the usual pictorial representations of the interacting atomic orbitals, taking particular note of the signs of the wave functions. For this purpose, we may use the simple "balloon" representations of atomic orbitals, such as those shown in Fig. 4.1. These are actually rough representations of 90–99% of the probability distribution, which as the product of the wave function and its complex conjugate (or simply the square, if the function is real) is inherently a positive number. The signs indicated in Fig. 4.1 are those of the wave function, Ψ, itself and are shown only as a convenience for understanding how wave functions combine in bonding interactions. Note that these are representations for the simplest examples of each type of orbital (e.g., $1s$, $2p$, $3d$). Orbitals of each type (same value of the quantum number l) with higher than the minimum allowed value of the principal quantum number n are spatially more extensive, have more nodes, and (except of the s orbitals) have more lobes. However, the symmetries of orbitals of the same type are identical, regardless of the value of n.

For example, compare the contour diagram for a hydrogen $2p$ orbital with that of a $3p$ orbital, as shown in Fig. 4.2. Both are symmetric with respect to a C_∞ axis and any vertical mirror plane (σ_v) passing through the lobes. As objects, the three-dimensional models of these would both belong to the point group $C_{\infty v}$. Likewise, all s orbitals belong to the full rotation group R_3.* All "cloverleaf" d orbitals (d_{xy}, d_{xz}, d_{yz}, $d_{x^2-y^2}$) belong to D_{2h} (assuming attention is paid to the alternations of wave function sign), and all d_{z^2} orbitals belong to $D_{\infty h}$. Therefore, for the purposes of analyzing symmetry aspects of orbital interaction and bonding, we usually may use the simplest example of an orbital of a given type, regardless of the actual orbital that may be involved.

Consider the kinds of orbital interactions that may occur with $S > 0$. Figure 4.3 (left side) shows six examples of such bonding interaction. As these examples show, bonding results when all interacting regions of the two orbitals overlap with the same wave function sign. If a wave function for the bond, Ψ, were constructed from the two atomic wave functions, it would have higher amplitude in the region around the two nuclei. Given that probability is proportional to $\Psi\Psi^*$, this implies that the electron density about the nuclei would also be higher.

---

*The infinite group R_3 is composed of all point group operations in all possible orders to C_∞ and S_∞. It is the point group of any spherical object. All other point groups are subgroups of R_3. No molecules have this symmetry; therefore, R_3 was not discussed in conjunction with point groups of molecules in Chapter 1.

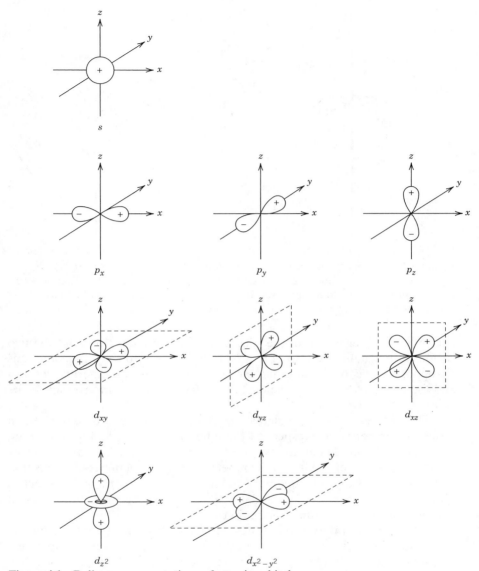

Figure 4.1 Balloon representations of atomic orbitals.

Figure 4.3 shows the two most frequently encountered kinds of positive overlap, *sigma* (σ) and *pi* (π). A sigma bonding interaction (e.g., Fig. 4.3 left, top three) is characterized by reenforcement along the internuclear axis. As these and other examples in Fig. 4.3 show, the interacting orbitals need not be the same type. A pi bonding interaction (e.g., Fig. 4.3 left, bottom three) is characterized by equal reenforcement above and below the internuclear

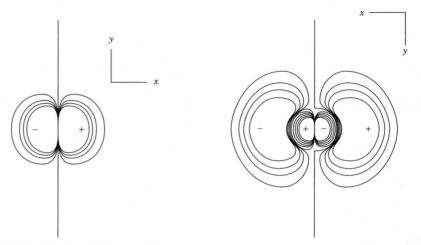

Contours at 0.0, 0.002, 0.004, 0.006, and 0.008   Contours at 0.0, 0.002, 0.004, 0.006 and 0.008

Figure 4.2   Contour diagrams of a $2p_x$ orbital (*left*) and a $3p_x$ orbital (*right*). Indicated values for contours are calculated values of ψ for hydrogen. [Reproduced with permission from A. Streitwieser, Jr. and P. H. Owens, *Orbital and Electron Density Diagrams*, Macmillan, New York, 1973.]

axis. Note that regardless of the specific interacting atomic orbitals, the two regions of maximum overlap of a pi bond have opposite wave function signs on opposite sides of the internuclear axis. As a result, the internuclear axis and the entire plane perpendicular to the plane of maximum overlap define a nodal surface ($P = 0$).

The difference between sigma and pi interactions can be described in terms of symmetry with respect to the internuclear axis. As Fig. 4.3 shows, sigma bonding interactions are symmetrical with respect to a C_2 axis collinear with the internuclear axis, and symmetrical to a σ_v plane containing that axis.* By contrast, *pi* bonding interactions are antisymmetric with respect to C_2 and σ_v (coplanar with the nodal plane); that is, the wave function sign changes with respect to these operations.

It follows from the overlap criterion of bonding that the interacting orbitals must also have these symmetries. Consequently, as a general approach in group theory, we often represent atomic orbitals capable of sigma bonding by vectors pointing from one atom toward another along the internuclear axis. Likewise, orbitals capable of pi interactions are represented by vectors perpendicular to the internuclear axis. Both of these vector representations express the symmetry of the interaction without the necessity of specifying which particular orbitals may be involved.

---

*In all cases of sigma overlap between s and p orbitals, the bonding interactions actually are symmetrical to C_∞ and an infinite number of σ_v planes intersecting along the C_∞ rotational axis. However, possible sigma bonding and antibonding interactions involving a single lobe on a d orbital (e.g., $d_{x^2-y^2} \pm p_x$ in Fig. 4.3) require the more restrictive symmetry definition.

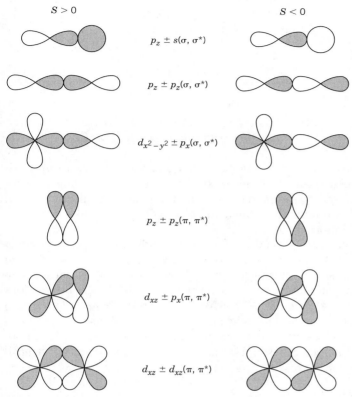

$S > 0$ $S < 0$

$p_z \pm s(\sigma, \sigma^*)$

$p_z \pm p_z(\sigma, \sigma^*)$

$d_{x^2 - y^2} \pm p_x(\sigma, \sigma^*)$

$p_z \pm p_z(\pi, \pi^*)$

$d_{xz} \pm p_x(\pi, \pi^*)$

$d_{xz} \pm d_{xz}(\pi, \pi^*)$

Figure 4.3 Examples of positive (*left*) and negative (*right*) orbital overlap. Shaded areas represent orbital lobes with positive wave function sign. (Axis orientations vary.)

The right side of Fig. 4.3 also shows examples of antibonding interactions. Note that the orbital combinations are the same as those shown for bonding interactions, except that the signs on one of the interacting orbitals have been reversed. Overlap with opposing signs creates destructive interference, resulting in an antibonding state ($S < 0$). Both sigma and pi antibonding interactions (σ^* and π^*) are possible. Their symmetries with respect to C_2 and σ_v are identical to those of the bonding interaction. The only difference is the mathematical sense in which the wave functions have been combined. In general, if two or more orbitals can be combined to define bonding combinations, then they must also be capable of forming an equal number of antibonding combinations of the same type (σ or π). In applied group theory, the vector representations of orbitals used to describe σ and π bonding combinations, as described above, also allow for σ^* and π^* antibonding combinations.

Figure 4.4 shows four examples of nonbonding interactions, for which $S = 0$. In all cases, any region of reenforcement (same wave function signs) is counterbalanced by an equal region of destructive interference (different

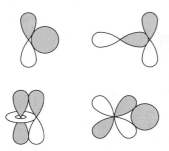

Figure 4.4   Examples of some overlaps that are required by symmetry to be zero ($S = 0$).

wave function signs). Such interactions neither add nor subtract from possible bonding between atoms. In terms of symmetry, the nonbonding state results from a lack of shared symmetry between interacting orbitals. In essence, an atomic orbital is nonbonding when there is no orbital of like symmetry available on a neighboring atom.

For completeness, we should also note one additional type of bonding and antibonding combination that can occur between d orbitals on adjacent atoms. If z is taken as the internuclear axis, pairs of $d_{x^2-y^2}$ or d_{xy} orbitals on adjacent atoms may interact with overlap on all four lobes of each. Interactions of this type result in a *delta bond* (δ) when $S > 0$ and a *delta antibond* (δ^*) when $S < 0$. Relative to the internuclear axis, both combinations are symmetric with respect to C_2, but antisymmetric with respect to C_4. This kind of interaction is postulated to describe metal–metal bonding in certain transition-metal cluster complexes.

4.2   Valence Bond Theory and Hybrid Orbitals

Historically there have been two major approaches for describing bonding on the basis of interactions of atomic orbitals: *Valence Bond Theory* and *Molecular Orbital Theory*. The valence bond (VB) approach, first described by Walter Heitler and Fritz London in 1927, is the older and in many ways simpler of the two. Both historically and conceptually, the VB model is an extension of the Lewis idea of bond formation through sharing of electron pairs. The important addition of the VB model is the identification of specific atomic orbitals on each atom that are assumed to be interacting to form bonds. Like the Lewis model, the electron sharing is presumed to be localized between pairs of atoms, so there is a strong correlation between orbital interactions and bonding linkages in a polyatomic molecule. Because of this, the VB approach tends to be more intuitively satisfying, even though it is often less successful in accounting for some of the quantitative details of molecular electronic structure.

In the VB approach, the electron density between bonding atoms can be defined on the basis of a new wave function, which in its simplest formulation is a product of the wave functions of the interacting orbitals. For example, consider two hydrogen atoms, each with a single electron, coming together to form a bond. The simplest product wave function for this interaction would have the form

$$\Psi = [1s_a(1)][1s_b(2)] \tag{4.7}$$

The subscripts a and b represent the two atoms with their associated electrons 1 and 2, respectively, each assigned to a $1s$ orbital on each atom. Actually, this trial wave function does not give very satisfactory agreement with the experimentally determined dependence of energy on internuclear separation (the Morse curve). However, it can be improved by modifications that account for the indistinguishability of electrons in a bond, the effective nuclear charge resulting from the two nuclei, and possible contributions to bonding from other orbitals (e.g., $2p_z$). With these refinements, the empirical wave function gives a satisfactory agreement with experimental results.

This general approach, which constructs localized bonds between pairs of atoms, can be extended to polyatomic molecules. However, in such cases, due consideration must be given to the geometrical orientations of the various atoms in the molecule—that is, the shape of the molecule. According to the Valence Shell Electron Pair Repulsion Theory (VSEPR theory) of Gillespie and Nyholm,* the shapes of molecules can be understood on the basis of minimization of electron coulombic repulsions and considerations of atomic sizes. The lengths and strengths of bonds are a consequence of these factors and the effective distribution of electron density throughout the molecule. In the VB approach, the total molecular electron density is partitioned into regions corresponding to bonds between pairs of atoms. Taking this approach, it follows that the electron density associated with a bond is the result of effective overlap between appropriately oriented atomic orbitals.

The need for appropriately oriented atomic orbitals consistent with the molecular shape presents a difficulty, since the conventional atomic orbitals ($1s$, $2s$, $2p$, etc.) generally do not have the correct geometries. This difficulty led Linus Pauling to postulate the formation of hybrid orbitals consistent with the geometrical requirements of molecules. For example, we can conceive the formation of such orbitals for carbon in CH_4 by a hypothetical process such as that illustrated in Fig. 4.5. From the ^3P ground state of carbon, arising from the configuration $1s^2 2s^2 2p^2$, energy would be required first to achieve the ^5S state, in which all four valence electrons are unpaired ($1s^2 2s^1 2p^3$). Additional energy would then be required to create four tetrahedrally directed sp^3 hybrid orbitals from the $2s$ and $2p$ orbitals. The resulting hypothetical state, V_4, would provide the carbon atom with four appropriately oriented orbitals, each with a single electron, to overlap with the four hydrogen $1s$ orbitals, leading to electron pairing and bond formation. The accompanying release of energy with bond formation, giving a lower total energy state, is seen as the impetus for hybrid formation.

*A discussion of VSEPR theory can be found in most introductory chemistry texts or the following: R. J. Gillespie and R. S. Nyholm, *Q. Rev.* **1957**, *XI*, 339; R. J. Gillespie, *J. Chem. Educ.* **1970**, *47*, 18; *ibid.* **1974**, *51*, 367; *ibid.*, **1992**, *69*, 116.

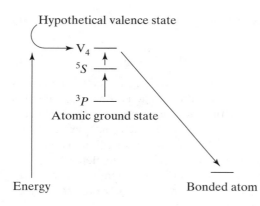

Hypothetical valence state

Atomic ground state

Energy

Bonded atom

Figure 4.5 Hypothetical energy process for the formation of carbon sp^3 hybrid orbitals.

It is important to realize that this hypothetical process is merely a convenient artifice for reconciling the formation of directed bonds with our notions of conventional atomic orbitals. The wave functions for the familiar atomic orbitals are but one set of solutions to the Schrödinger equation, and other solutions may be found as sums of these. In the case of tetrahedrally bonded carbon, combinations of $2s$, $2p_x$, $2p_y$, and $2p_z$ lead to the following four wave functions:

$$\Psi_1 = \frac{1}{2}\ (s + p_x + p_y + p_z) \tag{4.8a}$$

$$\Psi_2 = \frac{1}{2}\ (s + p_x - p_y - p_z) \tag{4.8b}$$

$$\Psi_3 = \frac{1}{2}\ (s - p_x + p_y - p_z) \tag{4.8c}$$

$$\Psi_4 = \frac{1}{2}\ (s - p_x - p_y + p_z) \tag{4.8d}$$

Each resulting orbital has the same shape (as shown in Fig. 4.6) and is oriented at 109° 28′ from any other orbital of the set. The wave functions for sp^3

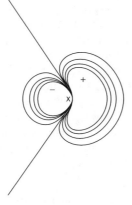

Figure 4.6   Contour diagram of a single sp^3 hybrid orbital. The two straight lines indicate the two-dimensional projection of a conical node. Indicated values for contours are calculated values of ψ. [Reproduced with permission from A. Streitwieser, Jr. and P. H. Owens, *Orbital and Electron Density Diagrams*, Macmillan, New York, 1973.]

Contours at 0.0, 0.002, 0.004, 0.006, and 0.008

hybrid orbitals, cast in this manner, are equally valid solutions to the Schrödinger equation for carbon as are the wave functions for the $2s$ and $2p$ orbitals themselves. The distinction between the hybrid orbitals and the standard atomic orbitals is merely a matter of convenience. The hybrid set is better for describing the tetrahedrally bonded carbon atom, while the standard set of $2s$ and $2p$ orbitals is better for describing the free atom. Indeed, as Linus Pauling pointed out, "if quantum theory had been developed by the chemist rather than the spectroscopist it is probable that the tetrahedral orbitals . . . would play the fundamental role in atomic theory, in place of the s and p orbitals."*

We have considered the case of sp^3 hybrid orbitals only because they are likely to be the most familiar. Actually, the sp^3 combination of standard atomic orbitals is not the only one that can be involved in tetrahedral bonding. We can identify all possible sets of orbitals that lead to tetrahedral geometry by using techniques of group theory. In approaching the problem, we must recognize the ways in which the standard atomic orbitals will transform in any point group. We have seen that s orbitals, which in isolation could be described by the full rotation group R_3, are symmetric with respect to all symmetry operations. As a result, in any molecular point group an s orbital at the center of the system will transform as the totally symmetric representation. The orientation of the three p orbitals along the cardinal axes of the coordinate system allows them to be represented by a unit vector along the same axis. As a result, p orbitals transform by the same species as the corresponding unit vectors (cf. Section 2.4), which are noted in the next-to-last column of each character table. Finally, the d orbitals transform according to the species of their corresponding direct products (cf. Section 2.4), as listed in the last column of each character table.

Disregarding our knowledge of sp^3 hybrid orbitals for the moment, suppose we wish to determine which specific atomic orbitals could be combined to form a set of four tetrahedrally directed hybrid orbitals. We will presume that such a set can be formed, and we will make it the basis for a representation in the point group of a tetrahedron, T_d. If we subject this basis set to the operations of T_d, we can deduce the characters of a reducible representation, which we shall call Γ_t. Reducing the representation Γ_t into its component irreducible representations will enable us to identify conventional orbitals with the requisite symmetries to form a tetrahedral set of hybrids.

We will represent the four equivalent hybrid orbitals by vectors, as shown in Fig. 4.7, labeled A, B, C, and D so that the effects of the symmetry operations of T_d can be followed. The base of each vector is at the center of the system and will not be shifted by any operation of the group. Therefore, we only need to consider the effects of the operations on the tips of the vectors. We will use 4×1 matrices to represent the positions before and after each

*Linus Pauling, *The Nature of the Chemical Bond*, 3rd ed., Cornell University Press, Ithaca, NY, 1960, p. 113.

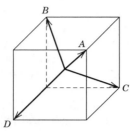

Figure 4.7   Vector basis for a representation of tetrahedral orbitals in T_d.

operation. The transformation effected by each operation will be represented by a 4×4 transformation matrix, whose trace will generate a character for the operation, which in turn will become part of our reducible representation Γ_t in T_d. The group T_d consists of 24 operations grouped into five classes. Recall that for any representation the characters for all operations in the same class must be the same. Therefore, we will only need to determine the effect produced by any one operation of each class in order to determine the character for all operations of the class. The effects of representative operations of each class are illustrated in Fig. 4.8.

For the identity operation, E, all vectors remain in their original positions. The effect, expressed in matrix notation, is

$$\begin{bmatrix} 1 & 0 & 0 & 0 \\ 0 & 1 & 0 & 0 \\ 0 & 0 & 1 & 0 \\ 0 & 0 & 0 & 1 \end{bmatrix} \begin{bmatrix} A \\ B \\ C \\ D \end{bmatrix} = \begin{bmatrix} A \\ B \\ C \\ D \end{bmatrix} \tag{4.9}$$

The trace of the 4×4 transformation matrix gives a character of 4 for the identity operation for our reducible representation.

For the C_3 operation, the projection illustrated in Fig. 4.8 shows the operation for the axis passing through A and the center of the system. Since A lies along the axis, its position is not changed by the C_3 rotation. Rotating in the clockwise manner shown, the positions of B, C, and D are interchanged as follows: B is replaced by D, C is replaced by B, and D is replaced by C. In matrix notation, this is

$$\begin{bmatrix} 1 & 0 & 0 & 0 \\ 0 & 0 & 0 & 1 \\ 0 & 1 & 0 & 0 \\ 0 & 0 & 1 & 0 \end{bmatrix} \begin{bmatrix} A \\ B \\ C \\ D \end{bmatrix} = \begin{bmatrix} A \\ D \\ B \\ C \end{bmatrix} \tag{4.10}$$

The resulting character from the transformation matrix is 1.

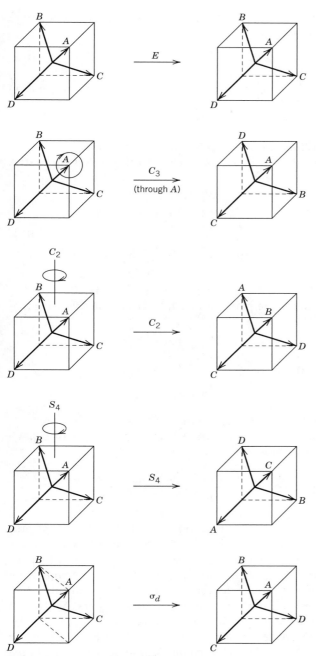

Figure 4.8 Effects of representative operations of T_d on the four-vector basis for a representation of tetrahedral hybrid orbitals.

The C_2 operation illustrated in Fig. 4.8 interchanges A and B and also C and D. In matrix notation, this is

$$\begin{bmatrix} 0 & 1 & 0 & 0 \\ 1 & 0 & 0 & 0 \\ 0 & 0 & 0 & 1 \\ 0 & 0 & 1 & 0 \end{bmatrix} \begin{bmatrix} A \\ B \\ C \\ D \end{bmatrix} = \begin{bmatrix} B \\ A \\ D \\ C \end{bmatrix} \qquad (4.11)$$

The transformation matrix in this case yields a character of 0.

The S_4 operation illustrated in Fig. 4.8 leads to the following matrix equation:

$$\begin{bmatrix} 0 & 0 & 1 & 0 \\ 0 & 0 & 0 & 1 \\ 0 & 1 & 0 & 0 \\ 1 & 0 & 0 & 0 \end{bmatrix} \begin{bmatrix} A \\ B \\ C \\ D \end{bmatrix} = \begin{bmatrix} C \\ D \\ B \\ A \end{bmatrix} \qquad (4.12)$$

This, too, yields a zero character for the transformation matrix.

Finally, the σ_d reflection in the plane of Fig. 4.8 leads to the following matrix equation:

$$\begin{bmatrix} 1 & 0 & 0 & 0 \\ 0 & 1 & 0 & 0 \\ 0 & 0 & 0 & 1 \\ 0 & 0 & 1 & 0 \end{bmatrix} \begin{bmatrix} A \\ B \\ C \\ D \end{bmatrix} = \begin{bmatrix} A \\ B \\ D \\ C \end{bmatrix} \qquad (4.13)$$

This gives a character of 2 for the transformation matrix.

Gathering all the characters from the transformation matrices of Eq. 4.9 through 4.13, we obtain the following representation:

T_d	E	$8C_3$	$3C_2$	$6S_4$	$6\sigma_d$
Γ_t	4	1	0	0	2

If we compare the characters of this representation with the actions that produced them, we can see a general result that will speed construction of reducible representations of this type in other cases. Notice that the character for each operation is equal to the number of vectors that are not shifted by the operation. In the present case, E leaves all four vectors nonshifted; C_3 leaves one vector, which lies on the axis, nonshifted; both C_2 and S_4 shift all vectors, since none lies on the axis; and a σ_d reflection leaves two vectors, which lie in the plane, nonshifted. Consequently, *to construct the reducible representation for any set of hybrid orbitals, count the number of vectors in the basis set that remain nonshifted by a representative operation of each class in the point group of the system. The number of nonshifted vectors is the character for the class in each case.*

The reduction can be accomplished by using Eq. (3.1) and the tabular method, explained in Section 3.1. The T_d character table needed to carry out this process can be found in Appendix A (or see Table 3.1). The work sheet for the reduction is shown below.

T_d	E	$8C_3$	$3C_2$	$6S_4$	$6\sigma_d$		
Γ_t	4	1	0	0	2	Σ	$\Sigma/24$
A_1	4	8	0	0	12	24	1
A_2	4	8	0	0	-12	0	0
E	8	-8	0	0	0	0	0
T_1	12	0	0	0	-12	0	0
T_2	12	0	0	0	12	24	1

From this we see that $\Gamma_t = A_1 + T_2$. Note that the dimension of Γ_t, $d_r = 4$, is satisfied by this sum of one nondegenerate representation and one triply degenerate representation. This reduction can be verified by summing the characters of A_1 and T_2:

T_d	E	$8C_3$	$3C_2$	$6S_4$	$6\sigma_d$
A_1	1	1	1	1	1
T_2	3	0	-1	-1	1
Γ_t	4	1	0	0	2

Our results show that a set of hybrid orbitals with tetrahedral geometry can be constructed by making suitable mathematical combinations of any orbital of A_1 symmetry with any set of three *degenerate* orbitals of T_2 symmetry. In other words, $\Gamma_t = A_1 + T_2$ is a kind of recipe for constructing the desired hybrids. We now need to determine which combinations of specific atomic orbitals fit this recipe.

Since s orbitals transform as the totally symmetric representation (A_1 in T_d), we know we can always include an s orbital as our A_1 orbital in the set of four used to construct the hybrids. To be sure that no other orbitals might be used as the A_1 choice (e.g., any individual p or d orbitals), we check the next-to-last and last columns of the T_d character table (cf. Appendix A). For A_1 there are no listings of vectors or direct products that correspond to atomic orbitals.* It appears, then, that any tetrahedral hybrid set must include an s orbital and no other as the A_1 contribution. This does not mean, however, that p or d orbitals cannot contribute to the hybrids, but if they do they can only contribute as part of the T_2 set.

To determine possible contributing orbitals for the triply degenerate set, we look at the vector and direct product listings for the species T_2. Here we find degenerate unit vectors (x, y, z), which in this context signify the three p orbitals p_x, p_y, and p_z. Likewise, we find the degenerate direct products $(xy,$

*Note that the notation $x^2 + y^2 + z^2$ does not correspond to any of the conventional d orbitals.

xz, yz), which signify a degenerate set of the three specific d orbitals, d_{xy}, d_{xz}, and d_{yz}. Since there are two sets of orbitals with T_2 symmetry, there are two possible choices of triply degenerate orbitals to include in constructing a hybrid set. We can construct one set of hybrids as a combination of s (A_1) with the three p orbitals p_x, p_y, and p_z (T_2) to obtain the familiar sp^3 hybrids, or we can construct another set as a combination of s (A_1) with the three d orbitals d_{xy}, d_{xz}, and d_{yz} (T_2) to obtain less familiar sd^3 hybrids. The wave functions for the sp^3 hybrids have been shown previously as Eq. (4.8a)–(4.8d). The wave functions of the sd^3 hybrids have the same mathematical form, except that the wave functions for the appropriate d orbitals replace those of the p orbitals in Eq. (4.8a)–(4.8d).

The question may arise, "Which hybrid set is correct?" The answer depends upon the energies of d orbitals for the central atom, and whether or not they are appropriate for significant participation in the bonding. For example, in carbon the energy of the d orbitals lies so much higher than that of the p orbitals that bonding in species such as CH_4 and CCl_4 is adequately described in the VB model by invoking sp^3 hybrids with virtually no sd^3 contribution. For molecules with main group central atoms, such as SO_4^{2-}, d orbital involvement is generally minimal.* However, if the central atom is a transition metal (as in CrO_4^{2-}), d orbital participation in the bonding may be significant.

Note that the $d_{x^2-y^2}$ and d_{z^2} orbitals cannot contribute to forming a tetrahedral set of hybrid orbitals, since in T_d they transform as the species E, rather than the required species A_1 or T_2. Also note that either hybrid set allowed by symmetry is comprised of four conventional atomic orbitals, the same as the number of hybrids being constructed. This correspondence between the number of atomic orbitals used and the number of hybrid orbitals formed is general.

The approach we have taken with tetrahedral hybrids can be used to deduce appropriate atomic orbital combinations for hybrids consistent with the other primary shapes predicted by VSEPR theory (e.g., linear, trigonal planar, trigonal bipyramidal, and octahedral). As an example, let us consider constructing a set of two linear hybrid orbitals on a central atom. Such a set might be used to describe bonding in a linear MX_2 or MXY species, such as BeH_2, CO_2, or $BeFCl$. Previous familiarity with hybrid orbitals should suggest sp hybrids as one possibility. The question here is whether there are other combinations of atomic orbitals that could be used, as well.

The linear geometry of two equivalent hybrid orbitals belongs to the point group $D_{\infty h}$. Accordingly, we should take two vectors directed 180° from one another as the basis for our representation in the group. This will yield a

---

*For further discussions of d orbital involvement in bonding with main-group elements see L. Suidan, J. K. Badenhoop, E. D. Glendening, and F. Weinhold, *J. Chem. Educ.*, **1995**, *72*, 583; and W. Kutzelnigg, *Angew. Chem. Int. Ed. Engl.*, **1984**, *23*, 272.

reducible representation Γ_l whose dimension is $d_r = 2$. As a representation in an infinite-order point group, Γ_l cannot be reduced by applying Eq. (3.1). However, with such a small dimension the reduction can be accomplished by inspection, and it should not be necessary to use the technique described in Section 3.4.

As suggested in our working of the tetrahedral example, we can generate the characters for all classes of the reducible representation simply by noting the numbers of vectors that remain nonshifted by any operation in each class. Our vector basis is simply the following two vectors:

$$\longleftarrow \cdot \longrightarrow$$

From the character table in Appendix A, we see that the listed operations of $D_{\infty h}$ are E, $2C_\infty^\Phi$, $\infty\sigma_v$, i, $2S_\infty^\Phi$, ∞C_2. The notations $2C_\infty^\Phi$ and $2S_\infty^\Phi$ are proper and improper rotations, respectively, of any order in both directions about the z axis, in which both vectors lie. The notation $\infty\sigma_v$ indicates the infinite number of mirror planes that intersect along z, and ∞C_2 indicates the infinite number of dihedral axes that are perpendicular to the z axis. The operations E, $2C_\infty^\Phi$, and $\infty\sigma_v$ leave both vectors nonshifted. All other operations shift both vectors. This leads to the following reducible representation:

$D_{\infty h}$	E	$2C_\infty^\Phi$	...	$\infty\sigma_v$	i	$2S_\infty^\Phi$	...	∞C_2
Γ_l	2	2	...	2	0	0	...	0

By inspection (cf. $D_{\infty h}$ character table in Appendix A), we can see that this decomposes as $\Gamma_l = \Sigma_g^+ + \Sigma_u^+$. The proof of this is shown by summing the characters of the Σ_g^+ and Σ_u^+ representations:

$D_{\infty h}$	E	$2C_\infty^\Phi$	...	$\infty\sigma_v$	i	$2S_\infty^\Phi$	...	∞C_2
Σ_g^+	1	1	...	1	1	1	...	1
Σ_u^+	1	1	...	1	-1	-1	...	-1
Γ_l	2	2	...	2	0	0	...	0

This result means that we can construct a set of linear hybrid orbitals by taking combinations of one orbital of Σ_g^+ symmetry with one orbital of Σ_u^+ symmetry. Once again, since Σ_g^+ is the totally symmetric representation in $D_{\infty h}$, we can use an s orbital as part of a hybrid set. Beyond this, the direct product listing z^2 in the character table (cf. Appendix A) suggests that a d_{z^2} orbital could be used in place of an s orbital in such a set.* Turning to possible orbitals of Σ_u^+ symmetry, the character table shows only a unit vector listing z, suggesting a p_z orbital. Thus, we can construct two sets of hybrid or-

---

*The direct product listing $x^2 + y^2$ for Σ_g^+ in $D_{\infty h}$ does not correspond to one of the conventional d orbitals and can be ignored.

bitals: the familiar sp set, made of s and p_z orbitals; and a dp set, made of d_{z^2} and p_z orbitals. Using the sp set as a model, the hybrid orbital wave functions have the form

$$\Psi_1 = \frac{1}{\sqrt{2}}(s + p_z) \tag{4.14a}$$

$$\Psi_2 = \frac{1}{\sqrt{2}}(s - p_z) \tag{4.14b}$$

The shape of each sp hybrid is very similar to that of a single sp^3 hybrid, as shown in Fig. 4.6, except that the sp hybrid has 50% p character (compared to 75% for sp^3), and the node is tighter about the lobe with the negative sign. The dp set has the same mathematical form as Eq. (4.14a) and (4.14b) with the substitution of d_{z^2} for s. As with the previously discussed tetrahedral hybrids, dp hybrids could make a contribution to a VB model of bonding in a molecule with a central atom in the third or higher periods of the periodic table.

4.3   *Localized and Delocalized Molecular Orbitals*

In the VB approach the interacting orbitals in a chemical bond may be viewed as retaining much of their atomic character. By contrast, the molecular orbital (MO) approach seeks to construct new wave functions that define unique orbitals for the bonded system. In principle, the task amounts to defining the Schrödinger wave equation for a system in which the nuclei of the individual atoms are treated as if they formed a polycentric nucleus embedded in an electron distribution that surrounds the entire molecule. Solutions to this wave equation define MOs with characteristic energies. If the relative energies of these MOs are known, the ground-state configuration of the molecule can be deduced by an aufbau process, much like that used for determining the ground state configurations of single atoms. As with the case of isolated atoms, the aufbau process obeys the Pauli Exclusion Principle and Hund's Rule of Maximum Multiplicity.

    In practice, the exact construction and solution of a molecular wave equation is not feasible except in the simplest diatomic cases. Therefore, it is customary to construct empirical wave functions as mathematical sums of wave functions on the various atoms of the molecule. This approach is called the *Linear Combination of Atomic Orbitals (LCAO)* method. In the case of a general diatomic molecule, AB, wave functions for the molecule take on the forms

$$\Psi_1 = a\psi + b\psi \tag{4.15a}$$

$$\Psi_2 = a\psi - b\psi \tag{4.15b}$$

in which a and b are constants, sometimes called *mixing constants*, that reflect the relative contributions of each wave function to the LCAO wave function. The atomic orbitals (AOs) used to construct these MOs (a) must have similar energies, (b) must overlap appreciably, and (c) must have the same symmetry with respect to the internuclear axis. The last stipulation, understandably, has significance for applying group theory to MO construction in polyatomic cases.

If Eq. (4.15a) results in reenforcement ($S > 0$), the resulting LCAO will define a bonding MO. If that is the case, then Eq. (4.15b) defines an antibonding MO ($S < 0$), which will have a node between the two nuclei. The antibonding LCAO-MO always has a higher (less favorable) energy than the corresponding bonding LCAO-MO.

Hybridization is not an essential feature of the MO approach. However, if we recall that hybrid orbitals arise from wave functions that are as legitimate solutions to an atom's Schrödinger equation as those that give rise to the conventional orbitals, it is reasonable to suppose that hybrid orbitals could be used as a starting point in an LCAO-MO treatment of bonding in polyatomic molecules. Assuming hybridization on the central atom, *a priori*, is most often useful when taking a *localized MO approach*. A localized MO approach is essentially an extension of the VB model, whereby molecular orbitals are defined for pairs of atoms that are connected by chemical bonds. The LCAO wave functions in this limited approach involve only two atoms and take on forms such as Eqs. (4.15a) and (4.15b).

As an illustration of a localized MO model, consider the bonding in gaseous BeH_2. The Lewis model for this species is simply H–Be–H, from which VSEPR theory predicts a linear structure ($D_{\infty h}$). The electron density about the molecule is represented by the contour diagram shown in Fig. 4.9. The VB approach implicitly partitions this distribution into two equivalent Be–H bonds, each formed by overlap of one sp hybrid orbital on the beryllium atom with a $1s$ orbital on a hydrogen atom. We can extend this model to become a localized MO model by defining wave functions between pairs of adjacent atoms with the following forms:

$$\sigma_1 = a[sp(1)_{Be}] + b[1s_{H'}] \tag{4.16a}$$

$$\sigma_2 = a[sp(2)_{Be}] + b[1s_{H''}] \tag{4.16b}$$

$$\sigma_3^* = a[sp(1)_{Be}] - b[1s_{H'}] \tag{4.16c}$$

$$\sigma_4^* = a[sp(2)_{Be}] - b[1s_{H''}] \tag{4.16d}$$

Here hybrid orbital $sp(1)$ on beryllium points toward hydrogen atom H′, and hybrid orbital $sp(2)$ points toward hydrogen atom H″. Equations (4.16a) and (4.16b) lead to sigma-bonding MOs, and Eqs. (4.16c) and (4.16d) lead to sigma-antibonding MOs (distinguished by superscript asterisk notation). The

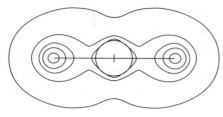

Contours at 0.01, 0.07, 0.13, 0.19, and 0.25

Figure 4.9   Electron density about BeH_2. Indicated values for contours are calculated values of ψ. [Reproduced with permission from A. Streitwieser, Jr. and P. H. Owens, *Orbital and Electron Density Diagrams*, Macmillan, New York, 1973.]

antibonding MOs σ_3^* and σ_4^* have no electrons in the ground state of BeH_2. Figure 4.10 shows a contour diagram for one of the localized sigma-bonding MOs. As might be expected, it bears great similarity to an *sp* hybrid, magnified by reenforcing overlap with the 1*s* orbital on the hydrogen atom.

On the basis of this localized MO model, we can construct a qualitative molecular orbital energy level diagram, such as shown in Fig. 4.11. Diagrams such as this show, by means of tie lines, which AOs are used in formulating bonding, antibonding, and nonbonding MOs by the LCAO approach. The ordering of levels of both AOs and MOs is meant to suggest the relative energy order of the orbitals, from lowest at the bottom (most stable) to highest at the top (least stable). The filling of electrons in the MOs follows in an aufbau manner, using the available electrons from the participating atoms.

The scheme shown in Fig. 4.11 includes the beryllium $2p_x$ and $2p_y$ orbitals, which do not have the appropriate symmetry for either bonding or antibonding overlap with the hydrogen 1*s* orbitals ($S=0$). As such, they form nonbonding π^n orbitals, which are essentially the 2*p* orbitals perturbed by the presence of the two hydrogen nuclei at relatively close proximity.* The beryllium 1*s* orbital and its two electrons are not shown in this scheme, because it is assumed that these are not involved with the bonding. As such they are considered to be nonbonding core electrons.

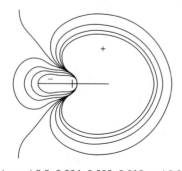

Contours at 0.0, 0.004, 0.008, 0.012, and 0.016

Figure 4.10   Contour diagram of a localized Be–H molecular orbital of BeH_2. Indicated values for contours are calculated values of ψ. [Reproduced with permission from A. Streitwieser, Jr. and P. H. Owens, *Orbital and Electron Density Diagrams*, Macmillan, New York, 1973.]

*These levels are designated π^n, rather than σ^n, because their symmetry would permit π-MO formation if similar orbitals were available on the pendant (outer) atoms.

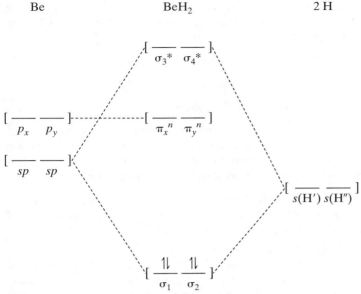

Be BeH_2 2 H

Figure 4.11 Qualitative localized MO energy level scheme for BeH_2.

The localized MO scheme for BeH_2 suggests that two pairs of electrons are localized in degenerate bonding MOs (i.e., orbitals with equivalent energies). This is, however, somewhat misleading. The equivalence of the bonding sigma MOs is artificial, since in setting up the problem we constrained the electrons to be localized in two equivalent regions. In other words, this result is an artifact of the way in which we chose to partition the total electron density shown in Fig. 4.9. Nonetheless, such localized MO models are useful for discussions of chemical bonds and accounting for bonding and nonbonding electron pairs. Moreover, it conforms to familiar notions of bonds and electron pairing growing out of classical Lewis models.

If we do not constrain electrons to localized bonds, *a priori*, we will obtain a *general* or *delocalized MO* model. This approach usually yields MOs with energies that are more consistent with electronic spectra and ionization energies. However, this approach requires that we abandon the VB notion of a chemical bond as the sharing of a pair of electrons by two adjacent atoms. In the general MO approach, MOs and the electrons associated with them typically extend across the molecule.

Hybridization is usually not a starting assumption for a general MO treatment. Rather, we seek to identify bonding, antibonding, and nonbonding combinations as LCAOs of ordinary AOs on all atoms of the molecule. For simple molecules of the type MX_n we can approach the problem by matching symmetries of the orbitals on the central M atom with those of mathematical combinations of orbitals on the outer X atoms, called *pendant atoms*. These mathematical constructs of pendant-atom AOs are called *symmetry-*

adapted linear combinations (SALCs). Generally, the SALCs do not have physical meaning outside their use in constructing MOs. When SALCs are employed, the resulting LCAO-MOs take on the form

$$\Psi_{MO} = a\psi_{AO}(M) \pm b\psi_{SALC}(nX) \qquad (4.17)$$

in which the SALCs have the general form

$$\psi_{SALC} = c_1\psi_1 \pm c_2\psi_2 \pm c_3\psi_3 \pm \cdots \pm c_n\psi_n \qquad (4.18)$$

We can use techniques of group theory to determine the symmetries of possible SALCs formed from pendant-atom AOs. Then we can determine which SALCs will combine with which AOs on the central atom to form MOs by the LCAO method. The process involves the following steps:

1. Use the directional properties of potentially bonding orbitals on the outer atoms (shown as vectors on a model) as a basis for a representation of the SALCs in the point group of the molecule.

2. Generate a reducible representation for all possible SALCs by noting whether vectors are shifted or nonshifted by each class of operations of the group. Each vector shifted through space contributes 0 to the character for the class. Each nonshifted vector contributes 1 to the character for the class. A vector shifted into the negative of itself (base nonshifted but tip pointing in the opposite direction) contributes -1 to the character for the class.

3. Decompose the representation into its component irreducible representations to determine the symmetry species of the SALCs. The number of SALCs, including members of degenerate sets, must equal the number of AOs taken as the basis for the representation.

4. Determine the symmetries of potentially bonding central-atom AOs by inspecting unit vector and direct product transformations listed in the character table of the group. Remember that an *s* orbital on a central atom always transforms as the totally symmetric representation of the group.

5. Central-atom AOs and pendant-atom SALCs with the same symmetry species will form both bonding and antibonding LCAO-MOs.

6. Central-atom AOs or pendant-atom SALCs with unique symmetry (no species match between AOs and SALCs) form nonbonding MOs.

To contrast this approach with the localized MO approach, let us generate the general MO model for BeH_2. Keeping in mind that the orbitals on the various atoms must have similar energies for effective overlap, we can confine our considerations, as before, to the $2s$ and $2p$ orbitals on the beryllium atom and the $1s$ orbitals on the two hydrogen atoms. The two hydrogen $1s$ orbitals can only form sigma interactions, so we will represent them as two vectors pointing towards the central beryllium atom:

$$\longrightarrow \quad \cdot \quad \longleftarrow$$

The hydrogen $1s$ wave functions, which these vectors represent, form a basis for a representation in $D_{\infty h}$. The SALCs that can be formed from them will combine with beryllium AOs of the appropriate symmetry to form LCAO-MOs.

Only E, $2C_\infty^\Phi$, and $\infty\sigma_v$ leave both vectors nonshifted. All other operations interchange the two vectors, shifting them through space. This gives the following reducible representation:

$D_{\infty h}$	E	$2C_\infty^\Phi$	...	$\infty\sigma_v$	i	$2S_\infty^\Phi$	...	∞C_2
Γ_{SALC}	2	2	...	2	0	0	...	0

Note that this representation is the same as Γ_l, which we generated for two linear hybrid orbitals.* Thus, as with Γ_l, Γ_{SALC} decomposes into the sum of the two irreducible representations Σ_g^+ and Σ_u^+.

The result $\Gamma_{\text{SALC}} = \Sigma_g^+ + \Sigma_u^+$ means that one of the possible SALCs has the symmetry Σ_g^+, and the other has the symmetry Σ_u^+. It is not difficult to deduce the forms of these SALCs, since they are combinations of only two AOs. The Σ_g^+ combination is totally symmetric to all operations of the group, including inversion, which makes it a *gerade* function (hence the g subscript in the Mulliken symbol). This can only occur if both $1s$ wave functions on the two hydrogen atoms are combined in a positive sense. The Σ_u^+ combination is *ungerade*, which implies a change of sign with the operation of inversion. This would occur if one $1s$ wave function were taken in the positive sense and the other were taken in the negative sense (i.e., a subtractive combination). The two normalized SALCs, then, must have the form

$$\Phi_g = \frac{1}{\sqrt{2}}(1s_{\text{H}'} + 1s_{\text{H}''}) \tag{4.19a}$$

$$\Phi_u = \frac{1}{\sqrt{2}}(1s_{\text{H}'} - 1s_{\text{H}''}) \tag{4.19b}$$

These SALCs are illustrated in Fig. 4.12.

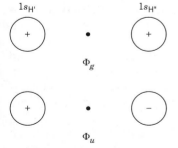

$1s_{\text{H}'}$ $1s_{\text{H}''}$

Φ_g

Φ_u

Figure 4.12   Symmetry-adapted linear combinations (SALCs) of hydrogen $1s$ orbitals for BeH_2.

*As we shall see in other cases, this occurs because there is no difference in symmetry between a set of vectors pointing out from the center of a system and a set of vectors with the same geometry pointing in toward the center of the system. Both give the same reducible representation.

In order for the SALCs to form bonding and antibonding combinations, AOs with the same symmetry properties must exist on the beryllium atom. The AOs on beryllium that we assume might be involved in bonding are $2s$ and $2p$, implying that $1s$ remains a nonbonding core orbital. As always, the s orbitals of the central atom transform as the totally symmetric representation, here Σ_g^+. The symmetries of p orbitals can be seen from the unit vector transformation properties listed in the $D_{\infty h}$ character table (cf. Appendix A). From those listings we see that p_z transforms as Σ_u^+ and that p_x and p_y transform as a degenerate pair by Π_u.

The Σ_g^+ symmetry of the $2s$ orbital on the beryllium atom matches that of the SALC Φ_g of the two hydrogen atoms. Following the form of Eq. (4.17), we can form bonding and antibonding MOs with the following wave functions:

$$\sigma_g = c_1 2s + c_2 \Phi_g \tag{4.20a}$$

$$\sigma_g{}^* = c_3 2s - c_4 \Phi_g \tag{4.20b}$$

The constants c_1, c_2, c_3, and c_4 are mixing constants that reflect the extent of interaction between the beryllium $2s$ orbitals and the hydrogen SALCs. The labels σ_g and σ_g^* follow customary practice, by which MOs are designated with lowercase equivalents to the Mulliken symbols. In similar manner, the $2p_z$ orbital matches the Σ_u^+ symmetry of the Φ_u SALC, giving MOs with the following wave functions:

$$\sigma_u = c_5 2p_z + c_6 \Phi_u \tag{4.20c}$$

$$\sigma_u^* = c_7 2p_z - c_8 \Phi_u \tag{4.20d}$$

Representations of these molecular orbitals are shown in Fig. 4.13. Note that these orbitals extend across the entire molecule and are not confined to individual bonds. Nonetheless, the bonding MOs are characterized by reenforcement in the region of each Be–H bond, while the antibonding MOs are characterized by a nodal plane passing between each Be–H pair.

The $2p_x$ and $2p_y$ orbitals on beryllium, which form a degenerate pair of Π_u symmetry, have no matching SALCs with the same symmetry. This indicates that they are incapable of forming either bonding or antibonding MOs. Therefore, they remain as a degenerate set of nonbonding orbitals, designated π^n.

Figure 4.14 shows a qualitative energy level scheme for the delocalized molecular orbitals of BeH_2. The σ_g-MO, which results from the most effective overlap and has no nodes, lies lower in energy than the σ_u-MO, which is interrupted at the beryllium atom by a nodal plane passing perpendicular to the plane of the projection in Fig. 4.13.

As with the localized MO scheme (Fig. 4.11), we fill this scheme in an aufbau manner with the four available valence electrons (ignoring the pair in the presumably nonbonding core $1s$ orbital on beryllium). This places two pairs in bonding sigma molecular orbitals, as does the localized model.

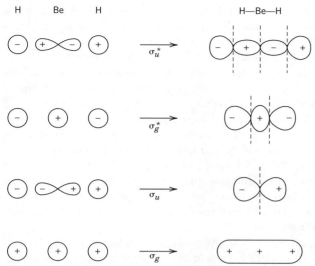

Figure 4.13   Linear combinations of atomic orbitals (LCAOs) and the resulting delocalized bonding (σ_g and σ_u) and antibonding (σ_g^* and σ_u^*) molecular orbitals (MOs) of BeH_2. Dashed lines indicate nodal planes perpendicular to the molecular axis. Orbital energy increases from bottom to top.

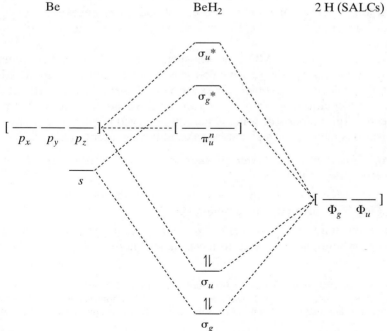

Figure 4.14   Qualitative delocalized MO energy level scheme for BeH_2.

However, the delocalized scheme shows that these two pairs do not have the same energy, a consequence of the distinct symmetries of the σ_g- and σ_u-MOs.* Both the delocalized model configuration $(\sigma_g)^2(\sigma_u)^2$ and the localized model configurations $(\sigma_1)^2(\sigma_2)^2$ give the same overall electron density for the molecule (cf. Fig. 4.9). The difference between the two models is essentially a matter of how that total electron density is partitioned. The delocalized model, however, gives a more realistic assessment of the relative energies of the bonding electrons. Without the artificial constraint of confinement to Be–H pairs, the two bonding electron pairs are suggested by symmetry to have different energies. Although data are lacking for BeH_2, energy distinctions such as this, suggested by the symmetry results of a delocalized approach, generally are consistent with experimental results (e.g., photoelectron spectroscopy). In other words, for BeH_2 we should not expect to find two pairs of electrons with exactly the same energy, contrary to the suggestion of the VB and localized MO models.

If electrons are delocalized across the molecule, as the general MO approach suggests, then the concept of bond order between two adjacent atoms in a molecule must be assessed in that context. In the case of BeH_2, the VB and localized MO models clearly indicate that both Be–H bonds are in every way equivalent, each being a single bond (bond order = B.O. = 1). The general MO approach in no way contradicts this. To arrive at the same conclusion, we must realize that the two pairs are in bonding MOs that extend equally across both Be–H bonds. This results in a total bond order of 2 over two identical Be–H bonds, or B.O. = 1 for each bond. In essence, each single bond results from sharing half the electron densities of both the σ_g- and σ_u-bonding MOs.

The ordering of levels in MO schemes such as Fig. 4.14 deserves some comment. Strictly speaking, the relative energies of MOs cannot be predicted without detailed calculations, subject to experimental verification. In the absence of such information, we can arrive at a tentative ordering of MOs for simple molecules on the basis of some generally observed results. The following generalizations may be used as guides to establishing a tentative ordering:

1. Bonding MOs always lie lower in energy than the antibonding MOs formed from the same AOs.

2. Nonbonding MOs tend to have energies between those of bonding and antibonding MOs formed from similar AOs.

3. Pi interactions tend to have less effective overlap than sigma interactions. Therefore, π-bonding MOs tend to have higher energies than σ-bonding

---

*It is possible under special circumstances for two orbitals of different symmetries to have virtually the same energy, a condition called *accidental degeneracy*. This, however, is more the exception than the rule, and one should generally assume that different symmetries imply different energies.

MOs formed from similar AOs. Likewise, π^* MOs tend to be less anti-bonding and have lower energies than σ^* MOs formed from similar AOs.

4. MO energies tend to rise as the number of nodes increases. Therefore, MOs with no nodes tend to lie lowest, and those with the greatest number of nodes tend to lie highest in energy.

5. Among σ-bonding MOs, those belonging to the totally symmetric representation tend to lie lowest.

As a summarizing example, consider the delocalized MO treatment of methane. It is reasonable to assume that the bonds are formed by interactions of the $1s$ orbitals on the four hydrogen atoms with $2s$ and $2p$ orbitals on the carbon atom. The carbon $1s$ electrons are assumed to be nonbonding. The four hydrogen atoms in their sigma interactions with carbon may be represented by a set of four tetrahedrally oriented vectors pointing toward the central atom (Fig. 4.15). These are taken as the basis for a representation in T_d to determine the symmetries of hydrogen SALCs. From a consideration of shifted and nonshifted vectors, the following reducible representation emerges:

T_d	E	$8C_3$	$3C_2$	$6S_4$	$6\sigma_d$
Γ_{SALC}	4	1	0	0	2

This representation is identical to Γ_t, which we generated for a set of four hybrid orbitals in Section 4.2. Therefore, as before, it decomposes as $\Gamma_{SALC} = A_1 + T_2$. This means that we can form one totally symmetric SALC and a set of three degenerate SALCs.

The totally symmetric SALC is formed by taking all four hydrogen $1s$ wave functions in a positive sense. Using labels A, B, C, and D to distinguish the hydrogen atoms, the normalized A_1 SALC can be written as

$$\Phi_1 = \frac{1}{2}\{1s_A + 1s_B + 1s_C + 1s_D\} \tag{4.21a}$$

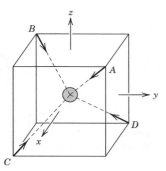

Figure 4.15 Vector basis for a representation of hydrogen SALCs of CH_4.

The three degenerate SALCs of T_2 are constructed by taking all possible combinations of two hydrogen wave functions in a positive sense with two in a negative sense. The normalized T_2 SALCs have the form

$$\Phi_2 = \frac{1}{2}\{1s_A + 1s_B - 1s_C - 1s_D\} \tag{4.21b}$$

$$\Phi_3 = \frac{1}{2}\{1s_A - 1s_B - 1s_C + 1s_D\} \tag{4.21c}$$

$$\Phi_4 = \frac{1}{2}\{1s_A - 1s_B + 1s_C - 1s_D\} \tag{4.21d}$$

On carbon, both the $1s$ and $2s$ orbitals have A_1 symmetry. If we assume that the $1s$ orbital is a nonbonding core orbital, then we may consider it as forming the nonbonding MO σ_1. The $2s$ orbital will form bonding and anti-bonding MOs with the A_1 SALC Φ_1. The resulting LCAO-MOs have the forms

$$\sigma_2 = c_1(2s) + c_2\Phi_1 \tag{4.22a}$$

$$\sigma_6^* = c_3(2s) - c_4\Phi_1 \tag{4.22b}$$

The form of the LCAO for σ_2 is shown in Fig. 4.16. The antibonding combination [Eq. (4.22b)] results from changing the wave function signs on the hydrogen atoms, which is identical to the negative of the Φ_1 SALC.

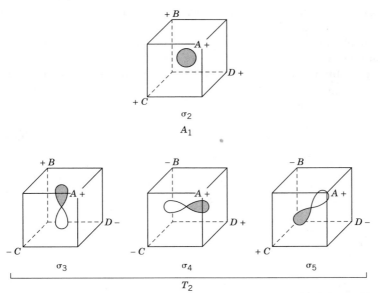

Figure 4.16   Representations of the bonding LCAOs of methane. The four hydrogen atom wave functions are labeled A, B, C, and D to correspond with the notation of Eqs. (4.21a)–(4.21d).

As the vector transformation properties in the T_d character table indicate, the three $2p$ orbitals have T_2 symmetry and will form bonding and antibonding combinations with the degenerate SALCs Φ_2, Φ_3, and Φ_4. With each SALC, the bonding LCAO is formed by matching a $2p$ orbital whose positive lobe is directed between two hydrogen wave functions with positive sign, and whose negative lobe is directed between two hydrogen wave functions with negative sign (cf. Fig. 4.16). The resulting wave functions for bonding and antibonding MOs are

$$\sigma_3 = c_5(2p_z) + c_6\Phi_2 \tag{4.22c}$$

$$\sigma_7^* = c_7(2p_z) - c_8\Phi_2 \tag{4.22d}$$

$$\sigma_4 = c_9(2p_y) + c_{10}\Phi_3 \tag{4.22e}$$

$$\sigma_8^* = c_{11}(2p_y) - c_{12}\Phi_3 \tag{4.22f}$$

$$\sigma_5 = c_{13}(2p_x) + c_{14}\Phi_4 \tag{4.22g}$$

$$\sigma_9^* = c_{15}(2p_x) - c_{16}\Phi_4 \tag{4.22h}$$

Bonding MOs σ_3, σ_4, and σ_5 are degenerate and necessarily have the same energy. Antibonding MOs σ_7^*, σ_8^*, and σ_9^* are likewise degenerate and have the highest energy of all the MOs we have defined for CH_4.

Figure 4.17 shows a qualitative MO scheme based on the LCAO-MOs defined by Eqs. (4.22a)–(4.22h). In the conventional manner, the energy levels are labeled with the appropriate Mulliken symbols, written in lowercase. This model contrasts noticeably with what might be expected from either a VB or localized MO model. Instead of four equal-energy electron pairs confined to four equivalent bonds, we see three pairs of electrons at one energy level and a single pair at a lower energy level. The total electron distribution predicted by this model is not significantly different from that predicted by the more localized models, and all approaches predict that all four C–H bonds are in every way equivalent. However, by the delocalized MO approach, the electron density of each C–H single bond is 75% from electrons in the degenerate t_2 MOs and 25% from electrons in the a_1 MO. The single-bond order can be rationalized by seeing that all four pairs are equally delocalized over the four C–H bonds, implying that each bond results from one-fourth of the four-pair density; that is, B.O. = 4/4 = 1.

The delocalized model of CH_4 predicts that there are two different energies of electrons with a population ratio of $3:1$. This is consistent with the observed *photoelectron spectrum*, which measures ionization energies of valence electrons by determining their kinetic energies after ejection by incident X-ray

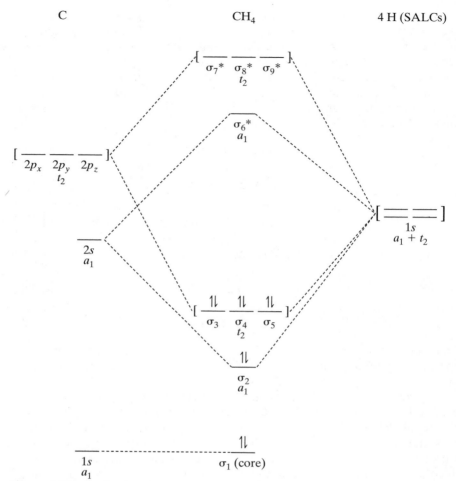

C CH_4 4 H (SALCs)

Figure 4.17 Qualitative delocalized MO energy level scheme for CH_4.

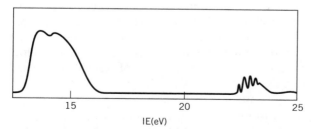

IE(eV)

Figure 4.18 Photoelectron spectrum of CH_4. [Adapted with permission of the Royal Society of Chemistry from A. W. Potts, T. A. Williams, and W. C. Price, *Faraday Disc. Chem. Soc.*, **1972**, *54*, 104.]

or UV radiation.* If the energy of the incident radiation, $h\nu$, is known and the kinetic energy, KE, of electrons ejected from a particular MO is measured, then the ionization energy of the electrons, IE, is given by

$$IE = h\nu - KE \tag{4.23}$$

The higher the ionization energy, the lower in energy lies the MO from which the electrons were ejected. The photoelectron spectrum of CH_4 (Fig. 4.18) shows a large band at approximately 13.5 eV and a smaller band at approximately 23.0 eV. The larger band corresponds to ejection of the three electron pairs in t_2 bonding MOs—namely, σ_3, σ_4, and σ_5. The smaller band arises from ejection of electrons from the lower-lying a_1 MO, σ_2, and therefore requires greater ionization energy. Both bands have considerable vibrational fine structure, characteristic of electrons ejected from bonding MOs.

Figure 4.18 does not show the band arising from the two electrons in the core level σ_1, which is essentially a $1s$ orbital on the carbon atom. This ionization energy is considerably higher (approximately 291 eV) than the energies required to eject electrons from the bonding MOs. X-ray photoelectron spectroscopy† is needed to observe the ionization band arising from σ_1.

The absence of four equal-energy pairs can be seen as a necessary consequence of methane's T_d symmetry. Since the highest-dimension irreducible representations of T_d are triply degenerate (T_1 and T_2), there can be no higher than threefold degeneracy among the MOs. In other words, the fourfold degeneracy among electron pairs that VB and localized MO models suggest is not allowed by the symmetry of the molecule. Therefore, while all models give comparable qualitative results for bond types and strengths, the delocalized approach generally gives more satisfactory predictions of electron energy levels, consistent with the molecular symmetry.

4.4 MX$_n$ Molecules with Pi-Bonding

The examples we have considered up to this point have only involved sigma interactions. Both BeH_2 and CH_4 have no pi-bonding interactions because the pendant hydrogen atoms' $2p$ orbitals lie so much higher in energy, and there are too few electrons for additional bonding beyond the sigma interac-

---

*Brief introductory treatments of photoelectron spectroscopy may be found in some advanced inorganic chemistry textbooks, such as G. L. Miessler and D. A. Tarr, *Inorganic Chemistry*, Prentice-Hall, Englewood Cliffs, NJ, 1991, p. 132*ff.*, and I. S. Butler and J. F. Harrod, *Inorganic Chemistry: Principles and Applications*, Benjamin/Cummings, Redwood City, CA, 1989, pp. 230–237. More extensive treatments include D. W. Turner, C. Baker, A. D. Baker, and C. R. Brundle, *Molecular Photoelectron Spectroscopy*, John Wiley & Sons, London, 1970; C. R. Brundle and A. D. Baker, eds., *Electron Spectroscopy: Theory, Techniques, and Applications*, Academic Press, London, 1977, Vols. I and II; and A. D. Baker and D. Betteridge, *Photoelectron Spectroscopy—Chemical and Analytical Aspects*, Pergamon, Oxford, 1972.
†U. Gelius, *J. Electronic Spectr.* **1974**, *5*, 985.

tions. If the pendant atoms are members of the second or higher periods of the periodic table, pi interactions with the p orbitals on the central atom may be possible.

As an example of a simple molecule with both sigma and pi bonding, consider carbon dioxide. Its 16 valence electrons result in the familiar Lewis structure,

$$\ddot{\text{O}}\!=\!\text{C}\!=\!\ddot{\text{O}}$$

As predicted by VSEPR theory, carbon dioxide is a linear molecule (point group $D_{\infty h}$). In the VB model, the carbon atom is assumed to have sp hybrid orbitals, each with two electrons, which form separate sigma bonds by overlap and electron-pair sharing with a $2p_z$ orbital on each oxygen. This leaves "empty" $2p_x$ and $2p_y$ orbitals on the carbon atom, which can form pi bonds by overlap with a $2p_x$ orbital on one oxygen atom and a $2p_y$ orbital on the other oxygen atom. In a "bookkeeping" sense, the oxygen $2p$ orbitals can be considered to provide the electron pairs for the resulting pi-bond formation. For example, if we assume that one oxygen atom (call it O_a) shares a pair from its $2p_x$ orbital by overlap with and donation to a $2p_x$ orbital on the carbon atom, then the other oxygen atom (call it O_b) shares a pair from its $2p_y$ orbital by overlap with and donation to the $2p_y$ orbital on the carbon atom. This simplistic model predicts two equivalent C=O double bonds ($\sigma + \pi$), with the maximum C–O_a pi overlap in the xz plane and the maximum C–O_b pi overlap in the yz plane. This leaves each oxygen atom with a nonbonding $2s$ and a nonbonding $2p$ orbital (either $2p_x$ or $2p_y$, depending on which was not used in pi-bond formation), each with two electrons.

A localized MO model follows directly from this VB model. Pairs of equivalent sigma-bonding MOs (σ_a and σ_b) would lie lowest in energy, with pairs of equivalent pi-bonding MOs (π_a and π_b) lying somewhat higher in energy. The highest-energy occupied MOs would be the nonbonding $2s$ and $2p$ orbitals localized on the oxygen atoms, designated σ_a^n and σ_b^n, and π_a^n and π_b^n, respectively. Unoccupied pairs of σ^* and π^* antibonding MOs would lie above these. The electronic configuration for this localized MO model is

$$[(\sigma_a)^2(\sigma_b)^2][(\pi_a)^2(\pi_b)^2][(\sigma_a^n)^2(\sigma_b^n)^2][(\pi_a^n)^2(\pi_b^n)^2]$$

On the basis of either the VB or localized MO models, we might expect each bond to have the typical C=O bond length of 124 pm. Instead, the observed length is 116 pm, suggesting a somewhat stronger bond. This is sometimes rationalized by admitting the following resonance forms to the Lewis description:

$$\ominus:\ddot{\text{O}}\!-\!\text{C}\!\equiv\!\text{O}:\oplus \quad \longleftrightarrow \quad \oplus:\text{O}\!\equiv\!\text{C}\!-\!\ddot{\text{O}}:\ominus$$

On the basis of our localized MO model, we might rationalize the shorter bonds by recognizing that it is equally probable that the maximum C–O_a pi interaction might be in the yz plane, while the maximum C–O_b pi interaction is in the xz plane, instead of the reversed orientations previously described.

This is tantamount to postulating delocalization of π electrons across the entire molecule. We can anticipate, then, that a general MO approach, which inherently assumes delocalization, might yield a more satisfying model.

In setting up a general MO treatment, we will assume initially that the oxygen $2s$ electrons do not participate in bond formation, as we did with the VB and localized MO approaches. Having for the moment ruled out interactions with carbon AOs, we can assume that the two SALCs that can be defined for these orbitals will be equivalent to two nonbonding σ^n MOs, whose forms are

$$\sigma_g^n\,(O_{2s}) = \frac{1}{\sqrt{2}}(2s_a + 2s_b) \tag{4.24a}$$

$$\sigma_u^n\,(O_{2s}) = \frac{1}{\sqrt{2}}(2s_a - 2s_b) \tag{4.24b}$$

These belong to the species Σ_g^+ and Σ_u^+, respectively, in $D_{\infty h}$. They are identical in form to the hydrogen SALCs of BeH_2 shown in Fig. 4.12.

Having accounted for the $2s$ orbitals, we can turn our attention to forming SALCs among the $2p$ orbitals on the two oxygen atoms. These will form MOs by combination with the $2s$ and $2p$ orbitals on the central carbon atom. Figure 4.19 shows a set of six vectors (three on each oxygen atom, for the $2p_x$, $2p_y$, and $2p_z$ orbitals), which we will take as the basis for a representation for oxygen SALCs. To avoid the problem of infinite order, we will generate the representation in the subgroup D_{2h}, rather than the true group of CO_2, $D_{\infty h}$, employing the correlation technique described in Section 3.4.

In D_{2h} we obtain the following reducible representation:

D_{2h}	E	$C_2(z)$	$C_2(y)$	$C_2(x)$	i	$\sigma(xy)$	$\sigma(xz)$	$\sigma(yz)$
Γ_{SALC}	6	-2	0	0	0	0	2	2

The character for the identity operation should be apparent, but the other nonzero characters may not be as obvious. In the case of $C_2(z)$, the operation shifts the two x vectors and the two y vectors into the negatives of themselves, contributing -4 to the overall character. The two z vectors, however, are not shifted by $C_2(z)$ and contribute $+2$ to the overall character. The character for $C_2(z)$, then, is the sum $-4 + 2 = -2$. In a similar manner, $\sigma(xz)$ leaves the two x vectors and the two z vectors nonshifted but reverses the di-

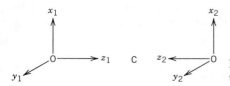

Figure 4.19 Vector basis for a representation of oxygen SALCs of CO_2.

rection of the y vectors, giving an overall character of $+4 + (-2) = +2$. In the case of $\sigma(yz)$, it is the x vectors that are reversed, while the y and z vectors remain nonshifted, giving an over character of $+2$.

The dimension of Γ_{SALC} is sufficiently large to warrant systematic reduction. Using Eq. (3.1) and the tabular method described in Section 3.1, we obtain the following work sheet:

D_{2h}	E	$C_2(z)$	$C_2(y)$	$C_2(x)$	i	$\sigma(xy)$	$\sigma(xz)$	$\sigma(yz)$		
Γ_{SALC}	6	-2	0	0	0	0	2	2	Σ	$\Sigma/8$
A_g	6	-2	0	0	0	0	2	2	8	1
B_{1g}	6	-2	0	0	0	0	-2	-2	0	0
B_{2g}	6	2	0	0	0	0	2	-2	8	1
B_{3g}	6	2	0	0	0	0	-2	2	8	1
A_u	6	-2	0	0	0	0	-2	-2	0	0
B_{1u}	6	-2	0	0	0	0	2	2	8	1
B_{2u}	6	2	0	0	0	0	-2	2	8	1
B_{3u}	6	2	0	0	0	0	2	-2	8	1

From this we see that $\Gamma_{SALC} = A_g + B_{2g} + B_{3g} + B_{1u} + B_{2u} + B_{3u}$ in D_{2h}. We can correlate these species with the equivalent species in $D_{\infty h}$ by using Table 3.9. The ascent from the working group D_{2h} to the actual group $D_{\infty h}$ reveals that the subgroup species B_{2g} and B_{3g} are degenerate as Π_g in the true group of the molecule. Likewise, the subgroup species B_{2u} and B_{3u} are degenerate as Π_u in $D_{\infty h}$. Thus, in $D_{\infty h}$, $\Gamma_{SALC} = \Sigma_g^+ + \Pi_g + \Sigma_u^+ + \Pi_u$. This means that our six SALCs will be composed of two that are nondegenerate and two pairs that will be doubly degenerate.

We must find the AOs on the central carbon atom that have the appropriate symmetry to form bonding and antibonding combinations with these SALCs. The carbon $2s$ orbital transforms as the totally symmetric representation, Σ_g^+. From the vector transformation properties listed in the $D_{\infty h}$ character table, we can conclude that the carbon $2p_z$ orbital transforms as Σ_u^+, and that the $2p_x$ and $2p_y$ orbitals transform degenerately as Π_u.

We are now ready to combine AOs with SALCs. As we proceed, it may be useful to compare results with those obtained for BeH_2 in Section 4.3. Since both BeH_2 and CO_2 are linear molecules, we should expect the symmetry species of the AOs on the central atom to be the same in both cases. In BeH_2, the $2s$ and $2p_z$ orbitals form sigma combinations with the hydrogen $1s$ SALCs of Σ_g^+ and Σ_u^+ symmetry, respectively. The $2p_x$ and $2p_y$ orbitals on beryllium remain nonbonding. In CO_2, the $2s$ and $2p_z$ orbitals likewise will have sigma interactions with the pendant atom SALCs of Σ_g^+ and Σ_u^+ symmetry, respectively, but here the SALCs are formed as combinations of oxygen $2p_z$ orbitals. The resulting LCAO-MOs have the following forms, where terms between the braces are expressions for the SALCs:

$$\sigma_g(s) = c_1 2s + c_2\left\{\frac{1}{\sqrt{2}}[2p_z(a) + 2p_z(b)]\right\} \qquad (4.25a)$$

$$\sigma_g^*(s) = c_3 2s - c_4 \left\{ \frac{1}{\sqrt{2}} [2p_z(a) + 2p_z(b)] \right\} \qquad (4.25b)$$

$$\sigma_u(z) = c_5 2p_z + c_6 \left\{ \frac{1}{\sqrt{2}} [2p_z(a) - 2p_z(b)] \right\} \qquad (4.26a)$$

$$\sigma_u^*(z) = c_7 2p_z - c_8 \left\{ \frac{1}{\sqrt{2}} [2p_z(a) - 2p_z(b)] \right\} \qquad (4.26b)$$

Unlike BeH$_2$, in CO$_2$ the degenerate $2p_x$ and $2p_y$ orbitals on carbon have matching symmetry SALCs with which they can form bonding and antibonding combinations. These SALCs are formed from matching pairs of $2p_x$ or $2p_y$ orbitals on the oxygen atoms. The LCAO-MOs so formed are degenerate (Π_u), like the AOs and SALCs used to form them. They have the following mathematical forms:

$$\pi_u(x) = c_9 2p_x + c_{10} \left\{ \frac{1}{\sqrt{2}} [2p_x(a) + 2p_x(b)] \right\} \qquad (4.27a)$$

$$\pi_u(y) = c_{11} 2p_y + c_{12} \left\{ \frac{1}{\sqrt{2}} [2p_y(a) + 2p_y(b)] \right\} \qquad (4.27b)$$

$$\pi_u^*(x) = c_{13} 2p_x - c_{14} \left\{ \frac{1}{\sqrt{2}} [2p_x(a) + 2p_x(b)] \right\} \qquad (4.27c)$$

$$\pi_u^*(y) = c_{15} 2p_y - c_{16} \left\{ \frac{1}{\sqrt{2}} [2p_y(a) + 2p_y(b)] \right\} \qquad (4.27d)$$

The LCAO-MOs defined by Eqs. (4.25)–(4.27) account for all matches between SALCs and AOs. However, as the reduction of Γ_{SALC} shows, there is a degenerate pair of SALCs of Π_g symmetry for which there are no matching AOs. These SALCs are formed as the negative combination of a pair of $2p_x$ orbitals and the negative combination of a pair of $2p_y$ orbitals on the oxygen atoms. Since they have no match among carbon AOs, they form two nonbonding π^n-MOs:

$$\pi_g^n(x) = \frac{1}{\sqrt{2}} [2p_x(a) - 2p_x(b)] \qquad (4.28a)$$

$$\pi_g^n(y) = \frac{1}{\sqrt{2}} [2p_y(a) - 2p_y(b)] \qquad (4.28b)$$

Figure 4.20 shows representations of the bonding, nonbonding, and antibonding π-MOs. A delocalized MO energy level scheme for CO$_2$, based on the assumptions of the foregoing development, is shown in Fig. 4.21. The energy level scheme shows four pairs of electrons in bonding MOs and no electrons in antibonding MOs. This means there are four bonds, just as the VB and localized MO models predict, but now the electrons are seen to be evenly distributed across the molecule (see Fig. 4.20). The pi-bonding system, consisting of $\pi_u(x)$ and $\pi_u(y)$ MOs, envelops the molecule. Although the individual π-MOs may be visualized as pairs of "sausages" with opposite wave function signs (see Fig. 4.20), together they form a cylindrical "sleeve" of electron density, partitioned along the internuclear axis by two orthogonal nodal planes, dihedral to the x and y axes. The delocalization of the pi-bonding system gives extra strength to the C–O bonds in CO_2, resulting in somewhat shorter C–O bond length.

Based on the MO scheme of Fig. 4.21, the electronic configuration of CO_2 could be written as

$$[\sigma_g^n]^2[\sigma_u^n]^2[\sigma_g(s)]^2[\sigma_u(z)]^2\{[\pi_u(x)]^2[\pi_u(y)]^2\}\{[\pi_g^n(x)]^2[\pi_g^n(y)]^2\}$$

or in simplified notation

$$(\sigma_g^n)^2(\sigma_u^n)^2[\sigma_g(s)]^2[\sigma_u(z)]^2[\pi_u(x, y)]^4[\pi_g^n(x, y)]^4.$$

From this we should expect the photoelectron spectrum to exhibit six bands, resulting from ionizations from each of the levels. As seen in Fig. 4.22, this is essentially the observed result, although the bands from the two lowest levels (highest ionization energy) are beyond the range of the ultraviolet photoelectron spectrometer used to obtain the spectrum. If our MO predictions are entirely correct, we also should expect vibrational fine structure on the second,

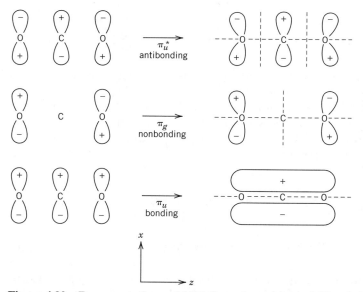

Figure 4.20 Representations of LCAOs and resulting π-MOs of CO_2, projected in the xz plane. Nodal planes perpendicular to the page are shown with dashed lines. Degenerate LCAOs and MOs similar to those shown lie in the yz plane.

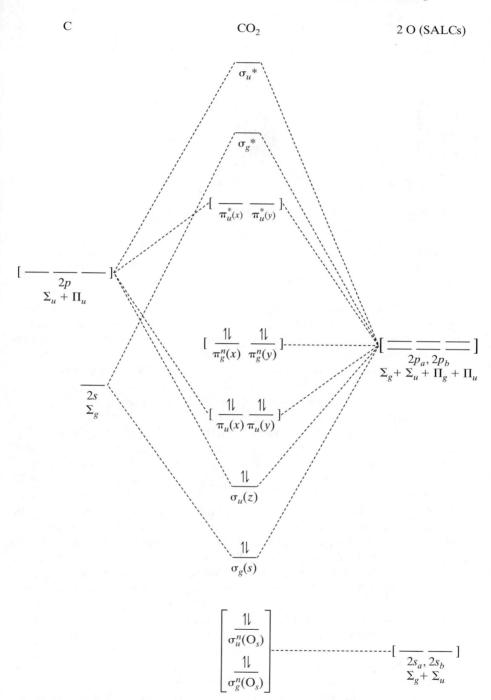

Figure 4.21 Qualitative delocalized MO energy level scheme (simplified) for CO_2.

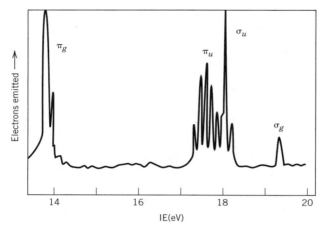

Figure 4.22   Photoelectron spectrum of CO_2. [Adapted with permission from D. W. Turner, C. Baker, A. D. Baker, and C. R. Rundle, *Molecular Photoelectron Spectroscopy*, Wiley-Interscience, London, 1970.]

third, and fourth bands (counting from low to high energy), which arise from the presumably bonding MOs $\pi_u(x, y)$, $\sigma_u(z)$, and $\sigma_g(s)$, respectively. However, only the second band, assigned to ionizations from $\pi_u(x, y)$, shows fine structure. The absence of fine structure on the third and fourth bands suggests that both $\sigma_g(s)$ and $\sigma_u(z)$ are nonbonding, contrary to our expectations from the MO scheme in Fig. 4.21. (The absence of fine structure on the first band, assigned to ionizations from the nonbonding $\pi_g^n(x, y)$ level, is consistent with Fig. 4.21.)

The nonbonding character of the $\sigma_g(s)$ and $\sigma_u(z)$ MOs results from s–p mixing. We initially assumed that only SALCs formed from the oxygen $2p_z$ orbitals would make effective bonding and antibonding combinations with the $2s$ and $2p_z$ orbitals on carbon. However, the oxygen SALCs formed from the $2s$ orbitals have the same symmetries (Σ_g^+ and Σ_u^+) as those formed from the $2p_z$ orbitals. On the basis of symmetry alone, the $2s$ SALCs are as capable of forming the appropriate bonding and antibonding combinations with the carbon $2s$ and $2p_z$ orbitals as are the $2p_z$ SALCs. Our assigning these SALCs as nonbonding levels, localized to the oxygen atoms, was based on the assumption that their energies were too different from those of the carbon AOs to have effective overlaps. To the contrary, as the photoelectron spectrum suggests, these SALCs appear to have significant interactions with carbon AOs, as do the $2p_z$ SALCs. Since both $2s$ and $2p_z$ SALCs have the same symmetry, they can mix in their interactions with the carbon AOs. This is equivalent to postulating $2s$–$2p$ mixing on the oxygen atoms. This mixing stabilizes the lower $\sigma_g(O_{2s})$ and $\sigma_u (O_{2s})$ levels (i.e., lowers their energy), making them bonding levels through overlap with carbon $2s$ and $2p_z$ orbitals. Consistent with this, we shall now designate these MOs $1\sigma_g$ and $1\sigma_u$. In the same manner, mixing destabilizes the higher $\sigma_g(s)$ and $\sigma_u(z)$ levels (i.e., raises their energy), thereby reducing the effectiveness of their overlap with carbon

$2s$ and $2p_z$ orbitals and making them nonbonding levels. In keeping with this, we shall now designate these MOs $2\sigma_g^n$ and $2\sigma_u^n$. Both the $\pi_u(x, y)$ and $\pi_g^n(x, y)$ MOs are unaffected by mixing, because they belong to different symmetry species. These mixing effects are represented in Fig. 4.23. Consistent

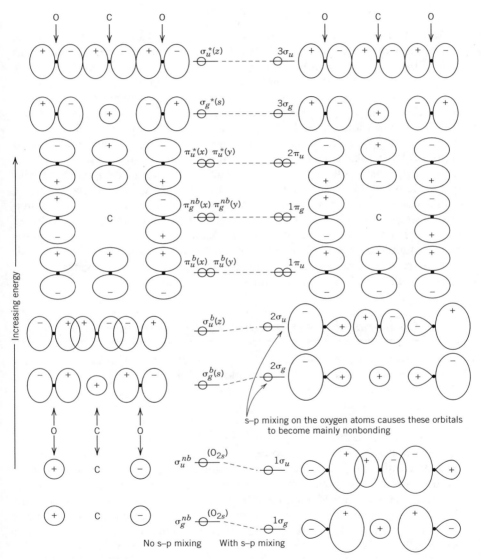

Figure 4.23 The effect of $s–p$ mixing on the shapes and energies of the molecular orbitals of CO_2. [Reproduced with permission from R. L. DeKock and H. B. Gray, *Chemical Structure and Bonding*, 2nd ed., University Science Books, Sausalito, CA, 1989.]

with the photoelectron spectrum, we may more accurately represent the electronic configuration of CO_2 as

$$[1\sigma_g]^2[1\sigma_u]^2[2\sigma_g^n]^2[2\sigma_u^n]^2\{[1\pi_u(x)]^2[1\pi_u(y)]^2\}\{[1\pi_g^n(x)]^2[1\pi_g^n(y)]^2\}$$

or in simplified notation

$$(1\sigma_g)^2(1\sigma_u)^2(2\sigma_g^n)^2(2\sigma_u^n)^2(1\pi_u)^4(1\pi_g^n)^4$$

Note that this configuration in no way alters our previous perceptions of the bond order or bond strength. We still have a total of four pairs of electrons delocalized in bonding MOs distributed across the two equivalent C–O bonds, resulting in an approximate bond order of 2 for each bond.

The case of CO_2 illustrates several points that are worth keeping in mind when constructing molecular energy level schemes for simple molecules, either with or without the aid of group theory. In general, it is convenient to make simplifying assumptions about which orbitals may or may not be most significantly involved in bonding and antibonding MO formation. However, the conclusions resulting from these assumptions should be reconciled with experimental data whenever possible. In particular, whenever two or more orbitals or SALCs have the same symmetries, they have the potential for mixing. In general, like-symmetry orbitals will repel one another, with the lower level becoming more stable (lower energy) and the upper level becoming less stable (higher energy). Whether or not such mixing occurs, or effectively alters the results from simpler assumptions that exclude mixing, depends upon the particular properties of the molecule in question.

4.5 Pi-Bonding in Aromatic Ring Systems

In terms of Lewis and VB models, benzene is represented as a resonance hybrid of the two well-known Kekulé canonical forms:

Each carbon atom is assumed to be sp^2-hybridized with the remaining p_z orbital available for pi interactions with similar orbitals on neighboring ring atoms. Since each p_z orbital is assigned one electron, the pi system consists of six electrons. In either Kekulé form, these add a total of three bonds beyond the six sigma bonds in the ring. The two resonance forms imply that these electrons are delocalized around the ring. Thus, each C–C bond has a bond order of $1\frac{1}{2}$.

If we consider developing a delocalized MO model with the aid of group theory, we immediately realize that, unlike the examples of MX_n molecules we have seen thus far, benzene has no central atom whose orbitals are to be matched with SALCs formed from combinations of orbitals on outer-lying

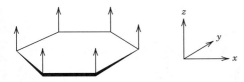

Figure 4.24 Vector basis for a representation of π-MOs of benzene.

atoms. Thus, we only need to consider interactions among the six $2p_z$ orbitals on the carbon atoms of the ring. Combinations among these six AOs form six π-MOs. The process is very much like that used to form SALCs in MX_n cases, and in fact the difference between SALCs and the π-LCAOs we shall form is merely a matter of semantics.

Figure 4.24 shows a set of six vectors, representing the six $2p_z$ orbitals on the ring carbon atoms, which may be taken as the basis for a representation in D_{6h}, the point group of benzene. By examining the effects of the operations of D_{6h} on these vectors, we arrive at the following reducible representation:

D_{6h}	E	$2C_6$	$2C_3$	C_2	$3C_2'$	$3C_2''$	i	$2S_3$	$2S_6$	σ_h	$3\sigma_d$	$3\sigma_v$
Γ_π	6	0	0	0	-2	0	0	0	0	-6	0	2

The elements for C_2' and σ_v pass through pairs of carbon atoms on opposite sides of the ring. In the case of C_2', the vectors at those atoms are transformed into the negatives of themselves, while all other vectors are transformed through space, resulting in an overall character of -2. A σ_v operation reflects two vectors into themselves and moves all other vectors through space, resulting in an overall character of 2. The σ_h operation transforms all vectors into their negatives, resulting in an overall character of -6. Except for identity, all other operations move all vectors through space, resulting in overall characters of zero. By applying Eq. (3.1), it can be shown that $\Gamma_\pi = B_{2g} + E_{1g} + A_{2u} + E_{2u}$. This indicates that the six π-LCAO-MOs consist of two that are nondegenerate (B_{2g} and A_{2u}) and two pairs that are doubly degenerate (E_{1g} and E_{2u}).

The forms of the pi-bonding LCAO-MOs are represented in Fig. 4.25. The positions of nodal planes, which result from alternations of wave function signs, are indicated for each MO in Fig. 4.26. The lowest-energy π-MO is formed by taking the totally positive combination of p_z orbitals on all six carbon atoms:

$$\pi_1 = \frac{1}{\sqrt{6}}(\phi_a + \phi_b + \phi_c + \phi_d + \phi_e + \phi_f) \qquad (4.29)$$

This combination, which has the full symmetry of the six vectors shown in Fig. 4.24, transforms as A_{2u}. Note that this MO is symmetric with respect to C_6, antisymmetric with respect to σ_h, and antisymmetric with respect to i, all of which is consistent with the Mulliken designation A_{2u} (cf. Figs. 4.25 and 4.26).

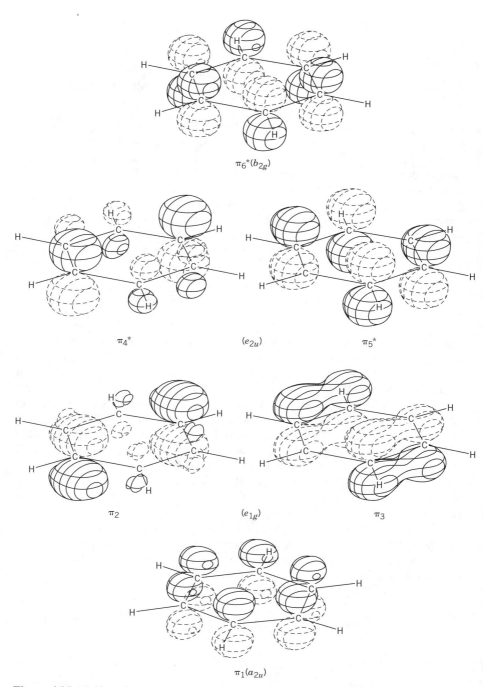

Figure 4.25   Delocalized π-MOs of benzene. Solid and dotted line contours represent positive and negative signs of the wave function, respectively. [Reproduced with permission from William L. Jorgensen and Lionel Salem, *The Organic Chemist's Book of Orbitals*, Academic Press, New York, 1973.]

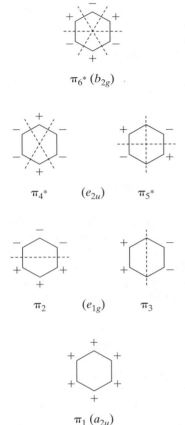

$\pi_6^*\ (b_{2g})$

π_4^* (e_{2u}) π_5^*

π_2 (e_{1g}) π_3

$\pi_1\ (a_{2u})$

Figure 4.26 Nodal planes (perpendicular to the ring plane) for the pi-bonding and anti-bonding MOs of benzene.

The next energy level consists of the degenerate pair of MOs π_2 and π_3, defined by the wave equations

$$\pi_2 = \frac{1}{2\sqrt{3}}(2\phi_a + \phi_b - \phi_c - 2\phi_d - \phi_e + \phi_f) \qquad (4.30a)$$

$$\pi_3 = \frac{1}{2}(\phi_b + \phi_c - \phi_e - \phi_f) \qquad (4.30b)$$

As Figs. 4.25 and 4.26 show, these are symmetric with respect to inversion and $C_2(z)$ rotation, which identifies them as the E_{1g} MOs expected from the reduction of Γ_π. The next highest MOs are a degenerate pair of antibonding MOs, π_4^* and π_5^*, whose wave functions have the form

$$\pi_4^* = \frac{1}{2\sqrt{3}}(2\phi_a - \phi_b - \phi_c + 2\phi_d - \phi_e - \phi_f) \qquad (4.31a)$$

$$\pi_5^* = \frac{1}{2}(-\phi_b + \phi_c - \phi_e + \phi_f) \qquad (4.31b)$$

These are the pair with E_{2u} symmetry. Note from Figs. 4.25 and 4.26 that these are antisymmetric with respect to inversion and symmetric with respect to $C_2(z)$, as expected for the species E_{2u}. The highest-energy MO is the antibonding π_6^*, formed by alternating signs on adjacent p_z wave functions around the ring:

$$\pi_6^* = \frac{1}{\sqrt{6}}(\phi_a - \phi_b + \phi_c - \phi_d + \phi_e - \phi_f) \qquad (4.32)$$

This combination has nodal planes perpendicular to the ring between each pair of carbon atoms (see Figs. 4.25 and 4.26). The alternation of signs on the p_z wave functions around the ring makes this combination antisymmetric with respect to C_6, antisymmetric with respect to σ_h, and symmetric with respect to i, consistent with the species B_{2g}.

The π-MO scheme for benzene is shown in Fig. 4.27. Note that the energies of the MOs rise with increasing numbers of nodes (cf. Fig. 4.26). As shown in Fig. 4.27, the three bonding orbitals are occupied with three pairs of electrons. The configuration $(\pi_1)^2(\pi_2)^2(\pi_3)^2$ adds a total of three bonds to the ring, in addition to the six from sigma bonds. Since these nine bonds are distributed among six C–C pairs, the average bond order for any C–C bond is $1\frac{1}{2}$, consistent with the VB model based on two resonance forms. However, with the general MO approach, the delocalization of pi electrons is a natural consequence of the model.

The π-MO scheme for benzene suggests that the six electrons are distributed in a $1:2$ ratio between two distinct orbital energies (a_{2u} and e_{1g}). The photoelectron spectrum of benzene is shown in Fig. 4.28. The assignment of all bands to specific σ and π MOs has been subject to some controversy,[*] since the bands for σ levels appear to overlap with that for the lowest π level. Nonetheless, the band at 9.25 eV (labeled π_A, π_B in Fig. 4.28), which shows considerable vibrational fine structure, can be assigned reliably to π_2 and π_3.

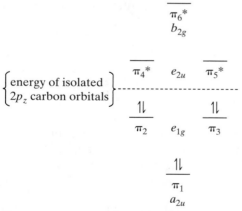

Figure 4.27   The π-MO energy level scheme for benzene.

*For a more complete discussion of the photoelectron spectrum of benzene see A. D. Baker and D. Betteridge, *Photoelectron Spectroscopy: Chemical and Analytical Aspects*, Pergamon Press, Oxford, 1972, p. 75*ff.*

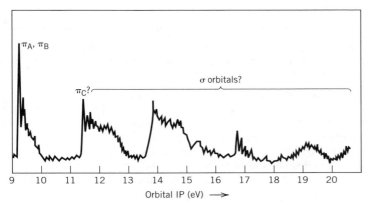

Figure 4.28 Photoelectron spectrum of benzene. [Adapted with permission from A. D. Baker and D. Betteridge, *Photoelectron Spectroscopy: Chemical and Analytical Aspects*, Pergamon Press, Oxford, 1972.]

The band labeled π_C in Fig. 4.28, with approximately half the intensity of the 9.25 eV band, is tentatively assigned to π_1. The bonding character of the MO giving rise to this band is evident from the vibrational fine structure.

You may have noticed that the pattern of the levels in Fig. 4.27 mimics the shape of the conjugated ring system being described—that is, a hexagon. This phenomenon, sometimes called the *polygon rule* or the *shadow method*, is common to all the π-MO schemes for other delocalized, single-ring structures. Thus, it is easy to write down the qualitative π-MO schemes from the ring geometries. In each case, the pattern of the π-MO scheme is laid out in the same geometrical arrangement as the polygon of the ring. The lowest level is a nondegenerate bonding MO (a point of the polygon) which belongs to the A representation by which z transforms in the molecule's point group. This symmetry species is antisymmetric to σ_h (and also i if inversion exits for the ring system). For ring systems with an even number of carbon atoms, the highest π^*-MO belongs to a B_2 species. In all cases, the doubly degenerate levels rise in energy in order of the numerical subscripts on their Mulliken symbols; that is, $E_1 < E_2$. Overall, the ordering of levels from lowest to highest energy corresponds with increasing numbers of nodes. The midline of the geometry separates π and π^* levels. For example, the π-MO energy level diagram for the cyclopentadienyl anion, $C_5H_5^-$ (D_{5h}), has a pentagonal pattern consisting of a nondegenerate, lowest-lying bonding MO (a_2''), two degenerate slightly bonding MOs of intermediate energy (e'), and two degenerate antibonding MOs with highest energy (e'').

The preceding description of π-bonding in single-ring systems is only qualitative. The Hückel approximation provides a computational method for determining the relative MO energies more quantitatively. Presentations of Hückel methodology in the context of symmetry arguments can be found in more advanced texts.*

*For example, see F. A. Cotton, *Chemical Applications of Group Theory*, 3rd ed., John Wiley & Sons, New York, 1990, Chapter 7.

Problems

4.1   Determine the sets of specific atomic orbitals that can be combined to form hybrid orbitals with the following geometries: (a) trigonal planar, (b) square planar, (c) trigonal bipyramidal, (d) octahedral.

4.2   Hybrid orbitals are usually discussed in the context of perfect geometries. Yet in many cases the molecular structures for which these hybrids are postulated deviate significantly from the ideal. For example, consider an ideally tetrahedral MX_4 molecule whose actual structure is slightly flattened along the z axis (a C_2 axis in both the ideal and distorted geometries). What sets of atomic orbitals could be combined to form hybrids with the appropriate orientations to describe bonding in such a molecule? How do these results compare with the sp^3 and sd^3 hybrids appropriate for perfectly tetrahedral geometry?

4.3   Borane, BH_3, is an unstable compound produced by thermal decomposition of $H_3B \cdot PF_3$. Although it has not been isolated and structurally characterized, it probably is trigonal planar.

(a)   Develop a general MO scheme for BH_3. Assume that only the boron $2s$ and $2p$ orbitals interact with the hydrogen $1s$ orbitals (i.e., the boron $1s$ orbital is nonbonding).

(b)   The photoelectron spectrum of BH_3 has not been observed. Nonetheless, if it could be taken, what would you expect it to look like, based on your MO scheme?

(c)   Compare and contrast the general MO description of BH_3 with a valence bond (VB) model and its related localized MO model.

4.4   Consider H_2O, a bent molecule for which $\angle$ H–O–H = 104.5°.

(a)   Develop a general MO scheme for H_2O. Assume that only the $2s$ and $2p$ orbitals of oxygen interact with the hydrogen $1s$ orbitals (i.e., the oxygen $1s$ orbital is nonbonding). The molecule's plane should be taken as the xz plane, the plane of σ_v in C_{2v}. [*Hint:* Two AOs match symmetry with one of the SALCs, thereby forming three MOs. Both AOs contribute in varying degrees to all three MOs.]

(b)   The photoelectron spectrum (P.E.S.) of H_2O has four bands (not including ionizations from the oxygen $1s$ core electrons) [cf. A. W. Potts and W. C. Price, *Proc. R. Soc. Lond.* **1972**, *A326*, 181–197]. The three highest-energy ionizations give bands with vibrational fine structure (although the highest-energy band has not been resolved, owing to instrumental limitations). The fourth band, from the least energetic ionization, shows no such fine structure. Explain these results on the basis of your MO scheme, modifying it if necessary to be consistent with the P.E.S. results.

(c)   The VB description of H_2O assumes sp^3 hybridization on the oxygen atom, resulting in two lone pairs protruding like "Mickey Mouse" ears from the back of the molecule. How well does this picture agree with the general MO model and P.E.S. results? Justify your answer by making sketches of the LCAO-MOs.

(d)   Addition of two pairs of electrons to the MO scheme for BeH_2 (Fig. 4.14) would adapt it to describe bonding in H_2O, if water were linear. Compare your MO scheme for bent H_2O with this hypothetical scheme for linear

H_2O. On the basis of these schemes alone (i.e., without using VSEPR arguments) give reasons why H_2O is bent rather than linear.

4.5 Consider NH_3, a pyramidal molecule, for which $\angle$ H–N–H = 106.6°.

(a) Develop a general MO scheme for NH_3. Assume that only the $2s$ and $2p$ orbitals of nitrogen interact with the hydrogen $1s$ orbitals (i.e., the nitrogen $1s$ orbital is nonbonding). [*Hint*: Two AOs match symmetry with one of the SALCs, thereby forming three MOs. Of these, the lowest energy MO is essentially a bonding combination formed between the hydrogen SALC and the $2s$ AO on nitrogen, and the highest MO is an antibonding combination formed from a mixture of the two AOs and the same-symmetry SALC.]

(b) The photoelectron spectrum (P.E.S.) of NH_3 has three bands (not including ionizations from the nitrogen $1s$ core electrons) [cf. A. W. Potts and W. C. Price, *Proc. R. Soc. Lond.* **1972**, *A326*, 181–197]. All three bands, even the lowest-energy band, show evidence of vibrational fine structure (although the highest-energy band has not been resolved, owing to instrumental limitations). Explain these results on the basis of your MO scheme, modifying it if necessary to be consistent with the P.E.S. results.

(c) The VB description of NH_3 assumes sp^3 hybridization on the nitrogen atom, resulting in a single lone pair at the apex of the molecule. The Lewis base character of NH_3 is attributed to this lone pair. How does this model compare with the MO description? Is the MO model consistent with the Lewis base character of NH_3? Explain.

(d) Addition of a pair of electrons to the MO scheme for BH_3 (see Problem 4.3) would adapt it to describe bonding in NH_3, if ammonia were trigonal planar. Compare your MO scheme for pyramidal NH_3 with the hypothetical scheme for trigonal planar NH_3. On the basis of these schemes alone (i.e., without using VSEPR arguments) explain why NH_3 is pyramidal rather than trigonal planar.

4.6 The allyl anion, $[H_2CCHCH_2]^-$, has a delocalized, open, three-center $p\pi$ system. Develop the MO scheme for this system, show the electron filling in the scheme, and sketch the forms of the LCAO-MOs. [*Hint*: Although it is customary to assume that p_z orbitals are involved in forming $p\pi$ orbitals, in this case you may prefer to assume that p_x orbitals are used, in keeping with the standard character table and conventions of defining z as the principal axis and the yz plane as the plane of the C–C–C chain. If you assume that p_z orbitals form the $p\pi$ orbitals and that the principal axis is x or y, you will need to alter the character table to reflect the switched axes.]

4.7 Although BH_3 is unstable (see Problem 4.3), the BX_3 trihalides (X = F, Cl, Br) are stable but reactive compounds that have been well characterized. A significant advantage of the BX_3 compounds is the potential for $p\pi$-bonding between the $2p_z$ orbital of boron and the np_z orbitals of the halogens ($n = 2, 3, 4$). This bonding is most significant for BF_3.

(a) Develop a molecular orbital scheme for the $p\pi$-MOs of BF_3.

(b) Although the $2s$ orbitals on fluorine may be assumed to be nonbonding, SALCs can still be formed among them. What are the symmetries of the three fluorine $2s$ SALCs?

(c)   Assume that each of the fluorine atoms uses a $2p$ orbital directed toward the central boron atom to form sigma interactions. What are the symmetries of the three SALCs that can be formed from the three sigma-symmetry $2p$ orbitals on the fluorine atoms?

(d)   In addition to the fluorine $2p$ orbitals engaged in pi and sigma interactions, there are three $2p$ orbitals lying in the plane of the molecule that may be assumed to be nonbonding. What are the symmetries of the three SALCs that can be formed from these $2p$ orbitals?

(e)   Using your results from parts (a) through (d), develop a complete molecular orbital scheme for BF_3. You may want to consider the P.E.S. of BF_3 to verify the order of occupied MOs [G. H. King, S. S. Krishnamurthy, M. F. Lappert, and J. B. Pedley, *Faraday Disc. Chem. Soc.*, **1972**, *54*, 70].

(f)   In part (d) we assumed that the fluorine in-plane $2p$ orbitals not engaged in sigma interactions were nonbonding. On the basis of symmetry, are bonding and antibonding interactions with boron precluded for these orbitals? If bonding is possible, what effects would it have on the MO scheme you developed in part (e)?

(g)   Compare your MO scheme for BF_3 with the MO scheme you developed for BH_3 in Problem 4.3. Are your MO descriptions consistent with the relative stabilities of the two compounds?

(h) BF_3 is a Lewis acid that readily forms adducts with Lewis bases; for example, $BF_3 + NH_3 \rightarrow F_3B \cdot NH_3$. Based on your MO schemes for both BF_3 and NH_3 (Problem 4.5), describe the likely mechanism by which the adduct $F_3B \cdot NH_3$ is formed.

4.8   Like benzene, the cyclobutadiene dianion, $C_4H_4^{2-}$, has six electrons in a delocalized π-system.

(a)   Using methods of group theory and the polygon rule, develop a qualitative π-MO scheme for the $C_4H_4^{2-}$ ion. Label each MO by bond type and Mulliken symbol, and show the filling of electrons in the scheme.

(b)   Sketch the LCAOs for the π-MOs.

(c)   Explain why $C_4H_4^{2-}$ would be expected to be much less stable than C_6H_6, despite both species having six electrons in π-MOs, in keeping with the Hückel $4n + 2$ criterion for aromaticity.

(d)   Neutral cyclobutadiene is a very unstable, nonaromatic species that appears to have a rectangular structure (D_{2h}) composed of alternating single and double bonds [cf. P. Reeves, T. Devon, and R. Pettit, *J. Am. Chem. Soc.* **1969**, *91*, 5890].

On the basis of your MO scheme, account for the lack of stability of square planar cyclobutadiene. Why is rectangular 1,3-cyclobutadiene a somewhat more stable structure?

4.9   Using methods of group theory and the polygon rule, develop qualitative π-MO schemes for the following cyclic $(CH)_n$ systems, assuming planar geometry (D_{nh}). Label each MO by bond type and Mulliken symbol. Show the filling of electrons in the scheme for the neutral molecule, the $+1$ cation, and the -1 anion. Discuss the relative stabilities of the neutral molecule, cation, and anion.
(a) C_3H_3, (b) C_5H_5, (c) C_7H_7 (use the C_7 character table).

4.10 Cyclopropane, C_3H_6, is a remarkably stable molecule (m.p. $-127.6°C$, b.p. $-32.7°C$), despite the extreme ring strain. Indeed, from a traditional VB approach, assuming sp^3-hybridized carbon atoms, one might wonder why it exists at all. A delocalized MO model suggests that the ring receives stabilization through $p\pi$-bonding, a result hardly expected for an alkane. Based on Gaussian orbital SCF calculations and the photoelectron spectrum [cf. D. W. Turner, C. Baker, A. D. Baker, and C. R. Brundle, *Molecular Photoelectron Spectroscopy,* Wiley-Interscience, London, 1970, p. 203*ff.*], the electronic structure of cyclopropane can be written as

$$(\sigma_{C-H})^2(\sigma_{C-H})^4(\pi_{C-C})^2(\sigma_{C-C})^2(\pi_{C-C})^4(\sigma_{C-H})^4$$

This configuration shows only the lowest-energy, filled MOs. There are also a number of higher-energy, unoccupied bonding, nonbonding, and antibonding MOs. The entire catalogue of MOs for cyclopropane can be deduced by breaking the problem up into three kinds of orbital interactions: (1) hydrogen $1s$ with carbon $2s$ and in-plane $2p$ interactions, leading to σ_{C-H}-MOs of all types; (2) in-plane carbon $2p$ interactions, leading to σ_{C-C} bonding and antibonding MOs; and (3) out-of-plane carbon $2p$ interactions, leading to π_{C-C} bonding and antibonding MOs. Proceeding through the following steps, determine the symmetries and bonding types of the MOs of cyclopropane.

(a) The most effective σ_{C-H} interactions are formed between the six hydrogen $1s$ orbitals and the three carbon $2s$ orbitals. Determine the symmetries of SALCs formed from the three carbon $2s$ orbitals. Then, determine the symmetries of SALCs formed from the six hydrogen $1s$ orbitals. By matching symmetries, give the Mulliken designations and possible bonding types for MOs that can be formed from carbon $2s$ with hydrogen $1s$ interactions.

(b) Less effective σ_{C-H} interactions are formed between the hydrogen $1s$ orbitals and the in-plane carbon $2p$ orbitals with the following orientations:

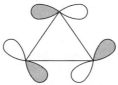

Determine the symmetries of SALCs formed from these three carbon $2p$ orbitals. By matching symmetries with the hydrogen $1s$ SALCs previously determined, give the Mulliken designations and possible bonding types for all MOs that can be formed.

(c) Ring sigma bonding results from in-plane carbon $2p$ orbitals with the following orientations:

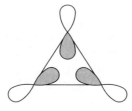

Determine the symmetries of SALCs formed from these three carbon $2p$ orbitals, and give the Mulliken designations and possible bonding types for all MOs that can be formed.

(d) Ring pi interactions result from out-of-plane carbon $2p$ orbitals with the following orientations:

Determine the symmetries of SALCs formed from these three carbon $2p$ orbitals, and give the Mulliken designations and possible bonding types for all MOs that can be formed.

(e) On the basis of your results from parts (a) through (d), assign Mulliken designations for the occupied orbitals in the electronic configuration given above.

(f) Make sketches of the LCAOs for the occupied bonding MOs.

4.11 Diborane, B_2H_6, has a structure with bridging hydrogen atoms.

$$\text{H}\underset{\text{H}}{\overset{\text{H}}{\diagdown}}\text{B}\underset{\text{H}}{\overset{\text{H}}{\diagup}}\text{B}\overset{\text{H}}{\underset{\text{H}}{}}$$

The terminal B–H bonds are conventional two-center, two-electron ($2c$–$2e$) covalent bonds, but each B–H–B bridge is an electron-deficient, three-center, two-electron ($3c$–$2e$) bond. Develop a general MO scheme in diborane's point group, D_{2h}, for the pair of bridge bonds, assuming that each boron atom is sp^3 hybridized. Although the choice of axes orientations in D_{2h} is arbitrary, a typical orientation would have the two boron atoms along the z axis and the hydrogen bridges lying in the xz plane. In setting up the problem, use four sp^3 hybrids, two on each boron atom, as the basis for a set of SALCs, and take the $1s$ orbitals on the two bridging hydrogen atoms as a basis for a separate set of SALCs.

4.12 The term *hypervalent* is sometimes used to describe molecules in which the central atom appears to exceed an octet. Examples include PF_5 and SF_4 (five electron pairs about the central atom), and SF_6 and XeF_4 (six electron pairs about the central atom). Although hypervalence has often been rationalized in terms of dsp^3 and d^2sp^3 hybridization schemes, most recent theoretical studies suggest that valence nd orbitals are not important for bonding in main-group elements [cf. L. Suidan, J. K. Badenhoop, E. D. Glendening, and F. Weinhold, *J. Chem. Educ.* **1995**, *72*, 583; D. L. Cooper, T. P. Cunningham, J. Gerratt, P. B. Karadakov, and M. Raimondi, *J. Am. Chem. Soc.* **1994**, *116*, 4414]. From the perspective of symmetry arguments, having nd orbitals with appropriate symmetry for bonding with pendant atom SALCs does not require that they be fully used. In the case of SF_6, *ab initio* SCF calculations suggest that the electron distribution on sulfur is approximately 32% in $3s$, 59% in $3p$, 8% in $3d$, and 1% in $4p$ [A. E. Reed and F. Weinhold, *J. Am. Chem. Soc.* **1986**, *108*, 3586].

(a) Ignoring any $4p$ participation, develop a qualitative sigma-only MO scheme for SF_6 that includes consideration of the symmetry properties of the $3d$

orbitals, but reflects their minimal involvement in the bonding. Assume that each fluorine uses a single $2p$ orbital directed at the central sulfur (i.e., assume that fluorine $2s$ and the remaining $2p$ orbitals are nonbonding).

(b)   Assuming that $3d$ orbital contributions to bonding can be ignored, what is the approximate S–F bond order implied by your MO scheme? How would you reconcile this with the observation that the S–F bond lengths (156 pm) are shorter than expected for a single bond?

(c)   An alternative approach to describe bonding in main group hypervalent molecules uses three-center, four-electron ($3c$–$4e$) bonds formed by overlap of p orbitals on the central and pendant atoms. [cf. R. E. Rundle, *J. Am. Chem. Soc.* **1963**, *85*, 112; G. C. Pimentel, *J. Chem. Phys.* **1951**, *19*, 446]. Thus, SF_6 is seen as three mutually perpendicular $3c$–$4e$ F–S–F bonds. What is the S–F bond order in these $3c$–$4e$ bonds? If the bonding in SF_6 were closer to this model, how would your MO scheme need to be modified to be consistent?

4.13   There is little evidence to support significant participation of $3d$ orbitals in the bonding of PF_5. [cf. D. L. Cooper, T. P. Cunningham, J. Gerratt, P. B. Karadakov, and M. Raimondi, *J. Am. Chem. Soc.* **1994**, *116*, 4414.] Develop a qualitative sigma-only MO scheme for PF_5, using only $3s$ and $3p$ orbitals on phosphorous and a single $2p$ orbital on each of the five fluorine atoms.

4.14   A satisfactory general MO description of the bonding in XeF_4 can be developed by considering only the $5s$ and $5p$ orbitals of Xe interacting with the $2p$ orbitals of the four F atoms [cf. K. O. Christe, E. C. Curtis, D. A. Dixon, H. P. Mercier, J. C. P. Sanders, and G. J. Schrobilgen, *J. Am. Chem. Soc.* **1991**, *113*, 3351]. The $2s$ orbitals of fluorine have considerably lower energy and can be assumed to be nonbonding. For the purposes of electron counting, Xe may be taken to have the valence configuration $5s^25p^2$, as Xe^{4+}. Fluorine may be taken to have the valence configuration $2p^6$, as F^-, distributed in the set of four atoms as follows: eight electrons in $2p_x$ orbitals directed at the central Xe; eight lone-pair electrons in $2p_y$ orbitals, orthogonal to the Xe–F bonds; eight electrons in $2p_z$ orbitals, perpendicular to the molecular plane. Use these three groupings as three separate bases for SALC representations, Γ_σ, $\Gamma_{\pi(\|)}$, and $\Gamma_{\pi(\perp)}$. By matching the symmetries of these SALCs with the symmetries of Xe AOs, develop a qualitative general MO scheme for XeF_4. For simplicity, all fluorine SALCs not involved in bonding and antibonding combinations with Xe orbitals may be grouped together in the center of the scheme, without attempting to sort out their relative energies. Actually, the highest occupied molecular orbital (HOMO) is the antibonding combination of in-plane fluorine $2p_y$ orbitals. The antibonding nondegenerate Xe–F σ^* and π^* levels lie below the HOMO, among other fluorine-only MOs.

4.15   X-ray structure analysis shows that $N(CH_3)_4^+XeF_5^-$ and related compounds contain the planar pentafluoroxenate(IV) anion, the first example of a pentagonal planar MX_5 species [cf. K. O. Christe, E. C. Curtis, D. A. Dixon, H. P. Mercier, J. C. P. Sanders, and G. J. Schrobilgen, *J. Am. Chem. Soc.* **1991**, *113*, 3351]. Taking an approach similar to that described for XeF_4 (Problem 4.14), develop a qualitative general MO scheme for the XeF_5^- ion. As with XeF_4, the HOMO is the antibonding combination of in-plane fluorine $2p_y$ orbitals. The antibonding nondegenerate Xe–F σ^* and π^* levels lie below the HOMO, among other fluorine-only MOs.

CHAPTER 5

Equations for Wave Functions

In Chapter 4 we formed wave functions for hybrid orbitals and molecular orbitals as linear combinations of specific atomic orbitals. In both cases the new functions conformed to the symmetry requirements of the system, and as such could be regarded as symmetry-adapted linear combinations (SALCs). In very simple cases, such as BeH_2 and CO_2, the explicit expressions for the SALCs could be deduced by inspection. But with even a relatively straightforward case like the pi molecular orbitals (π-MOs) of benzene, we found some SALCs whose mathematical forms were not intuitively obvious. It would be useful for cases like these to have a systematic way of generating the mathematical expressions for the SALCs. In this chapter we consider methodologies for satisfying this need. The projection operator approach, on which we will focus most of our attention, is a traditional and generally applicable method. However, as we shall see, the applicability of projection operators has limits, and one must bring a certain amount of "chemical intuition" to the analysis in certain cases.

5.1 Formulating SALCs with Projection Operators

The *projection operator* has been called a "function generating machine," because it generates algebraic equations more or less automatically. In our case, the functions we wish to obtain are linear combinations of atomic wave functions (e.g., pendant atom SALCs, π-MOs of ring systems) formulated from a collection of specific functions that form a *basis set*. As such, the functions we seek may in general be regarded as SALCs. As we have seen, each SALC must have the symmetry of an irreducible representation within the reducible representation for the problem under consideration. To generate a SALC belonging to one of these symmetry species, a projection operator for the particular irreducible representation is applied to one function in the basis set. As a result, the operator projects out the full linear combination for the SALC in terms of all the basis functions in the set. If we wish to obtain all the allowed SALCs, we must construct projection operators for each symmetry species comprising the reducible representation. The particular function to which we apply each projection operator can be chosen within the basis set more or less arbitrarily, so long as all the functions of the set are related to each other by symmetry operations of the group. We can be this arbitrary because the projection operator "knows" the existence of the other basis func-

138

tions and how they are mathematically related to the reference function. This is a consequence of the symmetry of the group and the symmetry relationships dictated by the irreducible representation to which the projection operator belongs.

The projection operator for any symmetry species can be constructed either in terms of the full operator matrices of the irreducible representation or in terms of its characters. For nondegenerate irreducible representations the two forms are equivalent. For doubly or triply degenerate representations the full matrix form has the advantage of generating the two or three degenerate functions in one mathematical process. However, to construct the projection operator in this form, we must know all the explicit operator matrices comprising the degenerate irreducible representation. Moreover, we must know the forms of these matrices for each and every operation, not just one representative operation of each class. This is necessary for degenerate irreducible representations, because all operations in a class usually have unique operator matrices, even though all have the same character. Listings of the full matrix forms of the irreducible representations are not generally available, and the effort of generating them does not justify the slight advantage obtained from producing the two or three degenerate functions directly. Using the projection operator in characters is less cumbersome and more direct, but only one function is generated immediately for doubly or triply degenerate irreducible representations. However, the individual degenerate functions usually can be generated either directly or by employing a variety of techniques, some of which we will illustrate. Consequently, we will only concern ourselves with the character form of the projection operator in this text.*

Suppose we wish to find the allowed SALCs constructed from a set of functions $\phi_1, \phi_2, \ldots, \phi_n$, which form the basis for a reducible representation of the group. We require that these basis functions be related to each other by the operations of the group. Thus, each function is interchanged with itself or other functions in the set in either a positive or negative sense through the effect of an operator, R, for each of the operations of the group. If we wish to construct the SALCs associated with the ith irreducible representation, S_i, we may apply the projection operator, P_i, to any one of the several basis functions, ϕ_t, according to the expression

$$S_i \propto P_i \phi_t = \frac{d_i}{h} \sum_R \chi_i^R R_j \phi_t \tag{5.1}$$

in which

d_i = dimension of the ith irreducible representation,

h = order of the group,

χ_i^R = each operation's character in the ith irreducible representation,

R_j = the operator for the jth operation of the group.

---

*For a derivation of both forms of the projection operator see F. A. Cotton, *Chemical Applications of Group Theory,* 3rd ed., John Wiley & Sons, New York, 1990, Chapter 6.

The result of the term $R_j\phi_t$ is the basis function, either in the positive or negative sense, obtained when the reference function ϕ_t is subjected to the action of the jth operation. Note that the summation is taken over all the individual operations, and not simply over each class of operations. This is necessary because each operation of a given class may transform ϕ_t into a different member of the basis set.

The results of Eq. (5.1) are not the final wave functions we seek. To obtain these, we normalize the functions generated from the projection operators, invoking the usual condition $N^2 \int \psi \psi^* \, d\tau = 1$, where N is the normalization constant. Therefore, we can routinely ignore the factor d_i/h of Eq. (5.1), which in all cases will be incorporated automatically within the normalization constant. In addition to normalization, we must insure that all the wave functions we generate meet the quantum mechanical requirement of *orthogonality*. Accordingly, for any two wave functions of the system we require that $\int \psi_i \psi_j \, d\tau = 0$ if $i \neq j$.

In carrying out normalization or testing for orthogonality on SALCs, we will be taking products of wave functions that are linear combinations of the basis functions. These products have the general form

$$(a_i\phi_i \pm a_{i+1}\phi_{i+1} \cdots \pm a_n\phi_n)(b_j\phi_j \pm b_{j+1}\phi_{j+1} \cdots \pm b_m\phi_m)$$

The basis functions composing these SALCs are presumably normalized and orthogonal. Thus, in carrying out the expansion of the products, terms of the type $\phi_i\phi_i$ or $\phi_j\phi_j$ will be unity and terms of the type $\phi_i\phi_j$ ($i \neq j$) will be zero. We can generalize these results in terms of the Kroneker delta function, δ_{ij}, by writing

$$\int \phi_i \phi_j \, d\tau = \delta_{ij} \tag{5.2}$$

where $\delta_{ij} = 1$ if $i = j$ and zero otherwise. As a result, we only need to concern ourselves with the nonvanishing $\phi_i\phi_i$ terms, since all the cross terms ($\phi_i\phi_j$) will be zero.

To illustrate the use of projection operators, let us formulate the σ-SALCs for the pendant atoms in an octahedral MX_6 molecule. By the procedures described in Chapter 4 we can readily determine that the six SALCs will have the symmetries $\Gamma_\sigma = A_{1g} + E_g + T_{1u}$. To find the mathematical forms of these SALCs we will construct projection operators in each of the three symmetry species. As we have noted, using Eq. (5.1) requires a term for each and every operation of the group. Since the order of O_h is 48, we might anticipate a rather cumbersome set of equations, each with 48 terms. However, we can cut the work at least in half by carrying out the process in the rotational subgroup O, for which $h = 24$. We choose O, rather than some lower-order subgroup, because it preserves the essential symmetry of the parent group, O_h, especially the degeneracies of its irreducible representations.*

*As we shall see in other cases, choosing a rotational subgroup of the molecule's true group is often a useful strategy for minimizing the labor of a problem. We can take this approach because an axial group's rotational subgroup either preserves the parent group's degeneracies or lifts them in a way that makes the correlation with the parent group species direct and unambiguous.

This makes it easy to correlate the results obtained in the subgroup with those that would be obtained in the full parent group. Note that in O we have the symmetry species $\Gamma_\sigma = A + E + T_1$, which have obvious correlations to the symmetry species in O_h. In actual practice, we would avoid further unnecessary work by realizing that the A SALC is totally symmetric. This means it must consist of the positive addition of all six sigma orbitals on the X atoms. Therefore, we really do not need to use a projection operator to obtain it. Nonetheless, as an introduction to the use of projection operators, we will carry out the process here.

Figure 5.1 shows the positions of the six X atoms whose sigma orbitals (labeled σ_1 through σ_6) point toward the central M atom. The corners of the surrounding cube have been labeled a through d to provide reference points for the orientations of the various symmetry elements and their operations in the group O. The listing $8C_3$ in the O character table (cf. Appendix A) refers to four C_3 operations and four C_3^2 operations performed about four axes that run along the cube diagonals. Thus, the C_3 axis we will label aa runs along the cube diagonal that connects the two corners labeled a. The $3C_2$, $3C_4$, and $3C_4^3$ operations are performed about axes that pass through pairs of *trans*-related positions and are so labeled. Thus, we will label the C_2 axis that passes through positions 1 and 2 as *12*. The $6C_2'$ axes lie in pairs in the three planes that intersect at the center of the octahedron. Relative to the reference cube, they pass through the midpoints on two opposite edges. We will label these C_2' axes according to the two-letter designations of the cube edges through which they pass. Thus, the C_2' axis that passes through the midpoints of the two ac edges will be labeled ac. In the case of threefold rotations we will take the clockwise sense viewed from the upper corners of the cube. The fourfold rotations will be taken in the clockwise sense viewed from the cube face of the lower numbered position (e.g., from the upper face $abcd$ for the axis *12*).

Having carefully defined the orientations and directions of the operations in this manner, we can proceed to determine their effects on an arbitrarily chosen reference function of the basis set of six sigma orbitals. Taking the reference basis function as σ_1, the operations of O effect the following transformations:

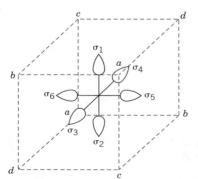

Figure 5.1   Orientation of the basis functions used to construct σ-SALCs for the six pendant atoms of an octahedral MX_6 molecule.

O	E	C_3	C_3	C_3	C_3	C_3^2	C_3^2	C_3^2	C_3^2	C_2	C_2	C_2
Label		aa	bb	cc	dd	aa	bb	cc	dd	12	34	56
$R_j\sigma_1$	σ_1	σ_5	σ_3	σ_6	σ_4	σ_3	σ_6	σ_4	σ_5	σ_1	σ_2	σ_2

	C_4	C_4	C_4	C_4^3	C_4^3	C_4^3	C_2'	C_2'	C_2'	C_2'	C_2'	C_2'
	12	34	56	12	34	56	ac	bd	ab	cd	ad	bc
	σ_1	σ_5	σ_4	σ_1	σ_6	σ_3	σ_2	σ_2	σ_3	σ_4	σ_5	σ_6

In the table above, the row labeled $R_j\sigma_1$ indicates that E transforms σ_1 into itself, $C_3(aa)$ transforms σ_1 into σ_5, $C_3(bb)$ transforms σ_1 into σ_3, and so forth.

We now can construct the projection operator $P(A)\sigma_1$ by multiplying each of the resulting basis functions shown above by the characters of the operations in the A representation. Since A is the totally symmetric representation, all the characters are $+1$ in this case. Adding lines to the preceding table to show the characters of A and the resulting products $\chi_i^R R_j\sigma_1$, we obtain the following results:

O	E	C_3	C_3	C_3	C_3	C_3^2	C_3^2	C_3^2	C_3^2	C_2	C_2	C_2
Label		aa	bb	cc	dd	aa	bb	cc	dd	12	34	56
$R_j\sigma_1$	σ_1	σ_5	σ_3	σ_6	σ_4	σ_3	σ_6	σ_4	σ_5	σ_1	σ_2	σ_2
A	1	1	1	1	1	1	1	1	1	1	1	1
$\chi_i^R R_j\sigma_1$	σ_1	σ_5	σ_3	σ_6	σ_4	σ_3	σ_6	σ_4	σ_5	σ_1	σ_2	σ_2

	C_4	C_4	C_4	C_4^3	C_4^3	C_4^3	C_2'	C_2'	C_2'	C_2'	C_2'	C_2'
	12	34	56	12	34	56	ac	bd	ab	cd	ad	bc
	σ_1	σ_5	σ_4	σ_1	σ_6	σ_3	σ_2	σ_2	σ_3	σ_4	σ_5	σ_6
	1	1	1	1	1	1	1	1	1	1	1	1
	σ_1	σ_5	σ_4	σ_1	σ_6	σ_3	σ_2	σ_2	σ_3	σ_4	σ_5	σ_6

Summing all the $\chi_i^R R_j\sigma_1$ terms gives

$$P(A)\sigma_1 \propto 4\sigma_1 + 4\sigma_2 + 4\sigma_3 + 4\sigma_4 + 4\sigma_5 + 4\sigma_6$$

$$\propto \sigma_1 + \sigma_2 + \sigma_3 + \sigma_4 + \sigma_5 + \sigma_6$$

which is the result we anticipated. Keeping in mind the general properties represented by Eq. (5.2), normalization of $P(A)\sigma_1$ gives

$$N^2 \int (\sigma_1 + \sigma_2 + \sigma_3 + \sigma_4 + \sigma_5 + \sigma_6)^2 \, d\tau$$
$$= N^2 \int (\sigma_1^2 + \sigma_2^2 + \sigma_3^2 + \sigma_4^2 + \sigma_5^2 + \sigma_6^2) \, d\tau$$
$$= N^2 (1 + 1 + 1 + 1 + 1 + 1) = 6N^2 \equiv 1$$
$$\Rightarrow N = 1/\sqrt{6}$$

Therefore, the normalized SALC is

$$\Sigma_1(A) = 1/\sqrt{6}(\sigma_1 + \sigma_2 + \sigma_3 + \sigma_4 + \sigma_5 + \sigma_6) \tag{5.3}$$

To obtain the first of the two degenerate E SALCs we take the results previously obtained for the transformations of σ_1 and multiply them by the characters of the E representation. From the O character table we see that there are zero characters for $6C_4$ and $6C_2'$. Consequently, we only need to consider the first 12 terms, shown in the upper half of our previous listing of basis function transformations. Thus, the essential elements of the $P(E)\sigma_1$ operator are given by the following:

O	E	C_3	C_3	C_3	C_3	C_3^2	C_3^2	C_3^2	C_3^2	C_2	C_2	C_2
Label		aa	bb	cc	dd	aa	bb	cc	dd	12	34	56
$R_j\sigma_1$	σ_1	σ_5	σ_3	σ_6	σ_4	σ_3	σ_6	σ_4	σ_5	σ_1	σ_2	σ_2
E	2	-1	-1	-1	-1	-1	-1	-1	-1	2	2	2
$\chi_i^R R_j\sigma_1$	$2\sigma_1$	$-\sigma_5$	$-\sigma_3$	$-\sigma_6$	$-\sigma_4$	$-\sigma_3$	$-\sigma_6$	$-\sigma_4$	$-\sigma_5$	$2\sigma_1$	$2\sigma_2$	$2\sigma_2$

Summing across all $\chi_i^R R_j\sigma_1$ gives

$$P(E)\sigma_1 \propto 4\sigma_1 + 4\sigma_2 - 2\sigma_3 - 2\sigma_4 - 2\sigma_5 - 2\sigma_6$$
$$\propto 2\sigma_1 + 2\sigma_2 - \sigma_3 - \sigma_4 - \sigma_5 - \sigma_6$$

which after normalization gives

$$\Sigma_2(E) = 1/(2\sqrt{3})(2\sigma_1 + 2\sigma_2 - \sigma_3 - \sigma_4 - \sigma_5 - \sigma_6) \tag{5.4}$$

We can demonstrate that this is orthogonal to our first function, $\Sigma_1(A)$, as follows:

$$\int (\sigma_1 + \sigma_2 + \sigma_3 + \sigma_4 + \sigma_5 + \sigma_6)(2\sigma_1 + 2\sigma_2 - \sigma_3 - \sigma_4 - \sigma_5 - \sigma_6) \, d\tau$$
$$= 2 + 2 - 1 - 1 - 1 - 1 = 0$$

However, $\Sigma_2(E)$ is only one of a degenerate pair. We must find the partner.

One way in which we can try to find the partner of $\Sigma_2(E)$ is to carry out the E projection on a different basis function, say, σ_3 instead of σ_1. Taking this approach, we obtain the following results:

O	E	C_3	C_3	C_3	C_3	C_3^2	C_3^2	C_3^2	C_3^2	C_2	C_2	C_2
Label		aa	bb	cc	dd	aa	bb	cc	dd	12	34	56
$R_j\sigma_3$	σ_3	σ_1	σ_6	σ_2	σ_6	σ_5	σ_1	σ_5	σ_2	σ_4	σ_3	σ_4
E	2	-1	-1	-1	-1	-1	-1	-1	-1	2	2	2
$\chi_i^R R_j\sigma_3$	$2\sigma_3$	$-\sigma_1$	$-\sigma_6$	$-\sigma_2$	$-\sigma_6$	$-\sigma_5$	$-\sigma_1$	$-\sigma_5$	$-\sigma_2$	$2\sigma_4$	$2\sigma_3$	$2\sigma_4$

This gives

$$P(E)\sigma_3 \propto -2\sigma_1 - 2\sigma_2 + 4\sigma_3 + 4\sigma_4 - 2\sigma_5 - 2\sigma_6$$
$$\propto -\sigma_1 - \sigma_2 + 2\sigma_3 + 2\sigma_4 - \sigma_5 - \sigma_6$$

On first encounter this may seem as good a result as that obtained from $P(E)\sigma_1$. After all, it is orthogonal to $\Sigma_1(A)$:

$$\int(\sigma_1 + \sigma_2 + \sigma_3 + \sigma_4 + \sigma_5 + \sigma_6)(-\sigma_1 - \sigma_2 + 2\sigma_3 + 2\sigma_4 - \sigma_5 - \sigma_6)\,d\tau$$
$$= -1 - 1 + 2 + 2 - 1 - 1 = 0$$

In fact, if this were our first result for E, it would be an acceptable wave function. However, if we accept our previous result for $\Sigma_2(E)$ from $P(E)\sigma_1$, our result from $P(E)\sigma_3$ cannot be an acceptable wave function, since it is not orthogonal with $\Sigma_2(E)$:

$$\int(2\sigma_1 + 2\sigma_2 - \sigma_3 - \sigma_4 - \sigma_5 - \sigma_6)(-\sigma_1 - \sigma_2 + 2\sigma_3 + 2\sigma_4 - \sigma_5 - \sigma_6)\,d\tau$$
$$= -2 - 2 - 2 - 2 + 1 + 1 = -6 \neq 0$$

Surely the two degenerate E functions must be orthogonal with each other, as well as with $\Sigma_1(A)$.

The problem we have just encountered—that two projection operators give acceptable functions in their own right but are not acceptable partners to each other—results from conflicting choices of axes. When we began with $P(E)\sigma_1$ we implicitly fixed the coordinate system. Let us say that the orientation was such that the z axis of the system passed through σ_1. If we then begin again with $P(E)\sigma_3$, we implicitly reorient the z axis to pass through σ_3. The choice of which orientation to use is completely arbitrary, since all of the basis functions are equivalent. However, once we have chosen one orientation, the degenerate partner must conform to that choice. In some cases, depending on the geometry of the system, the shift of the axis system when operating on a different basis function may be of no consequence, and the partner function can be generated directly, in either its positive or negative form. In other cases, choosing a different basis function may generate the same SALC as the first choice, in either its positive or negative form. In still other cases, operating on a different basis function may generate a projected function that is a linear combination of the first SALC and the partner SALC or SALCs.

Our expression from $P(E)\sigma_3$ is neither a partner to $\Sigma_2(E)$ nor the negative of $\Sigma_2(E)$ itself. Instead, it would appear to be a linear combination of $\Sigma_2(E)$ and the partner function we seek. This being the case, we need to find the appropriate combination of $P(E)\sigma_1$, the operator expression that gave us $\Sigma_2(E)$, and the expression for $P(E)\sigma_3$. In other words, the function we seek has the form $aP(E)\sigma_1 + bP(E)\sigma_3$, where a and b are small positive or negative integers. The correct values of a and b are those that yield a function that is orthogonal to $\Sigma_2(E)$, as well as $\Sigma_1(A)$. With a little trial-and-error manipulation, we can obtain the missing partner as $P(E)\sigma_1 + 2P(E)\sigma_3$:

$$
\begin{array}{c}
2\sigma_1 + 2\sigma_2 - \sigma_3 - \sigma_4 - \sigma_5 - \sigma_6 \\
\underline{-2\sigma_1 - 2\sigma_2 + 4\sigma_3 + 4\sigma_4 - 2\sigma_5 - 2\sigma_6} \\
3\sigma_3 + 3\sigma_4 - 3\sigma_5 - 3\sigma_6 \\
\propto \sigma_3 + \sigma_4 - \sigma_5 - \sigma_6
\end{array}
$$

This result is orthogonal to $\Sigma_2(E)$,

$$
\int(2\sigma_1 + 2\sigma_2 - \sigma_3 - \sigma_4 - \sigma_5 - \sigma_6)(\sigma_3 + \sigma_4 - \sigma_5 - \sigma_6)\,d\tau
$$
$$
= 0 + 0 - 1 - 1 + 1 + 1 = 0
$$

and also to $\Sigma_1(A)$,

$$
\int(\sigma_1 + \sigma_2 + \sigma_3 + \sigma_4 + \sigma_5 + \sigma_6)(\sigma_3 + \sigma_4 - \sigma_5 - \sigma_6)\,d\tau
$$
$$
= 0 + 0 + 1 + 1 - 1 - 1 = 0
$$

The normalized partner wave function, then, is

$$
\Sigma_3(E) = \frac{1}{2}(\sigma_3 + \sigma_4 - \sigma_5 - \sigma_6) \tag{5.5}
$$

Another approach, which leads to the same result, is based on the following general property of degenerate functions: *The effect of any group operation on a wave function of a degenerate set is to transform the function into the positive or negative of itself, a partner, or a linear combination of itself and its partner or partners.* For the purpose of finding the partner in the present case, we will want to pick an operation that is unlikely to transform $\Sigma_2(E)$ into itself in either a positive or negative sense. Any one of the C_3 operations would be good candidates for this task. Let us look at the effect of performing the $C_3(aa)$ rotation on $\Sigma_2(E)$. This operation effects the following transformations on the individual basis functions: $\sigma_1 \to \sigma_5$, $\sigma_2 \to \sigma_6$, $\sigma_3 \to \sigma_1$, $\sigma_4 \to \sigma_2$, $\sigma_5 \to \sigma_3$, $\sigma_6 \to \sigma_4$. Thus $\Sigma_2(E)$ is transformed as

$$
(2\sigma_1 + 2\sigma_2 - \sigma_3 - \sigma_4 - \sigma_5 - \sigma_6) \to (-\sigma_1 - \sigma_2 - \sigma_3 - \sigma_4 + 2\sigma_5 + 2\sigma_6)
$$

However, this is not orthogonal to $\Sigma_2(E)$:

$$
\int(2\sigma_1 + 2\sigma_2 - \sigma_3 - \sigma_4 - \sigma_5 - \sigma_6)(-\sigma_1 - \sigma_2 - \sigma_3 - \sigma_4 + 2\sigma_5 + 2\sigma_6)\,d\tau
$$
$$
= -2 - 2 + 1 + 1 - 2 - 2 = -6 \neq 0
$$

This suggests that the new function is a combination of $\Sigma_2(E)$ and the partner we seek, $\Sigma_3(E)$. As before, we can find $\Sigma_3(E)$ by trying various combinations of the new function and $\Sigma_2(E)$, again using orthogonality as the test for a valid partner. In this manner we can find the form of $\Sigma_3(E)$ by adding two times the negative of the new function to the negative of $\Sigma_2(E)$:

$$
\begin{array}{r}
-2\sigma_1 - 2\sigma_2 + \sigma_3 + \sigma_4 + \sigma_5 + \sigma_6 \\
+2\sigma_1 + 2\sigma_2 + 2\sigma_3 + 2\sigma_4 - 4\sigma_5 - 4\sigma_6 \\
\hline
3\sigma_3 + 3\sigma_4 - 3\sigma_5 - 3\sigma_6 \\
\propto \sigma_3 + \sigma_4 - \sigma_5 - \sigma_6
\end{array}
$$

This is the same result as we obtained previously, which on normalization gives $\Sigma_3(E)$, as shown in Eq. (5.5). Note that if we had subtracted two times the new function from $\Sigma_2(E)$ we would have obtained the negative of this result, which is merely the same as $\Sigma_3(E)$ taken in the negative. Our preference for the form of $\Sigma_3(E)$ shown in Eq. (5.5), rather than its negative, is an arbitrary choice. The resulting function is orthogonal either way.* However, any other addition or subtraction of $\Sigma_2(E)$ and the expression we obtained by performing $C_3(aa)$ on $\Sigma_2(E)$ leads to a result that is not orthogonal. Consequently, the choice of how to manipulate the two expressions to obtain the partner function $\Sigma_3(E)$ is dictated by the orthogonality requirement. Quite simply, we do whatever it takes to get to a function that passes this test.

We have now seen two ways of obtaining a partner function: (1) Apply the projection operator for the degenerate representation to a different basis function than that used to generate the first SALC of the degenerate set, and (2) subject the first obtained SALC to an appropriate symmetry operation of the group. Either approach gives the desired result, but performing a group operation on the first SALC is clearly less work. Which operation to choose for the job is not particularly important, so long as it does not merely transform the first SALC into the positive or negative of itself. However, if one happens to make the wrong choice, choosing another operation or possibly the same operation about a differently oriented symmetry element usually will give either the positive or negative of the partner function or a function from which the partner can be obtained by suitable addition or subtraction with the original SALC.

In some cases, the partner SALCs are not difficult to deduce once the first function has been obtained. The three T_1 σ-SALCs for the six pendant atoms of an octahedral MX_6 molecule are a case in point. In similar manner to our procedure for the two E SALCs, we can find the first of three degenerate T_1 SALCs by applying the appropriate projection operator to one of the six basis functions. As before, we will use σ_1. Since the character for $8C_3$ in the T_1 representation is zero, we can skip the eight terms after the first term for identity. Thus, we have the following results:

*Note that if two functions ψ_a and ψ_b are orthogonal, such that $\int \psi_a \psi_b \, d\tau = 0$, then their negatives are also orthogonal, since it must be that $\int (-\psi_a)\psi_b \, d\tau = \int \psi_a(-\psi_b) \, d\tau = \int (-\psi_a)(-\psi_b) \, d\tau = 0$.

O	E	C_3	C_3	C_3	C_3	C_3^2	C_3^2	C_3^2	C_3^2	C_2	C_2	C_2
Label		aa	bb	cc	dd	aa	bb	cc	dd	12	34	56
$R_j\sigma_1$	σ_1	σ_5	σ_3	σ_6	σ_4	σ_3	σ_6	σ_4	σ_5	σ_1	σ_2	σ_2
T_1	3	0	0	0	0	0	0	0	0	-1	-1	-1
$\chi_i^R R_j\sigma_1$	$3\sigma_1$									$-\sigma_1$	$-\sigma_2$	$-\sigma_2$

C_4	C_4	C_4	C_4^3	C_4^3	C_4^3	C_2'	C_2'	C_2'	C_2'	C_2'	C_2'
12	34	56	12	34	56	ac	bd	ab	cd	ad	bc
σ_1	σ_5	σ_4	σ_1	σ_6	σ_3	σ_2	σ_2	σ_3	σ_4	σ_5	σ_6
1	1	1	1	1	1	-1	-1	-1	-1	-1	-1
σ_1	σ_5	σ_4	σ_1	σ_6	σ_3	$-\sigma_2$	$-\sigma_2$	$-\sigma_3$	$-\sigma_4$	$-\sigma_5$	$-\sigma_6$

Summing across the $\chi_i^R R_j\sigma_1$ results, we obtain the expression

$$P(T_1)\sigma_1 \propto 4\sigma_1 - 4\sigma_2 \propto \sigma_1 - \sigma_2$$

We can readily show that this is orthogonal to the previous three functions for A and E, and on normalization we obtain the SALC

$$\Sigma_4(T_1) = 1/\sqrt{2}(\sigma_1 - \sigma_2) \qquad (5.6)$$

In this case the companion functions are not difficult to discern from the geometry of the system. We see that $\Sigma_4(T_1)$ is the combination of two basis functions from pairs of *trans*-related pendant atoms. We can conclude from this that the other two functions must involve the same kind of combination with the remaining pairs. Thus, we readily obtain the companion SALCs

$$\Sigma_5(T_1) = 1/\sqrt{2}(\sigma_3 - \sigma_4) \qquad (5.7)$$

$$\Sigma_6(T_1) = 1/\sqrt{2}(\sigma_5 - \sigma_6) \qquad (5.8)$$

If a more analytical approach is needed, note that applying C_3 and C_3^2 about the aa axis, for example, transforms $\Sigma_4(T_1)$ into these two functions.

In some cases, the character form of the projection operator of a degenerate irreducible representation does not yield a single SALC, but rather a combination of SALCs. This occurs in the case of the σ-SALCs for the pendant atoms of a tetrahedral MX$_4$ molecule. Let us consider the use of projection operators to generate the expressions for the hydrogen SALCs we used in constructing the LCAO-MOs for methane [Eqs. (4.21a)–(4.21d) in Section 4.3]. As previously shown, the symmetry of the four hydrogen SALCs is $\Gamma = A_1 + T_2$. Consistent with our previous notation (cf. Fig. 4.16) we will label the four equivalent hydrogen atoms as shown in Fig. 5.2. Here we will

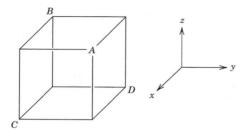

Figure 5.2   Orientation of the $1s$ basis functions used to construct SALCs for the four hydrogen atoms of methane.

simply indicate the functions as s, rather than $1s$, to avoid confusion in the projection operator expressions.

The group T_d has 24 operations ($h = 24$), which means that the projection operators will have as many terms. As before, we can avoid such cumbersome expressions by working in the rotational subgroup of the molecule's actual point group. In this case, the subgroup is T (cf. Appendix A for the character table), which has half as many operations. We are only interested in the two species A_1 and T_2 of T_d, which correspond to A and T in the group T. We really do not need to use projection operators to deduce the mathematical form of the A SALC, since this is the totally symmetric representation. Clearly, combining all four $1s$ wave functions with positive signs, as shown in Eq. (4.21a), is the only way to conform to the complete symmetry of the group. This only leaves the problem of determining the forms of the three degenerate T SALCs. From the character table for the group T, we see that the characters for both $4C_3$ and $4C_3^2$ are zero, so we can skip the eight terms associated with these operations. Thus, of the 12 terms for the projection operator for the representation T in the group T, only the four terms for the operations E, $C_2(x)$, $C_2(y)$, and $C_2(z)$ are nonzero. Considering only these nonzero terms, we obtain the following results for the T projection from the reference basis function s_A:

T	E	$\cdots$	$C_2(x)$	$C_2(y)$	$C_2(z)$
$R_j s_A$	s_A	$\cdots$	s_C	s_D	s_B
T	3	$\cdots$	-1	-1	-1
$\chi_i^R R_j s_A$	$3s_A$	$\cdots$	$-s_C$	$-s_D$	$-s_B$

This gives

$$P(T)s_A \propto 3s_A - s_C - s_D - s_B$$

This function is orthogonal to the A SALC, $\Phi_1 = \dfrac{1}{2}(s_A + s_B + s_C + s_D)$, and could be normalized to give the function

$$\Phi(T) = 1/(2\sqrt{3})(3s_A - s_C - s_D - s_B) \qquad (5.9)$$

However, this makes no sense as an individual function when we realize that the three T SALCs must overlap with the three degenerate $2p$ orbitals on the central carbon atom. As shown in Fig. 4.16, the lobes of each $2p$ orbital point toward opposite faces of the reference cube of the tetrahedron. For example, the positively signed lobe of the $2p_x$ orbital points toward the cube face of the A and C hydrogen atoms, and its negatively signed lobe points toward the face of the B and D hydrogen atoms. For bonding to occur, the signs on the hydrogen wave functions in the SALC must match the signs on the central atom with which they overlap. Thus, the SALC that matches the $2p_x$ orbital must use the A and C hydrogen $1s$ functions in a positive sense and the B and D hydrogen $1s$ functions in a negative sense. Similar matches must occur for the $2p_y$ and $2p_z$ orbitals. From these considerations we might surmise that the function projected by $P(T)s_A$ is a sum of the three SALCs we seek. In this case we can see that the following three functions add to give the overall expression for $P(T)s_A$:

$$P(T^z)s_A \propto s_A + s_B - s_C - s_D \tag{5.10a}$$

$$P(T^y)s_A \propto s_A - s_B - s_C + s_D \tag{5.10b}$$

$$P(T^x)s_A \propto s_A - s_B + s_C - s_D \tag{5.10c}$$

$$P(T)s_A \propto 3s_A - s_B - s_C - s_D$$

With normalization, $P(T^z)s_A$, $P(T^y)s_A$, and $P(T^x)s_A$ become Φ_2, Φ_3, and Φ_4, respectively, as shown in Eqs. (4.21b)–(4.21d).

As the preceding examples suggest, projection operators in character form do not automatically generate sets of degenerate SALCs. Moreover, it is difficult to generalize a procedure by which all members of a degenerate set may be most efficiently extracted from the initial projection. The best process in each case depends upon the peculiarities of the system under study. Both the geometry of the molecule and the form of the function that the projection operator first generates will dictate what method is most expedient. In the following section we will show yet another strategy for obtaining the members of degenerate sets. Although we will illustrate this for the π-SALCs of a conjugated ring system, the technique can be applied in other cases, as appropriate.

5.2 SALCs of Pi Systems

Projection operators can be used to generate π-SALCs just as effectively as they can be used to generate σ-SALCs. The π-functions might be pendant atom SALCs to be combined with π-symmetry orbitals on a central atom or group of atoms, or they might be π-MOs in a conjugated ring system. The approach is basically the same in both cases. As an illustration, let us use projection operators to obtain the six π-MO wave functions for benzene, which were presented without derivation in Section 4.5 [Eqs. (4.29)–(4.32)]. Taking the six p_z orbitals on the carbon atoms of the ring as basis functions for a representation in D_{6h} we found that the resulting reducible representation breaks down as $\Gamma_\pi = A_{2u} + B_{2g} + E_{1g} + E_{2u}$. Thus, we will need to construct projection operators for these symmetry species to form two nondegenerate MOs and two pairs of degenerate MOs. For reference we designate the carbon atoms alphabetically in a clockwise manner about the ring, as shown in Fig. 5.3. The six p_z basis functions of the ring (ϕ_a, ϕ_b, ϕ_c, ϕ_d, ϕ_e, ϕ_f) have the directional sense previously shown in Fig. 4.24.

We could construct our projection operators in D_{6h}, for which $h = 24$, but a more careful examination of the problem shows some ways in which the labor of the process can be minimized. In particular, consider the characters of the irreducible representations that comprise Γ_π.

D_{6h}	E	$2C_6$	$2C_3$	C_2	$3C_2'$	$3C_2''$	i	$2S_3$	$2S_6$	σ_h	$3\sigma_d$	$3\sigma_v$
A_{2u}	1	1	1	1	-1	-1	-1	-1	-1	-1	1	1
B_{2g}	1	-1	1	-1	-1	1	1	-1	1	-1	-1	1
E_{1g}	2	1	-1	-2	0	0	2	1	-1	-2	0	0
E_{2u}	2	-1	-1	2	0	0	-2	1	1	-2	0	0

The operations and characters in the box are those of the rotational subgroup C_6. In the cases of the two doubly degenerate species, the characters are the same as those in C_6 if the complex conjugate paired irreducible representations comprising E_1 and E_2 in that group are added together. (See Section 3.2.) Note that the characters for the subgroup operations are sufficient to differentiate between the four symmetry species of Γ_π. This suggests that we can save a great deal of labor by applying projection operators in the subgroup C_6, for which $h = 6$, rather than the full group D_{6h}. Examining the characters within the box in the table above and comparing them with the characters for the representations of C_6, we see that the correlation of species from D_{6h} to C_6 is as follows: $A_{2u} \rightarrow A$; $B_{2g} \rightarrow B$; $E_{1g} \rightarrow E_1$; $E_{2u} \rightarrow E_2$. Thus, in C_6 the necessary projection operators are those for A, B, E_1, and E_2. However, as the table for the group C_6 (cf. Appendix A) shows, the E_1 and E_2 representations are actually complex-conjugate pairs of irreducible representations, involving the imaginary integer $i = \sqrt{-1}$. This might seem like an inconvenience, but in this case it provides an advantage. The descent in symmetry from D_{6h} to C_6 lifts the degeneracy of the doubly degenerate representations, allowing us to construct separate projection operators for each of the two complex-conjugate representations. Thus, we can obtain the two companion functions for each degenerate pair relatively directly.

Figure 5.3   Position labels for the six $2p_z$ basis functions used to form π-MOs of benzene.

We begin with the A-symmetry function, which corresponds to A_{2u} in D_{6h}. In the working subgroup C_6, A is the totally symmetric representation. Therefore, we expect the result $\Pi(A) = 1/\sqrt{6}(\phi_a + \phi_b + \phi_c + \phi_d + \phi_e + \phi_f)$. Indeed, the projection operator $P(A)\phi_a$ in C_6 yields this result, as follows:

C_6	E	C_6	C_3	C_2	C_3^2	C_6^5
$R_j\phi_a$	ϕ_a	ϕ_b	ϕ_c	ϕ_d	ϕ_e	ϕ_f
A	1	1	1	1	1	1
$\chi_i R_j\phi_a$	ϕ_a	ϕ_b	ϕ_c	ϕ_d	ϕ_e	ϕ_f

$$\Rightarrow \Pi(A) = 1/\sqrt{6}(\phi_a + \phi_b + \phi_c + \phi_d + \phi_e + \phi_f) = \pi_1 \qquad (5.11)$$

This is identical to Eq. (4.29), the function for the lowest energy bonding MO, π_1.

The SALC for the B representation, which corresponds to B_{2g} in D_{6h}, is obtained with equal alacrity:

C_6	E	C_6	C_3	C_2	C_3^2	C_6^5
$R_j\phi_a$	ϕ_a	ϕ_b	ϕ_c	ϕ_d	ϕ_e	ϕ_f
B	1	-1	1	-1	1	-1
$\chi_i R_j\phi_a$	ϕ_a	$-\phi_b$	ϕ_c	$-\phi_d$	ϕ_e	$-\phi_f$

$$\Rightarrow \Pi(B) = 1/\sqrt{6}(\phi_a - \phi_b + \phi_c - \phi_d + \phi_e - \phi_f) = \pi_6{}^* \qquad (5.12)$$

This is identical to Eq. (4.32), the function for the highest energy antibonding MO, π_6^*.

Closer examination of the results for A and B reveals yet another simplification we can employ with conjugated ring systems such as benzene. Note that the factors for the various ϕ's in Eqs. (5.11) and (5.12) are the same as the characters of the A and B representations, respectively. This occurs because the effect of the rotations in the group C_6 is to carry the reference function ϕ_a into all the basis functions around the ring in succession. Hence, when we multiply by the characters of any irreducible representation, the resulting projected function takes on the form

$$\chi_1^i\phi_a + \chi_2^i\phi_b + \chi_3^i\phi_c + \chi_4^i\phi_d + \chi_5^i\phi_e + \chi_6^i\phi_f$$

where the coefficients $\chi_1^i, \chi_2^i, \ldots, \chi_6^i$ are the six successive characters of the ith irreducible representation of C_6. Thus, we can write down the expressions for the six π-SALCs simply by inspecting the characters in C_6 for each of the irreducible representations of Γ_π. The characters in each case are the factors for the ϕ's, taken in order, in the SALCs prior to normalization. Applying this method to the two doubly degenerate symmetry species, E_1 and E_2, gives the following four functions:

$$P(E_1^a)\phi_a \propto (\phi_a + \epsilon\phi_b - \epsilon^*\phi_c - \phi_d - \epsilon\phi_e + \epsilon^*\phi_f) \qquad (5.13a)$$

$$P(E_1^b)\phi_a \propto (\phi_a + \epsilon^*\phi_b - \epsilon\phi_c - \phi_d - \epsilon^*\phi_e + \epsilon\phi_f) \qquad (5.13b)$$

$$P(E_2^a)\phi_a \propto (\phi_a - \epsilon^*\phi_b - \epsilon\phi_c + \phi_d - \epsilon^*\phi_e - \epsilon\phi_f) \qquad (5.14a)$$

$$P(E_2^b)\phi_a \propto (\phi_a - \epsilon\phi_b - \epsilon^*\phi_c + \phi_d - \epsilon\phi_e - \epsilon^*\phi_f) \qquad (5.14b)$$

Note that by working in the rotational subgroup C_6, rather than the actual group D_{6h}, we have bypassed the problem of obtaining two degenerate functions from a single projection operator. In C_6 we obtain two separate equations for each pair of degenerate SALCs in D_{6h}. However, as they stand, the expressions are imaginary. Naturally, we would prefer to have real functions. To obtain real-number expressions we take the positive and negative sums of the complex conjugate pairs of functions in each case. For E_1, by adding Eqs. (5.13a) and (5.13b), we obtain

$$P(E_1^a)\phi_a + P(E_1^b)\phi_a$$
$$\propto \{2\phi_a + (\epsilon + \epsilon^*)\phi_b - (\epsilon + \epsilon^*)\phi_c - 2\phi_d - (\epsilon + \epsilon^*)\phi_e + (\epsilon + \epsilon^*)\phi_f\}$$

where

$$\epsilon + \epsilon^* = \left(\cos\frac{2\pi}{6} + i\sin\frac{2\pi}{6}\right) + \left(\cos\frac{2\pi}{6} - i\sin\frac{2\pi}{6}\right)$$
$$= 2\cos\frac{2\pi}{6} = 2(\tfrac{1}{2}) = 1$$

Thus, we have

$$P(E_1^a)\phi_a + P(E_1^b)\phi_a \propto 2\phi_a + \phi_b - \phi_c - 2\phi_d - \phi_e + \phi_f$$

which after normalization gives

$$\Pi(E_1^a) = 1/(2\sqrt{3})\{2\phi_a + \phi_b - \phi_c - 2\phi_d - \phi_e + \phi_f\} = \pi_2 \qquad (5.15a)$$

This result is identical to Eq. (4.30a).

By subtracting Eq. (5.13b) from Eq. (5.13a) we obtain

$$P(E_1^a)\phi_a - P(E_1^b)\phi_a$$
$$\propto \{0 + (\epsilon - \epsilon^*)\phi_b + (\epsilon - \epsilon^*)\phi_c + 0 - (\epsilon - \epsilon^*)\phi_e - (\epsilon - \epsilon^*)\phi_f\}$$

where

$$\epsilon - \epsilon^* = \left(\cos\frac{2\pi}{6} + i\sin\frac{2\pi}{6}\right) - \left(\cos\frac{2\pi}{6} - i\sin\frac{2\pi}{6}\right)$$
$$= 2i\sin\frac{2\pi}{6} = 2i\left(\frac{\sqrt{3}}{2}\right) = i\sqrt{3}$$

Thus, we have

$$P(E_1^a)\phi_a - P(E_1^b)\phi_a \propto i\sqrt{3}(\phi_b + \phi_c - \phi_e - \phi_f)$$

Although $i\sqrt{3}$ is imaginary, it is a nonzero constant, which can be factored out prior to normalization. Accordingly, after normalization we obtain

$$\Pi(E_1^\beta) = \frac{1}{2}\{\phi_b + \phi_c - \phi_e - \phi_f\} = \pi_3 \qquad (5.15b)$$

which is identical to Eq. (4.30b).

Following the same procedure with Eqs. (5.14a) and (5.14b) for E_2 we obtain

$$P(E_2^a)\phi_a + P(E_2^b)\phi_a$$
$$\propto \{2\phi_a - (\epsilon + \epsilon^*)\phi_b - (\epsilon + \epsilon^*)\phi_c + 2\phi_d - (\epsilon + \epsilon^*)\phi_e - (\epsilon + \epsilon^*)\phi_f\}$$

and

$$P(E_2^a)\phi_a - P(E_2^b)\phi_a$$
$$\propto \{0 - (\epsilon - \epsilon^*)\phi_b + (\epsilon - \epsilon^*)\phi_c + 0 - (\epsilon - \epsilon^*)\phi_e + (\epsilon - \epsilon^*)\phi_f\}$$
$$\propto -\phi_b + \phi_c - \phi_e + \phi_f$$

After normalization these yield

$$\Pi(E_2^a) = 1/(2\sqrt{3})\{2\phi_a - \phi_b - \phi_c + 2\phi_d - \phi_e - \phi_f\} = \pi_4^* \qquad (5.16a)$$

and

$$\Pi(E_2^\beta) = \frac{1}{2}\{-\phi_b + \phi_c - \phi_e + \phi_f\} = \pi_5^* \qquad (5.16b)$$

which are identical to Eqs. (4.31a) and (4.31b).

The procedure we have followed for benzene can be extended to become a general method for generating π-LCAO-MOs for similar conjugated ring systems:

1. Write down an initial set of SALCs by inspecting the character table C_n, which is a subgroup of the molecule's point group D_{nh}. These SALCs will have the form $\chi_1^i\phi_1 + \chi_2^i\phi_2 + \cdots + \chi_n^i\phi_n$, where $\chi_1^i, \chi_2^i, \cdots \chi_n^i$ are the characters of the ith irreducible representation of the representation Γ_π in the group C_n.

2. Make real functions for pairs of complex conjugate SALCs by adding and subtracting the imaginary functions. Factor out any overall coefficients containing i prior to normalization.

3. Normalize the functions.

This method also can be applied to obtain the π-SALCs of pendant atoms in planar MX_n molecules with D_{nh} symmetry.

5.3 Formulating Hybrid Orbitals

As we have seen in the preceding sections, the formation of SALCs involves combining basis functions from the various atoms of the molecule into suitable LCAO functions that conform to the symmetry species of the reducible representation. When we apply the projection operators for each symmetry species to a member of this basis set, we project the desired function for the whole system. The basis functions are the known quantities, and the SALCs are the solutions to the molecular orbital problem. In this sense, the unknown whole (the SALC) is defined in terms of its known parts (the basis functions). When we seek to construct hybrid orbitals as LCAOs, we must realize that the process is actually the inverse. As we saw in Section 4.2, we begin with the hybrids, with their desired geometrical orientations, as our basis set and generate a reducible representation, whose component irreducible representations indicate the symmetries of the various atomic orbitals that may be combined on the central atom. Here, the whole (the set of hybrids) is the known quantity and the component parts (the atomic orbitals comprising them) are the unknowns to be determined. As a result, if we were to apply projection operators to the hybrids that form the basis set, we would obtain expressions for each of the component conventional atomic orbitals (s, p, d) as SALCs of the various hybrid functions. This is hardly the result we wish. Rather, we would hope to be able to formulate the hybrids in terms of the conventional atomic orbitals. Nonetheless, proceeding in this seemingly backwards manner can get us to the goal we seek. Once we have obtained expressions for the conventional orbitals in terms of the hybrids, we can take advantage of the properties of matrices and their inverses to obtain the desired expressions for the hybrids as SALCs of the conventional orbitals with very little additional effort.

Let us illustrate this approach by developing the expressions for the four tetrahedral sp^3 hybrids. As we saw in Section 4.2, taking a set of hybrid orbitals with tetrahedral orientation as the basis for a reducible representation in the group T_d gives $\Gamma_t = A_1 + T_2$, which indicates that the s orbital and the degenerate set of p orbitals may be combined to form four hybrid wave functions, Ψ_A, Ψ_B, Ψ_C, and Ψ_D, as shown in Fig. 5.4. We could apply projection operators for A_1 and T_2 (or A and T in the rotational subgroup T) to any one of these hybrids, say Ψ_A, and project functions for the s and p orbitals as linear combinations of the four hybrid orbital functions, which form the basis set. Actually, we already carried out the equivalent of this process when we developed the four hydrogen SALCs of methane in Section 5.1. In that case we took four vectors directed inward, toward the center of the system, as our

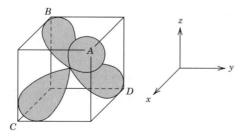

Figure 5.4   Orientations of the sp^3 functions Ψ_A, Ψ_B, Ψ_C, and Ψ_D.

basis set, whereas now we take four vectors directed outward from the center of the system. In terms of symmetry, this is a difference without distinction. Consequently, we can use the results we obtained previously for the hydrogen SALCs [the expression for A and the three expressions for T, Eqs. (5.10a)–(5.10c)] to obtain the following normalized expressions for the four conventional atomic orbitals in terms of the four hybrids:

$$s = \frac{1}{2}(\Psi_A + \Psi_B + \Psi_C + \Psi_D)$$

$$p_x = \frac{1}{2}(\Psi_A - \Psi_B + \Psi_C - \Psi_D)$$

$$p_y = \frac{1}{2}(\Psi_A - \Psi_B - \Psi_C + \Psi_D)$$

$$p_z = \frac{1}{2}(\Psi_A + \Psi_B - \Psi_C - \Psi_D)$$

We can write these equations in matrix form as

$$
\begin{bmatrix} s \\ p_x \\ p_y \\ p_z \end{bmatrix} =
\begin{bmatrix}
\frac{1}{2} & \frac{1}{2} & \frac{1}{2} & \frac{1}{2} \\
\frac{1}{2} & -\frac{1}{2} & \frac{1}{2} & -\frac{1}{2} \\
\frac{1}{2} & -\frac{1}{2} & -\frac{1}{2} & \frac{1}{2} \\
\frac{1}{2} & \frac{1}{2} & -\frac{1}{2} & -\frac{1}{2}
\end{bmatrix}
\begin{bmatrix} \Psi_A \\ \Psi_B \\ \Psi_C \\ \Psi_D \end{bmatrix}
\tag{5.17}
$$

The transformation of the Ψ's for the hybrids into the conventional atomic orbital wave functions is effected by the 4×4 matrix, which we shall call $\mathbf{A}$. Now, what we really seek is an equation for the inverse of the transformation expressed by Eq. (5.17); that is,

$$
\begin{bmatrix} \Psi_A \\ \Psi_B \\ \Psi_C \\ \Psi_D \end{bmatrix} =
\begin{bmatrix}
b_{11} & b_{12} & b_{13} & b_{14} \\
b_{21} & b_{22} & b_{23} & b_{24} \\
b_{31} & b_{32} & b_{33} & b_{34} \\
b_{41} & b_{42} & b_{43} & b_{44}
\end{bmatrix}
\begin{bmatrix} s \\ p_x \\ p_y \\ p_z \end{bmatrix}
\tag{5.18}
$$

What we need to find are the b_{ij} elements of the $\mathbf{B}$ matrix of Eq. (5.18). Since Eqs. (5.17) and (5.18) are the inverse transformations, it follows that the matrices $\mathbf{A}$ and $\mathbf{B}$ are the *inverses** of each other; that is, $\mathbf{AB} = \mathbf{AA}^{-1} = \mathbf{B}^{-1}\mathbf{B} = \mathbf{E}$.

*We denote the inverse of a matrix $\mathbf{M}$ as $\mathbf{M}^{-1}$ and require that $\mathbf{M}^{-1}\mathbf{M} = \mathbf{MM}^{-1} = \mathbf{E}$, where $\mathbf{E}$ is the *identity matrix*. $\mathbf{E}$ is a diagonal matrix consisting of all 1's along the trace and 0's everywhere else. In general, the elements of $\mathbf{E}$ are given by the Kronecker delta expression $e_{ij} = \delta_{ij}$ (0 when $i \neq j$, and 1 when $i = j$).

However, **A** and **B** are *orthogonal matrices,* which means that the inverse of one is the *transpose* of the other. As such, the elements in each successive row of **B** are the elements from each successive column of **A**, and vice versa; that is, $(\mathbf{B})_{ij} = b_{ij} = a_{ji}$. Using the transposed numeric coefficients from Eq. (5.17) as the coefficients b_{ij} in Eq. (5.18), we obtain

$$
\begin{bmatrix} \Psi_A \\ \Psi_B \\ \Psi_C \\ \Psi_D \end{bmatrix} =
\begin{bmatrix}
\frac{1}{2} & \frac{1}{2} & \frac{1}{2} & \frac{1}{2} \\
\frac{1}{2} & -\frac{1}{2} & -\frac{1}{2} & \frac{1}{2} \\
\frac{1}{2} & \frac{1}{2} & -\frac{1}{2} & -\frac{1}{2} \\
\frac{1}{2} & -\frac{1}{2} & \frac{1}{2} & -\frac{1}{2}
\end{bmatrix}
\begin{bmatrix} s \\ p_x \\ p_y \\ p_z \end{bmatrix}
\tag{5.19}
$$

which gives the following four equations for the individual sp^3 hybrid orbitals:

$$\Psi_A = \frac{1}{2}(s + p_x + p_y + p_z) \tag{5.20a}$$

$$\Psi_B = \frac{1}{2}(s - p_x - p_y + p_z) \tag{5.20b}$$

$$\Psi_C = \frac{1}{2}(s + p_x - p_y - p_z) \tag{5.20c}$$

$$\Psi_D = \frac{1}{2}(s - p_x + p_y - p_z) \tag{5.20d}$$

The procedure we have just seen can be extended to obtain the equations for other sets of equivalent hybrid orbitals.

1. Taking the n hybrid orbitals as a basis set, construct and decompose a reducible representation Γ_{hyb} to identify the appropriate conventional orbitals to be combined.

2. Using the hybrids themselves or an equivalent set of pendant atom sigma orbitals as the basis set, apply the projection operators for each of the irreducible representations comprising Γ_{hyb} to a representative function of the set to obtain expressions for the conventional orbitals as LCAOs of the hybrids. Normalize all functions.

3. Combine the equations obtained in step 2 into a single matrix equation, using the coefficients to form the $n \times n$ transformation matrix **A**. Take the transpose of **A** to form the transformation matrix **B**.

4. Write a matrix equation for the hybrids by applying the **B** matrix to a column matrix of the conventional orbitals, written in the same order as in the previous matrix equation. Expand the matrix equation to obtain a set of n equations, one for each hybrid orbital.

5.4 *Systems with Nonequivalent Positions*

The molecular systems we have considered thus far have had geometries in which all positions are symmetrically equivalent. In other words, every position can be generated by applying the symmetry operations of the group to any one of the positions. As a result, if we apply a projection operator to a function at any position, the SALC we produce is an expression in terms of functions at all positions. However, there are many molecular geometries in which there are two or more distinct kinds of positions that cannot be interchanged by any operation of the group. Among the idealized geometries of MX_n-type molecules the trigonal bipyramid (*tbp*) is probably the most familiar example.

Let us consider constructing pendant atom σ-SALCs for a *tbp* MX_5 molecule (e.g., PF_5). We can then use the expressions for these SALCs to formulate equations for dsp^3 hybrids on the central M atom, in the manner described in Section 5.3. Figure 5.5 shows the labeling of sigma functions and hybrid orbitals we will use in carrying out these tasks. Taking the pendant atom sigma functions as the basis set, we can readily show that the reducible representation for the SALCs in D_{3h} is $\Gamma_\sigma = 2A_1' + A_2'' + E'$. As previously noted, the symmetry of the SALCs is equivalent to that of a set of hybrids with the same geometry, so these are also the symmetry species of dsp^3 hybrids on the central M atom. Now, for a trigonal bipyramid we know that no operation of D_{3h} can convert one of the equatorial positions (1, 2, and 3) into either of the axial positions (4 and 5), and vice versa. Consequently, applying a projection operator to an equatorial reference basis function can only yield a SALC in terms of σ_1, σ_2, and σ_3. Likewise applying a projection operator to an axial reference function can only yield a SALC in terms of σ_4 and σ_5. Thus, if we choose to use projection operators, we are forced to break up the problem into two parts. In doing so, we recognize that Γ_σ can be seen as the sum of $\Gamma_{eq} = A_1' + E'$ and $\Gamma_{ax} = A_1' + A_2''$ for the equatorial and axial sets of positions, respectively. Proceeding on this basis we can formulate separate sets of SALCs for the two kinds of positions.

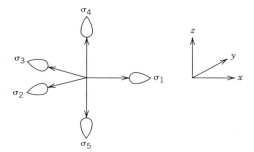

Figure 5.5 Positions of a trigonal bipyramid.

The SALCs for the equatorial set can be generated most readily with projection operators if we address the problem in the rotational subgroup C_3. As usual, we can write down the SALC for the totally symmetric representation (A_1' in D_{3h}, A in C_3) directly:

$$\Phi_{eq}(A_1') = 1/\sqrt{3}(\sigma_1 + \sigma_2 + \sigma_3) \tag{5.21}$$

Operating on the reference function σ_1, we can obtain functions for the two degenerate SALCs (E' in D_{3h}, E in C_3) by inspection of the C_3 character table:

$$P(E^a)\sigma_1 \propto \sigma_1 + \epsilon\sigma_2 + \epsilon^*\sigma_3$$
$$P(E^b)\sigma_1 \propto \sigma_1 + \epsilon^*\sigma_2 + \epsilon\sigma_3$$

By addition and subtraction, these two expressions lead to two real functions:

$$\{P(E^a)\sigma_1 + P(E^b)\sigma_1\} \propto 2\sigma_1 - \sigma_2 - \sigma_3$$

$$\Rightarrow \Phi_{eq}(E'^a) = 1/\sqrt{6}(2\sigma_1 - \sigma_2 - \sigma_3) \tag{5.22a}$$

$$\{P(E^a)\sigma_1 - P(E^b)\sigma_1\} \propto \sigma_2 - \sigma_3$$

$$\Rightarrow \Phi_{eq}(E'^b) = 1/\sqrt{2}(\sigma_2 - \sigma_3) \tag{5.22b}$$

The two SALCs for the axial positions can only involve positive and negative combinations of σ_4 and σ_5. Therefore, by inspection, they are

$$\Phi_{ax}(A_1') = 1/\sqrt{2}(\sigma_4 + \sigma_5) \tag{5.23}$$

$$\Phi_{ax}(A_2'') = 1/\sqrt{2}(\sigma_4 - \sigma_5) \tag{5.24}$$

If we consider the atomic orbitals with which these SALCs might interact in an MX_5 molecule, we can readily see the limitations of the approach we have taken in constructing them. The AOs on the central metal atom with matching symmetries are

$$s = A_1', \qquad (p_x, p_y) = E', \qquad p_z = A_2'', \qquad d_{z^2} = A_1'$$

We see from this that both A_1' SALCs have the appropriate symmetry to form bonding and antibonding combinations with both the s and d_{z^2} orbitals on the central atom. However, the two A_1' SALCs we have just formed are either confined to the xy plane [Eq. (5.21)] or to the z axis [Eq. (5.23)]. As a simplifying assumption, then, we might match the $\Phi_{eq}(A_1')$ SALC [Eq. (5.21)] with the s orbital and the $\Phi_{ax}(A_1')$ SALC [Eq. (5.23)] with the d_{z^2} orbital on the central M atom. Using the d_{z^2} orbital for the axial positions has some topological justification, inasmuch as the axial positions are usually longer than the equatorial positions in nonmetal MX_5 *tbp* species. Nonetheless, excluding s orbital involvement with the $\Phi_{ax}(A_1')$ SALC and d_{z^2} orbital involvement with the $\Phi_{eq}(A_1')$ SALC implies a model that must be regarded as a limiting case. The segregation of the two A_1' SALCs into either equatorial or axial combinations is an artifact of the projection operator method in this case.

Using the procedure described in Section 5.3, we may take the forms of Eqs. (5.21)–(5.24) to write the matrix equation for the five conventional orbitals on M ($s, p_x, p_y, p_z, d_{z^2}$) as linear combinations of five dsp^3 hybrid orbitals. Using Ψ to represent the hybrid functions, we have

$$
\begin{bmatrix} s \\ p_x \\ p_y \\ p_z \\ d_{z^2} \end{bmatrix}
=
\begin{bmatrix}
1/\sqrt{3} & 1/\sqrt{3} & 1/\sqrt{3} & 0 & 0 \\
2/\sqrt{6} & -1/\sqrt{6} & -1/\sqrt{6} & 0 & 0 \\
0 & 1/\sqrt{2} & -1/\sqrt{2} & 0 & 0 \\
0 & 0 & 0 & 1/\sqrt{2} & -1/\sqrt{2} \\
0 & 0 & 0 & 1/\sqrt{2} & 1/\sqrt{2}
\end{bmatrix}
\begin{bmatrix} \Psi_1 \\ \Psi_2 \\ \Psi_3 \\ \Psi_4 \\ \Psi_5 \end{bmatrix}
$$

Inverting the transformation matrix gives

$$
\begin{bmatrix} \Psi_1 \\ \Psi_2 \\ \Psi_3 \\ \Psi_4 \\ \Psi_5 \end{bmatrix}
=
\begin{bmatrix}
1/\sqrt{3} & 2/\sqrt{6} & 0 & 0 & 0 \\
1/\sqrt{3} & -1/\sqrt{6} & 1/\sqrt{2} & 0 & 0 \\
1/\sqrt{3} & -1/\sqrt{6} & -1/\sqrt{2} & 0 & 0 \\
0 & 0 & 0 & 1/\sqrt{2} & 1/\sqrt{2} \\
0 & 0 & 0 & -1/\sqrt{2} & 1/\sqrt{2}
\end{bmatrix}
\begin{bmatrix} s \\ p_x \\ p_y \\ p_z \\ d_{z^2} \end{bmatrix}
$$

From this we obtain five expressions for the hybrid orbitals:

$$\Psi_1 = 1/\sqrt{3}(s) + 2/\sqrt{6}(p_x) \tag{5.25a}$$

$$\Psi_2 = 1/\sqrt{3}(s) - 1/\sqrt{6}(p_x) + 1/\sqrt{2}(p_y) \tag{5.25b}$$

$$\Psi_3 = 1/\sqrt{3}(s) - 1/\sqrt{6}(p_x) - 1/\sqrt{2}(p_y) \tag{5.25c}$$

$$\Psi_4 = 1/\sqrt{2}(d_{z^2} + p_z) \tag{5.25d}$$

$$\Psi_5 = 1/\sqrt{2}(d_{z^2} - p_z) \tag{5.25e}$$

Like the SALCs we obtained previously, these must be regarded as a limiting case, since the axial hybrids exclude contributions from the s orbital, and the equatorial hybrids exclude contributions from the d_{z^2} orbital.

The inability of projection operators to produce linear combinations of both equivalent and nonequivalent functions limits us to obtaining artificially segregated results. Ideally, the set of σ-SALCs for a *tbp* molecule should have the form

$$\Phi_i = N(c_{i1}\sigma_1 \pm c_{i2}\sigma_2 \pm c_{i3}\sigma_3 \pm c_{i4}\sigma_4 \pm c_{i5}\sigma_5), \qquad i = 1, 2, 3, 4, 5$$

where the coefficients c_{ij} are nonzero, except as required by symmetry. In other words, we should not exclude any pendant atom functions *a priori*. Actually, we can obtain a set of more inclusive equations by taking a *pictorial approach*. Very simply, we determine the form of each SALC by sketching the pendant atom orbitals and assigning their wave function signs to conform with those of the matching central atom AO. These assignments give an idea of how all the pendant atom functions must be combined in either a positive or negative sense in each SALC. The relative contribution of each pendant atom to the SALC can often be deduced from the drawings or from results obtained by other means, such as projection operator methods. In this manner, we arrive at the drawings shown in Fig. 5.6 for the *tbp* case.

We see from Fig. 5.6 that neither E' SALC makes use of σ_4 and σ_5. These positions fall within the nodal planes of the matching p_x and p_y orbitals on M, thereby precluding effective overlap. Hence, by symmetry the expression for Φ_2 must have zero coefficients for c_{24} and c_{25}, and the expression for Φ_3 must have zero coefficients for c_{34} and c_{35}. Similarly, Φ_3 does not make use of σ_1, since it, too, falls within the nodal plane of the central p_y orbital. Thus, $c_{31} = 0$ in the expression for Φ_3. From this we conclude that the two E' SALCs, Φ_2 and Φ_3, have the same forms as the SALCs we formulated by projection operators [Eqs. (5.22a) and (5.22b)]. In the same manner, Φ_4 does not use any

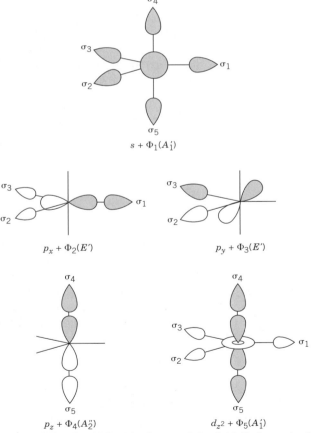

Figure 5.6 Identifying the forms of the SALCs for a *tbp* MX$_5$ molecule by matching wave function signs with central atom AOs with the appropriate symmetries. (Positive regions are indicated with shading.)

of the functions in the equatorial plane, because that is the nodal plane of the p_z orbital. Thus, for Φ_4 the coefficients c_{41}, c_{42}, and c_{43} must be zero, and the expression is identical to that obtained previously [Eq. (5.24)]. For these three SALCs—which involve combinations of equatorial or axial functions, but not both—the restrictions imposed by the projection operator approach are not inconsistent with the physical restrictions of the molecular topography. Therefore, they are complete and correct as originally formulated with projection operators. This is not the case, however, with the two A_1' SALCs.

We see from Fig. 5.6 that it is possible to form two A_1' SALCs that include all five pendant atom functions, taken in suitable combinations so as to form bonding and antibonding combinations with the s and d_{z^2} orbitals. However, since the equatorial and axial distances are generally not the same in *tbp* structures, the coefficients for the two kinds of positions must in general be different. However, given A_1' symmetry, we can say with confidence that $c_{i1} = c_{i2} = c_{i3}$ and $c_{i4} = c_{i5}$ for a perfect *tbp* structure. Now, we must realize that the equatorial and axial coefficients will not be fixed values, since the ratio of equatorial to axial bond lengths is highly variable among *tbp* molecules. Therefore, we cannot hope to formulate generally applicable equations for Φ_1 and Φ_5 with fixed values for the axial and equatorial coefficients. Nonetheless, we can propose expressions for an idealized case: a *tbp* structure in which the axial and equatorial bonds are equal. Although this is not the typical configuration, it is approximated in some cases. For example, the $[CdCl_5]^{3-}$ ion in $[Co(NH_3)_6][CdCl_5]$ has a bond length ratio $Cd\text{-}Cl_{eq}/Cd\text{-}Cl_{ax}$ of 1.015.* In such a structure, all five pendant atoms would have equal overlap with the central s orbital, resulting in an expression for the Φ_1 SALC in which all coefficients are equal. In the case of Φ_5, the signs for the equatorial functions must be opposite those of the axial functions, as shown in Fig. 5.6. If we assume equal contributions from equatorial and axial positions, the SALC should have the form $\sigma_4 + \sigma_5 - 2/3(\sigma_1 + \sigma_2 + \sigma_3)$. Including these two results with the three previously obtained, we can write the following five, orthogonal, normalized SALCs for the idealized equal-bond *tbp* case:

$$\Phi_1 = 1/\sqrt{5}(\sigma_1 + \sigma_2 + \sigma_3 + \sigma_4 + \sigma_5) \qquad (5.26a)$$

$$\Phi_2 = 1/\sqrt{6}(2\sigma_1 - \sigma_2 - \sigma_3) \qquad (5.26b)$$

$$\Phi_3 = 1/\sqrt{2}(\sigma_2 - \sigma_3) \qquad (5.26c)$$

$$\Phi_4 = 1/\sqrt{2}(\sigma_4 - \sigma_5) \qquad (5.26d)$$

$$\Phi_5 = 3/\sqrt{30}\{\sigma_4 + \sigma_5 - 2/3(\sigma_1 + \sigma_2 + \sigma_3)\} \qquad (5.26e)$$

For a real *tbp* molecule with unequal equatorial and axial bonds, the coefficients of Φ_1 and Φ_5 would need to be appropriately adjusted. In making such adjustments, normalization must be maintained, so that in each equation $\Sigma c_{ij}^2 = 1$.

---

*T. V. Long II, A. W. Herlinger, E. F. Epstein, and I. Bernal, *Inorg. Chem.* **1970**, *9*, 459.

As the *tbp* case illustrates, the functions generated by projection operators for molecules with nonequivalent positions are limited by the inability of the point group operations to generate all positions. In such cases, partitioning the problem into SALCs for equivalent sets may offer a reasonable, if limited, first model. However, it must be recognized that the functions so obtained are segregated into SALCs involving only equivalent basis functions, whether or not that is appropriate to the case at hand. When necessary, a better model may be achieved by modifying the projection operator results on the basis of a pictorial approach. Regardless of the method used, the set of equations obtained cannot be satisfactory unless all SALCs have been normalized and are orthogonal to each other.

Problems

5.1 Consider the case of σ-SALCs for the six pendant atoms in an octahedral MX_6 molecule, as described in Section 5.1. Given $\Sigma_2(E)$ [Eq. (5.4)] from the projection operator $P(E)\sigma_1$, obtain the functions for $P(E)\sigma_2$, $P(E)\sigma_4$, and $P(E)\sigma_5$. How are these functions related to $P(E)\sigma_1$? Can the companion function to $\Sigma_2(E)$ be obtained from any of these projections? If so, show how.

5.2 Consider the case of σ-SALCs for the six pendant atoms in an octahedral MX_6 molecule, as described in Section 5.1. Determine the effect of each of the three C_4 rotations on $\Sigma_2(E)$ [Eq. (5.4)]. Can the companion function to $\Sigma_2(E)$ be obtained from any of these results? If so, show how.

5.3 Show that the three expressions for the triply degenerate T_1 σ-SALCs of an octahedral MX_6 molecule [Eqs. (5.6)–(5.8)] are orthogonal to each other and to the A and E SALCs [Eqs. (5.3)–(5.5)].

5.4 Consider the hydrogen SALCs for methane. Show that the projection $P(T_2)s_A$ in the full group T_d gives the same result as that obtained from $P(T)s_A$ in the rotational subgroup T, as shown in Eq. (5.9).

5.5 Using projection operators, derive normalized functions for the four hydrogen SALCs of the following molecules: (a) $H_2C{=}CH_2$, (b) $H_2C{=}C{=}CH_2$.

5.6 Derive equations for the normalized π-MOs of the following conjugated ring systems, assuming a regular, planar structure in each case: (a) C_3H_3, (b) C_4H_4, (c) C_5H_5.

5.7 Consider a planar MX_3 molecule, for which X is an atom capable of sigma-, out-of-plane pi-, and in-plane pi-bonding with the central M atom. Derive equations for the normalized SALCs of the X atoms for these three modes of bonding.

5.8 As in Problem 5.7, derive equations for the normalized SALCs of the X atoms for a planar MX_4 molecule.

5.9 Using the procedure described in Section 5.3, derive expressions for the individual hybrid orbitals comprising the following sets: (a) trigonal planar sp^2, (b) square planar dsp^2, (c) octahedral d^2sp^3.

5.10 Given Eqs. (5.26a)–(5.26e), obtain expressions for the five dsp^3 hybrid orbitals of a *tbp* MX_5 molecule. Under what circumstances might such a set of hybrids be proposed to account for the bonding in a *tbp* molecule?

5.11 Derive a set of equations for the five-pendant-atom σ-SALCs of a square pyramidal (sp) MX_5 molecule, assuming that the valence orbitals on the central atom may be described as $d_{x^2-y^2}sp^3$ hybrids. How would your equations need to be modified if the d_{z^2} orbital were used instead of the p_z orbital?

5.12 As an alternative to the projection operator method, S. K. Dhar [*J. Coord. Chem.*, **1993**, *29*, 17] has proposed a nonrigorous method of constructing SALCs for MX_n molecules, based on the sum of the projections of the MX axes on the reference axes of the valence orbitals of the central atom. For the problem of obtaining SALCs for the *tbp* case, compare and contrast this approach to the methods presented in Section 5.4. What are the underlying assumptions in Dhar's method in this case? What are the advantages and disadvantages of this approach in general?

CHAPTER 6

Vibrational Spectroscopy

Theoretical chemists were aware of the power of group theory to handle problems in quantum mechanics as early as the late 1920s through the work of scientists such as Herman Weyl,* but most experimentalists had little need and even less interest in the subject. This changed after World War II with the development of commercial Raman spectrometers and most especially infrared spectrophotometers. Soon infrared spectroscopy became a standard laboratory procedure, and with this grew a need to understand the theoretical underpinnings of the technique. Two texts from the post-war period dealing with vibrational spectroscopy, which remain standard references today, are probably more responsible than any others for making chemists in general aware of the applicability of symmetry and group theory to practical problems of the experimentalist. The first of these, Gerhard Herzberg's 1945 work *Infrared and Raman Spectra of Polyatomic Molecules,*[†] contains spectroscopic data and assignments for nearly all simple compounds studied prior to its publication. Moreover, it introduced chemists to symmetry arguments, group theory techniques, and force constant calculations, while establishing the notation and nomenclature for the spectra of polyatomic molecules that has become standard today. Following Herzberg's book by a decade, Wilson, Decius, and Cross' *Molecular Vibrations*[‡] introduced many chemists to matrix techniques for carrying out force constant calculations and furthered understanding of the applicability of symmetry and group theory methods, including analyzing spectra of related compounds through group–subgroup relationships. As a result, many chemists who received their training after publication of these works first encountered symmetry and group theory in connection with their application to vibrational spectroscopy.

As with other applications of symmetry and group theory, these techniques reach their greatest utility when applied to the analysis of relatively small molecules in either the gas or liquid phases. In such cases, the observed spectroscopic frequencies can be assigned to specific vibrational motions involving all the atoms in the molecule. As the size of the molecule increases,

*Weyl's classic 1928 work *Gruppentheorie und Quantenmechanik* is available as a reprint of the 1931 English translation, *Group Theory and Quantum Mechanics,* Dover Publications, New York, 1950.

†G. Herzberg, *Molecular Spectra and Molecular Structure. II. Infrared and Raman Spectra of Polyatomic Molecules,* Van Nostrand, Princeton, NJ, 1945. A corrected reprint edition of this book has been published by Krieger Publishers, Melbourne, FL.

‡E. B. Wilson, Jr., J. C. Decius, and P. C. Cross, *Molecular Vibrations: The Theory of Infrared and Raman Vibrational Spectra,* McGraw-Hill, New York, 1955. A paperback reprint of this book has been published by Dover Publications, New York.

and particularly if its symmetry declines, many of these vibrations have very similar frequencies and are no longer individually distinguishable. At this level, spectroscopic assignment is usually confined to identifying frequencies associated primarily with specific chemical structures (e.g., functional groups). We will, therefore, confine our discussions to smaller molecules, where the power of symmetry and group theory is greatest. These are also the kinds of molecules for which detailed force constant determinations are most tractable, although we will not consider such calculations in this text.

6.1   Vibrational Modes and Their Symmetries

The individual atoms of a molecule are constantly in motion over the entire range of real temperatures above absolute zero. These individual atomic motions result in three kinds of molecular motions: *vibration, translation,* and *rotation.* To illustrate, we will first consider the simplest of molecules, a diatomic molecule AB.

First, suppose the A and B atoms move apart from their equilibrium internuclear distance, r_e (Fig. 6.1*a*), so as to stretch the chemical bond (Fig. 6.1*b*). As they move apart, away from the equilibrium position, they will experience a restoring force, F, in opposition to the motion. If we assume that the molecule follows classical mechanics, the restoring force will be proportional to the displacement from the equilibrium distance, Δr, and vary according to Hooke's law:

$$F = -k\Delta r \qquad (6.1)$$

where k is the *force constant.* At some point the restoring force will cause the two atoms momentarily to arrest their travel away from each other, after which they will reverse their motions and begin to travel toward each other. As they approach one another they will pass through the equilibrium internuclear distance and continue to move together, until their mutual repulsions arrest them at a minimum separation (Fig. 6.1*c*) and drive them back in the op-

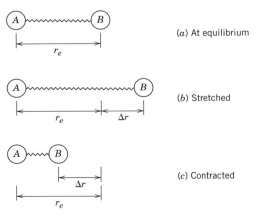

(*a*) At equilibrium

(*b*) Stretched

(*c*) Contracted

Figure 6.1   Motions of a harmonic oscillator, AB.

posite direction. This periodic series of motions constitutes one cycle of the *vibrational mode* of the molecule as a *harmonic oscillator.* If the molecule were to obey classical mechanics, the potential energy of the system throughout the vibrational cycle would vary parabolically as a function of displacement:

$$V = \frac{1}{2}k(\Delta r)^2 \tag{6.2}$$

However, unlike a classical mechanical system, the energy of a real vibrating molecule is subject to quantum mechanical restrictions. As such, the oscillating molecule can assume only certain values of vibrational energy. From the Schrödinger equation for a harmonic oscillator, the allowed energy levels are given by

$$E(\mathbf{v}) = h\nu(\mathbf{v} + \frac{1}{2}) \tag{6.3}$$

where $\mathbf{v}$ is the *vibrational quantum number,* whose values may be $\mathbf{v} = 0, 1, 2,...$; ν is the vibrational frequency in Hertz; and h is Planck's constant. By long-standing practice, vibrational spectroscopists usually quote frequencies in units of cm^{-1}, called *wavenumbers,* defined as $\tilde{\nu} = 1/\lambda$, where λ is the wavelength in centimeters.* In keeping with this, it is more convenient to define the energy of the system in wavenumber units, called *term values, T.* The term value is defined as $T = E/hc$, where c is the speed of light *in vacuo* in cm·s^{-1}. Thus, Eq. (6.3) becomes

$$T(\mathbf{v}) = \tilde{\nu}(\mathbf{v} + \frac{1}{2}) \tag{6.4}$$

Equation (6.4) suggests a model in which we have a series of equally spaced energy levels, as shown in Fig. 6.2. The minimum energy of the system, called the *vibrational ground state,* is attained when $\mathbf{v} = 0$. Note that it does not lie at the minimum of the parabola defined for the classical oscillator [Eq. (6.2)]. The classical minimum refers to the hypothetical condition of a quiescent molecule with its atoms at the equilibrium internuclear separation. The difference between this hypothetical minimum and the actual minimum energy of a vibrating molecule in its ground state ($\mathbf{v} = 0$) is called the *zero point energy.* Notice that for increasing values of $\mathbf{v}$ and corresponding higher values of vibrational energy the internuclear separation becomes greater at the extreme stretch and less at the extreme compression of the vibration. Hence, increasing vibrational energy occurs with higher vibrational amplitude. The vibrational frequency of the harmonic oscillator in any state is related to the force constant by

$$\nu = \frac{1}{2\pi}\sqrt{\left(\frac{k}{\mu}\right)} \tag{6.5}$$

*Strictly speaking, wavenumbers (cm^{-1}) are not frequency units, but rather reciprocal wavelength units. However, $\nu = c/\lambda = c\tilde{\nu}$, where c is the speed of light *in vacuo,* so wavenumbers are directly proportional to frequency. Thus, "wavenumber" is routinely used as if it were a frequency unit.

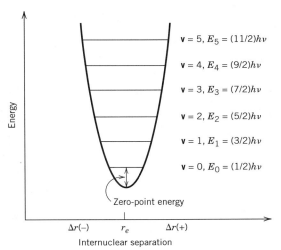

Figure 6.2   Energy levels of a harmonic oscillator.

where μ is the reduced mass given by $\mu = m_A m_B/(m_A + m_B)$, in which m_A and m_B are the masses of the individual atoms. Note that ν does not depend on the value of the quantum number **v**. In other words, the molecule's vibrational frequency is the same in all states, even though the energy E (or T) changes with **v** [Eq. (6.3)].

The basic quantum mechanical selection rule for a harmonic oscillator limits changes in vibrational energy to transitions between adjacent states; that is, $\Delta \mathbf{v} = \pm 1$. Applying Eq. (6.4) to any two states **v** and **v** + 1, we see that the energy separation between successive levels is $\Delta T = \tilde{\nu}$. In other words, for a harmonic oscillator the energy of the transition in wavenumbers is the same as the molecule's vibrational frequency, $\tilde{\nu}$.

Real molecules are not perfect harmonic oscillators. The variation of the potential energy of the system with internuclear separation usually is not a symmetric parabola, but rather tends to have the skewed appearance of a *Morse curve,* as shown in Fig. 6.3. This kind of potential energy dependence describes the behavior of an *anharmonic oscillator.* By solving the Schrödinger equation for the Morse potential, energy states for the anharmonic oscillator as term values are given by the equation

$$T(\mathbf{v}) = \tilde{\nu}\left(\mathbf{v} + \frac{1}{2}\right) - \tilde{\nu} x_e\left(\mathbf{v} + \frac{1}{2}\right)^2 + \cdots \tag{6.6}$$

where x_e is the *anharmonicity constant.* The higher terms in Eq. (6.6) are usually small and are routinely omitted. The separation between any two successive energy states, then, is given by

$$\Delta T = \tilde{\nu} - 2x_e \tilde{\nu}(\mathbf{v} + 1) \qquad (\Delta \mathbf{v} = \pm 1) \tag{6.7}$$

where **v** is taken as the value of the vibrational quantum number for the lower energy state. With rare exceptions, x_e is a positive number, so the separation between successive states becomes progressively smaller as **v** increases.

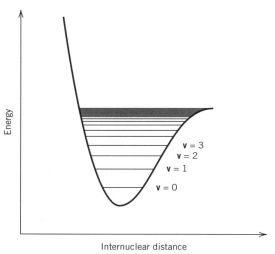

Energy

v = 3
v = 2
v = 1

v = 0

Internuclear distance

Figure 6.3 Energy levels of an anharmonic oscillator (Morse potential curve). Note the successively smaller separations between energy states as **v** increases (bottom to top).

The frequency recorded for a molecular vibration by infrared or Raman spectroscopy corresponds to the energy difference between two vibrational states. A transition between the state **v** = 0 and the state **v** = 1 defines a *vibrational fundamental*. Infrared absorption involves exciting a molecule from a lower vibrational state, usually the ground state **v** = 0, to a higher state for which **v** > 0.* For a fundamental, then, this transition is **v** = 0 → **v** = 1. With Raman scattering, $\Delta v = +1$ for the Stokes fundamentals (**v** = 0 → **v** = 1) and $\Delta v = -1$ for the anti-Stokes fundamentals (**v** = 1 → **v** = 0). However, in most Raman work, only the Stokes frequencies are sought, since they are significantly more intense. Thus, in most cases the observed spectroscopic frequency of the fundamental corresponds to a vibrational transition for which $\Delta v = +1$. This frequency, expressed in wavenumbers (cm^{-1}), is ΔT in Eq. (6.7). By this, taking **v** = 0, we can see that the spectroscopically observed frequency for a fundamental is $\Delta T = \tilde{v} - 2x_e\tilde{v}$. This frequency will more closely approximate the molecular vibrational frequency, $\tilde{v}$, when the anharmonicity, x_e, is extremely small. At the limit of a true harmonic oscillator ($x_e = 0$), the observed spectroscopic frequency is the same as the molecule's vibrational frequency. However, for many real diatomic molecules, the anharmonicity is appreciable and causes a significant difference between the spectroscopically observed frequency and the molecule's vibrational frequency. Looking ahead to polyatomic molecules, with which we will be primarily concerned, anhar-

*For most molecules at room temperature the thermal distribution of energies highly favors the ground vibrational state. For example, for H ^{35}Cl at room temperature it can be shown that the fraction of the molecular population with **v** = 1 to that with **v** = 0 is 8.9×10^{-7}. Thus, virtually all molecules are in the vibrational ground state.

monicity is usually ignored in routine work. Ignoring anharmonicity in these cases is often more a practical matter, since including quantitative allowances for it would add an extreme level of complication to the analysis of the spectra. Thus, to a certain degree of approximation, the spectroscopically observed frequencies for most polyatomic molecules are routinely taken as equal to the molecular vibrational frequencies.*

We have taken some care here to describe the atomic motions leading to vibration, because these are the kinds of motions that can be directly measured with infrared and Raman spectroscopy. However, the atoms of a diatomic molecule also can move in ways that are not vibrational. Suppose both A and B move in parallel in the same direction, resulting in a translation of the entire molecule through space. This is not a periodic motion, so it has no interaction with electromagnetic radiation; that is, it cannot be detected by infrared or Raman spectroscopy. Since any motion in space can be resolved into projections along x, y, and z of a Cartesian coordinate system, we can see that for every molecule there must exist three translations, T_x, T_y, and T_z (Fig. 6.4). Similarly, suppose A and B move in opposite directions perpendicular to the bond axis. This will cause the molecule to tumble or rotate. Unlike translation, rotation can be detected spectroscopically, because it occurs with a repeating periodic cycle. However, the frequencies of rotations lie in the microwave region and are not *directly* observed in the frequency range of most vibrational spectroscopy.† From the standpoint of vibrational analysis, rotations constitute a nonvibrational oscillation. For a diatomic molecule,

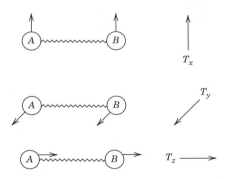

Figure 6.4 Movement of A and B atoms resulting in translations of the molecule AB.

*The effects of anharmonicity in the vibrations of polyatomic molecules become apparent in the spectroscopic frequencies of overtones and combinations, as discussed in Section 6.5.

†With samples in the gas phase, the presence of quantized rotational states can often be seen as bands of fine structure (P and R branches in the infrared spectrum; O and S branches in the Raman spectrum) on both sides of the frequency for the pure vibration (Q branch). They can also be observed directly in the Raman spectrum as very low frequency bands on both sides (Stokes and anti-Stokes) of the exciting frequency (Rayleigh line). In many cases the Q branch, for the vibration without rotation, is not spectroscopically observable, being forbidden on the basis of quantum mechanical considerations. Further detail of the theory of vibrational–rotational spectra, which is covered in most physical chemistry texts, is not essential to the purposes of this chapter. See, for example, P. Atkins, *Physical Chemistry*, 5th ed., W. H. Freeman, New York, 1994, Chapter 16.

we can define rotational motions about the x and y axes (designated R_x and R_y), but we cannot define such motion about the z axis (the bond axis), as shown in Fig. 6.5. Spinning about the z axis does not cause the molecule itself to move. Looking beyond the diatomic case, we can see that this is true for any linear molecule, regardless of the number of atoms composing it. However, for a nonlinear molecule the rotation about z does cause the molecule to tumble (cf. Fig. 6.6a,b). Thus, for all nonlinear molecules there are three rotations, R_x, R_y, and R_z.

Let us now consider the motions of a polyatomic molecule composed of n atoms. The motions of each atom can be resolved into components along

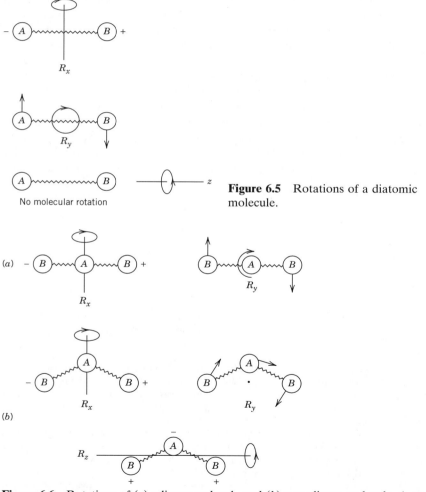

Figure 6.5 Rotations of a diatomic molecule.

Figure 6.6 Rotations of (a) a linear molecule and (b) a nonlinear molecule. A nonstandard axis choice has been made for the nonlinear molecule to facilitate comparison.

the three directions of a Cartesian coordinate system. Therefore, any molecule composed of n atoms possesses $3n$ *degrees of freedom* of motion. As with diatomic molecules, these $3n$ degrees of freedom include vibrations, translations, and rotations. The vibrational motions of the atoms can always be resolved into fundamental vibrational motions for the entire molecule, called *normal modes* of vibration. The number of these normal modes will be $3n$ minus the number of nonvibrational motions (i.e., translations and rotations). As we have seen, a linear molecule has three translations and two rotations. Subtracting these from the $3n$ degrees of freedom, we see that *a linear molecule possesses $3n - 5$ vibrational modes*. A nonlinear molecule has three translations and three rotations. Therefore, *a nonlinear molecule possesses $3n - 6$ normal modes*. Like the diatomic case, each normal mode of vibration has a characteristic frequency and can assume a series of quantized energies. The frequencies recorded by infrared and Raman spectroscopy arise from transitions between these states. A transition of the type $\mathbf{v} = 0 \rightarrow \mathbf{v} = 1$ defines a fundamental of the normal mode.

Although we have seen why a linear molecule has one less rotation than a nonlinear molecule, you may be wondering how this necessarily implies that a linear molecule has one more vibrational mode than a nonlinear one with the same number of atoms. The answer is that one of the rotational modes of a nonlinear molecule becomes a bending vibrational mode when the molecule is made linear. This is most easily seen for a triatomic molecule, as shown in Fig. 6.7.

Relative to the overall molecular symmetry, all of the $3n$ degrees of freedom—normal modes of vibration, translations, and rotations—have symmetry relationships consistent with the irreducible representations (species) of the molecule's point group. In other words, we can catalogue all degrees of freedom according to the appropriate Mulliken symbols for their corresponding irreducible representations. To do this we locate a set of three vectors along Cartesian coordinates at each atom, representing the three degrees of freedom of that atom. We then make the entire set of $3n$ vectors a basis for a representation in the molecule's point group. The reducible representa-

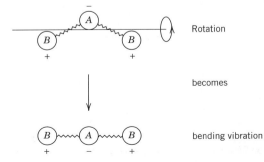

Rotation

becomes

bending vibration

Figure 6.7 One rotation of a nonlinear triatomic molecule becomes a bending vibrational mode if the molecule is made linear.

tion is then reduced into its component irreducible representations, which correspond to the symmetries of the $3n$ degrees of freedom. The symmetries of the translations and rotations can be identified from the character table from the listed transformation properties for unit vectors and rotational vectors, respectively. These are subtracted from the collection of symmetry species comprising the overall reducible representation, leaving the collection of Mulliken symbols for the $3n - 6$ normal modes ($3n - 5$ normal modes for linear molecules).

To illustrate, consider the bent molecule SO_2 (C_{2v}). Figure 6.8 shows the vector basis for the reducible representation of the $3n$ degrees of freedom, Γ_{3n}. We will need 9×9 transformation matrices to show the effects of the operations of C_{2v} on the nine individual vectors. By extrapolation, we can see that for any molecule we would need $3n \times 3n$ matrices. Clearly, carrying out this kind of mathematical work even for small molecules would be cumbersome, so we will look for shortcuts to the full-matrix approach as we proceed with this example.

For SO_2, the identity operation, E, leaves all vectors in place. In full-matrix notation this is

$$
\begin{bmatrix}
1 & 0 & 0 & 0 & 0 & 0 & 0 & 0 & 0 \\
0 & 1 & 0 & 0 & 0 & 0 & 0 & 0 & 0 \\
0 & 0 & 1 & 0 & 0 & 0 & 0 & 0 & 0 \\
0 & 0 & 0 & 1 & 0 & 0 & 0 & 0 & 0 \\
0 & 0 & 0 & 0 & 1 & 0 & 0 & 0 & 0 \\
0 & 0 & 0 & 0 & 0 & 1 & 0 & 0 & 0 \\
0 & 0 & 0 & 0 & 0 & 0 & 1 & 0 & 0 \\
0 & 0 & 0 & 0 & 0 & 0 & 0 & 1 & 0 \\
0 & 0 & 0 & 0 & 0 & 0 & 0 & 0 & 1
\end{bmatrix}
\begin{bmatrix}
x_1 \\ y_1 \\ z_1 \\ x_2 \\ y_2 \\ z_2 \\ x_3 \\ y_3 \\ z_3
\end{bmatrix}
=
\begin{bmatrix}
x_1 \\ y_1 \\ z_1 \\ x_2 \\ y_2 \\ z_2 \\ x_3 \\ y_3 \\ z_3
\end{bmatrix}
\tag{6.8}
$$

From the trace of the transformation matrix we obtain a character, χ_E, of 9, which is the value of $3n$ for SO_2. We can see from this that for any molecule of n atoms it must be that $\chi_E = 3n$.

The C_2 rotation exchanges the two oxygen atoms and reverses the sense of all x and y vectors. In full-matrix notation this is

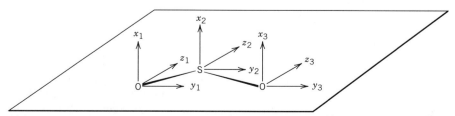

Figure 6.8 Orientations of vectors for SO_2 to form the basis for a representation of the $3n$ degrees of freedom. The molecule lies in the yz plane.

$$
\begin{bmatrix}
0 & 0 & 0 & 0 & 0 & 0 & -1 & 0 & 0 \\
0 & 0 & 0 & 0 & 0 & 0 & 0 & -1 & 0 \\
0 & 0 & 0 & 0 & 0 & 0 & 0 & 0 & 1 \\
0 & 0 & 0 & -1 & 0 & 0 & 0 & 0 & 0 \\
0 & 0 & 0 & 0 & -1 & 0 & 0 & 0 & 0 \\
0 & 0 & 0 & 0 & 0 & 1 & 0 & 0 & 0 \\
-1 & 0 & 0 & 0 & 0 & 0 & 0 & 0 & 0 \\
0 & -1 & 0 & 0 & 0 & 0 & 0 & 0 & 0 \\
0 & 0 & 1 & 0 & 0 & 0 & 0 & 0 & 0
\end{bmatrix}
\begin{bmatrix}
x_1 \\ y_1 \\ z_1 \\ x_2 \\ y_2 \\ z_2 \\ x_3 \\ y_3 \\ z_3
\end{bmatrix}
=
\begin{bmatrix}
-x_3 \\ -y_3 \\ z_3 \\ -x_2 \\ -y_2 \\ z_2 \\ -x_1 \\ -y_1 \\ z_1
\end{bmatrix}
\qquad (6.9)
$$

This gives a character of -1 for the C_2 operation. Note that only the central 3×3 block matrix, associated with the vector transformations of the nonshifted S atom, contributes to the overall character of the full 9×9 matrix.

The $\sigma(xz)$ operation exchanges the two oxygen atoms and reverses the sense of all y vectors. In full-matrix notation this is

$$
\begin{bmatrix}
0 & 0 & 0 & 0 & 0 & 0 & 1 & 0 & 0 \\
0 & 0 & 0 & 0 & 0 & 0 & 0 & -1 & 0 \\
0 & 0 & 0 & 0 & 0 & 0 & 0 & 0 & 1 \\
0 & 0 & 0 & 1 & 0 & 0 & 0 & 0 & 0 \\
0 & 0 & 0 & 0 & -1 & 0 & 0 & 0 & 0 \\
0 & 0 & 0 & 0 & 0 & 1 & 0 & 0 & 0 \\
1 & 0 & 0 & 0 & 0 & 0 & 0 & 0 & 0 \\
0 & -1 & 0 & 0 & 0 & 0 & 0 & 0 & 0 \\
0 & 0 & 1 & 0 & 0 & 0 & 0 & 0 & 0
\end{bmatrix}
\begin{bmatrix}
x_1 \\ y_1 \\ z_1 \\ x_2 \\ y_2 \\ z_2 \\ x_3 \\ y_3 \\ z_3
\end{bmatrix}
=
\begin{bmatrix}
x_3 \\ -y_3 \\ z_3 \\ x_2 \\ -y_2 \\ z_2 \\ x_1 \\ -y_1 \\ z_1
\end{bmatrix}
\qquad (6.10)
$$

This gives a character of 1 for the $\sigma(xz)$ operation. Again, only the central 3×3 block matrix for the nonshifted S atom contributes to the character.

Finally, the $\sigma(yz)$ operation leaves all atoms in place but reverses the sense of all x vectors. In full-matrix notation this is

$$
\begin{bmatrix}
-1 & 0 & 0 & 0 & 0 & 0 & 0 & 0 & 0 \\
0 & 1 & 0 & 0 & 0 & 0 & 0 & 0 & 0 \\
0 & 0 & 1 & 0 & 0 & 0 & 0 & 0 & 0 \\
0 & 0 & 0 & -1 & 0 & 0 & 0 & 0 & 0 \\
0 & 0 & 0 & 0 & 1 & 0 & 0 & 0 & 0 \\
0 & 0 & 0 & 0 & 0 & 1 & 0 & 0 & 0 \\
0 & 0 & 0 & 0 & 0 & 0 & -1 & 0 & 0 \\
0 & 0 & 0 & 0 & 0 & 0 & 0 & 1 & 0 \\
0 & 0 & 0 & 0 & 0 & 0 & 0 & 0 & 1
\end{bmatrix}
\begin{bmatrix}
x_1 \\ y_1 \\ z_1 \\ x_2 \\ y_2 \\ z_2 \\ x_3 \\ y_3 \\ z_3
\end{bmatrix}
=
\begin{bmatrix}
-x_1 \\ y_1 \\ z_1 \\ -x_2 \\ y_2 \\ z_2 \\ -x_3 \\ y_3 \\ z_3
\end{bmatrix}
\qquad (6.11)
$$

This gives a character of 3 for the $\sigma(yz)$ operation. Here, all atoms remain nonshifted, and all their associated 3×3 matrices contribute to the overall character.

We see from Eqs. (6.8) through (6.11) that each 9×9 operator matrix contains three 3×3 block operation matrices, one for each atom. For any operation R_i, all the individual-atom operator matrices (3×3 matrices) have identical form and therefore have the same character, χ_i. Only individual-atom block matrices that lie along the diagonal of the full operator matrix contribute to the overall character for the operation, χ_R. Moreover, we see that only those 3×3 block matrices for atoms that are *not shifted* by the operation contribute to the character of the overall matrix. *Thus, to find χ_R, the character for the overall operation, count the number of atoms that remain nonshifted by the operation, N_i, and multiply by the contribution per nonshifted atom, χ_i; that is,*

$$\chi_R = N_i \chi_i \qquad (6.12)$$

where χ_i is equivalent to the character of the 3×3 block matrices of which the full matrix for the operation is composed. For the operations of C_{2v} the contributions per nonshifted atom for each operation, obtained from the characters of the individual-atom block matrices for each operation, are

$$\chi_E = 3, \qquad \chi_{C_2} = -1, \qquad \chi_{\sigma(xz)} = 1, \qquad \chi_{\sigma(yz)} = 1$$

Note that the contribution per nonshifted atom for both reflections is the same, even though the individual 3×3 matrices are different. This is an example of a general result: *The contribution per nonshifted atom for a particular operation is the same regardless of the orientation of its associated symmetry element.* Moreover, the value of the contribution per nonshifted atom for a particular operation is the same in any point group in which that kind of operation is found.

The values of χ_i are important enough for this application that they have been tabulated in some texts.* Actually, such tables are not needed, since the values can be obtained easily from any character table, such as the one being used for the molecule under consideration. From an examination of the individual-atom transformation matrices in Eq. (6.8) through (6.11) we can see that the individual elements c_{11}, c_{22}, c_{33} (along the trace of the 3×3 matrix) indicate the effects of the operation on the coordinates x, y, z, respectively, of the atom on which they operate. Therefore, just like the general vector $\mathbf{v}$ we considered in Section 2.3, the values for these elements will be identical to the characters for the operation in the irreducible representations by which the unit vectors x, y, and z transform. To find a needed value of χ_i, then, simply add up the characters in the character table under the operation for the irreducible representations by which all three unit vectors transform. If two or three unit vectors transform *nondegenerately* by the same species (e.g., x and y in C_{2h}, and x, y, z in C_i), double or triple the character for that species, as the case may be, as the contributions of the associated vectors for each operation. If two or three unit vectors transform *degenerately* by the same

---

*For example, see E. B. Wilson, Jr., J. C. Decius, and P. C. Cross, *Molecular Vibrations: The Theory of Infrared and Raman Vibrational Spectra*, McGraw-Hill, New York, 1955, Table 6-1, p. 105.

species (e.g., (x, y) in D_{4h}, and (x, y, z) in T_d), use the listed character for that species as the contributions of the associated vectors for each operation. To illustrate, for C_{2v} z transforms as A_1, x transforms as B_1, and y transforms as B_2. Using the characters from these three species we obtain the following values of χ_i for the four operations of the group:

C_{2v}	E	C_2	$\sigma(xz)$	$\sigma(yz)$	
A_1	1	1	1	1	z
B_1	1	−1	1	−1	x
B_2	1	−1	−1	1	y
χ_i	3	−1	1	1	

Using Eq. (6.12), we obtain the following characters for the reducible representation Γ_{3n}:

C_{2v}	E	C_2	$\sigma(xz)$	$\sigma(yz)$
N_i	3	1	1	3
χ_i	3	−1	1	1
Γ_{3n}	9	−1	1	3

From the character for E we see that the dimension of the representation is consistent with $3n = 9$. Note that the individual values of χ_R of which our Γ_{3n} is composed are the same as those from the traces of the full 9×9 operation matrices in Eqs. (6.11)–(6.14).

We will use the tabular method to reduce Γ_{3n} (cf. Section 3.1). The following work sheet incorporates the information needed to generate the reducible representation, and it also notes the transformation properties of the unit vectors and rotational vectors from the character table (cf. Appendix A).

C_{2v}	E	C_2	$\sigma(xz)$	$\sigma(yz)$			
N_i	3	1	1	3			
χ_i	3	−1	1	1			
Γ_{3n}	9	−1	1	3	Σ	n_i	
A_1	9	−1	1	3	12	3	z
A_2	9	−1	−1	−3	4	1	R_z
B_1	9	1	1	−3	8	2	x, R_y
B_2	9	1	−1	3	12	3	y, R_x

This shows that $\Gamma_{3n} = 3A_1 + A_2 + 2B_1 + 3B_2$. Consistent with Eq. (2.10), we have found nine nondegenerate species for the nine degrees of freedom possible for SO_2.

Our interest is in sulfur dioxide's $3n - 6 = 3$ normal modes of vibration. We can find the symmetry species of the genuine normal modes of vibration by identifying and removing from Γ_{3n} the species for the three translations and three rotations the molecule possesses; that is,

$$\Gamma_{3n-6} = \Gamma_{3n} - \Gamma_{\text{trans}} - \Gamma_{\text{rot}} \qquad (6.13)$$

The species comprising Γ_{trans} and Γ_{rot} are the same as those of the three unit-vector and three rotational-vector transformations, respectively. From the C_{2v} character table, as noted in the worksheet above, we deduce that the translations are $\Gamma_{\text{trans}} = A_1 + B_1 + B_2$, corresponding to T_z, T_x, and T_y, respectively. Likewise, the rotations are $\Gamma_{\text{rot}} = A_2 + B_1 + B_2$, corresponding to R_z, R_y, and R_x, respectively. Applying Eq. (6.13) we have $\Gamma_{3n-6} = 2A_1 + B_2$.

This result tells us the symmetry of the motions that constitute the three normal modes of SO_2. Two of the modes preserve the symmetry relationships implied by the A_1 representation, which is the totally symmetric representation of the group. This means that the motions of the atoms in these two normal modes maintain the complete set of symmetry relationships of the group C_{2v} through all phases of the vibrational cycle. The third mode belongs to the species B_2. From the characters of B_2 (or from the symmetry and antisymmetry implied by the Mulliken symbol) we know that this mode is antisymmetric with respect to both C_2 rotation and $\sigma(xz)$ reflection.

With only three atoms in the molecule, it is not difficult to deduce the motions of these three normal modes. For the A_1 modes, both oxygen atoms must be moving in phase in the yz plane.* If they were to move out-of-phase, C_2 and $\sigma(xz)$ would not be maintained. If they were to move out of the yz plane, they would cause molecular rotation (R_y if in the same direction, R_z if in opposite directions). With these restrictions, we conclude that there are only two types of totally symmetric motions. The two A_1 modes must be (a) an in-phase stretching and contracting motion primarily along the bond axes (*symmetric stretching* mode) and (b) a back-and-forth bending motion of the O–S–O bond angle (*bending* mode). Likewise, the B_2 mode must be confined to motions in the yz plane, but the oxygen atoms must be moving in opposite phases so as to be antisymmetric with respect to C_2 and $\sigma(xz)$. The motion implied by these restrictions involves contraction of one S–O bond while the other S–O bond is stretching, and vice versa. This is an *antisymmetric stretching* mode.† The forms of all three normal modes are depicted in Fig. 6.9.

In addition to Mulliken symbol designations, normal modes are usually identified by frequency numbers as ν_1, ν_2, ν_3,...(cf. Fig. 6.9). For SO_2 the observed frequencies are $\nu_1 = 1151$ cm^{-1}, $\nu_2 = 519$ cm^{-1}, and $\nu_3 = 1361$ cm^{-1}. The numbering is often assigned systematically in descending order of symmetry species (as listed down the left column of the appropriate character

---

*The sulfur atom must also move in opposition to the oxygen atoms to maintain the position of the center of gravity of the molecule. Otherwise, translation would result. Nonetheless, for the present purposes, it is easier to envision the normal modes primarily in terms of the oxygen motions.

†Often such modes are called *asymmetric,* which literally means lacking symmetry. This, of course, is not the case. Current practice favors the term antisymmetric, to indicate that the symmetry of the motion is in some respects contrary to the complete symmetry of the molecule.

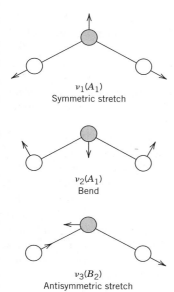

$\nu_1(A_1)$
Symmetric stretch

$\nu_2(A_1)$
Bend

$\nu_3(B_2)$
Antisymmetric stretch

Figure 6.9 Normal modes of SO_2. Arrow lengths have been exaggerated for clarity.

table), and among modes of the same symmetry species in descending order of vibrational frequency.* In the case of SO_2, for example, the two totally symmetric A_1 modes are assigned as ν_1 and ν_2. Stretching modes tend to have higher frequencies than bending modes, as observed in the case of SO_2, so the symmetric stretch is assigned as ν_1 and the bending mode is assigned as ν_2. The B_2 antisymmetric stretch, being lower in symmetry, is assigned as ν_3, despite its having a higher frequency than either ν_1 or ν_2. The relative ordering of observed frequencies tends to be the same for various molecules with the same structure, so the assignment of modes to frequency numbers often carries over from one compound to another. For example, H_2O is a bent triatomic molecule, like SO_2, so its three normal modes are numerically assigned in the same way [$\nu_1(A_1) = 3657$ cm^{-1}, $\nu_2(A_1) = 1595$ cm^{-1}, $\nu_3(B_2) = 3756$ cm^{-1}].†

*For historical reasons, many of which are no longer evident, the numbering for certain well-known structure types does not follow this convention. For example, the frequencies of octahedral XY_6 are traditionally numbered ν_1 (A_{1g}), ν_2 (E_g), ν_3 (T_{1u}), ν_4 (T_{1u}), ν_5 (T_{2g}), ν_6 (T_{2u}), in keeping with the older form of the O_h character table (e.g., see G. Herzberg, *Molecular Spectra and Molecular Structure. II. Infrared and Raman Spectra of Polyatomic Molecules,* Van Nostrand, Princeton, NJ, 1945, p. 123). Today's character tables list *gerade* species first, then *ungerade* species. If the numbering had been based on this ordering, the T_{2g} frequency would be ν_3. The reasoning behind the frequency numbering of some other structures (e.g., XY_4 square planar, ZXY_2 planar, ethane-type X_2Y_6, and nonlinear X_2Y_2) is not as apparent. Except for these cases, we will use systematic numbering by symmetry species.

†The frequencies quoted for both SO_2 and H_2O are for samples in the gas phase. The actual spectroscopic bands show rotational fine structure (P, Q, R branches), subject to quantum mechanical restrictions. In the case of the A_1 modes infrared bands for the pure vibrations (Q branches) are forbidden.

For simple molecules such as SO_2 and H_2O there is a one-to-one correspondence between the number of fundamental frequencies and the number of normal modes, because the low molecular symmetry does not allow degeneracy. If the molecule belongs to a point group that has doubly or triply degenerate irreducible representations, some vibrational modes may be degenerate and therefore have identical frequencies. In these cases the number of frequencies (ν_1, ν_2, ν_3, ...) will be less than the $3n - 6$ (or $3n - 5$, if linear) normal modes.

As an example of a molecule with degenerate normal modes, let us consider the symmetry of the normal modes of a tetrahedral XY_4 molecule (e.g., CH_4, CCl_4, SO_4^{2-}). Using the approach we detailed for SO_2, we will count the number of atoms that are nonshifted by each operation of the group (T_d) and multiply by the contribution per nonshifted atom for the operation. In the case of T_d the unit vectors x, y, and z transform degenerately as T_2, so the contributions per nonshifted atom for the operations are identical to their characters for the T_2 representation (cf. character table for T_d in Appendix A). There is a single character for all operations of a class, and likewise a single value for the contribution per nonshifted atom for all members of the class. Since there are five classes in T_d, we only need to consider five representative operations, one from each class, to generate the reducible representation.

Proceeding in this manner, we note that E leaves all five atoms nonshifted, C_3 (and likewise C_3^2) leaves the X and Y atoms on the chosen axis nonshifted, C_2 and S_4 (and likewise S_4^3) leave only the central X atom nonshifted, and any σ_d leaves the X and two Y atoms in the plane nonshifted. From this we obtain the following work sheet.

T_d	E	$8C_3$	$3C_2$	$6S_4$	$6\sigma_d$			
N_i	5	2	1	1	3			
χ_i	3	0	-1	-1	1			
Γ_{3n}	15	0	-1	-1	3	Σ	n_i	
A_1	15	0	-3	-6	18	24	1	
A_2	15	0	-3	6	-18	0	0	
E	30	0	-6	0	0	24	1	
T_1	45	0	3	-6	-18	24	1	(R_x, R_y, R_z)
T_2	45	0	3	6	18	72	3	(x, y, z)

This shows that $\Gamma_{3n} = A_1 + E + T_1 + 3T_2$. From the listed vector transformation properties, $\Gamma_{trans} = T_2$, and $\Gamma_{rot} = T_1$. Subtracting the translations and rotations we obtain $\Gamma_{3n-6} = A_1 + E + 2T_2$. Thus, we predict that there should be four frequencies: $\nu_1(A_1)$, $\nu_2(E)$, $\nu_3(T_2)$, $\nu_4(T_2)$.

How can there be only four frequencies if by $3n - 6$ we expect nine normal modes? One pair of modes, $\nu_2(E)$, is doubly degenerate and gives rise to a single frequency. Three other modes are triply degenerate and give rise to their own single frequency, $\nu_3(T_2)$, while another three modes are triply degenerate among themselves and give rise to a different single frequency, $\nu_4(T_2)$.

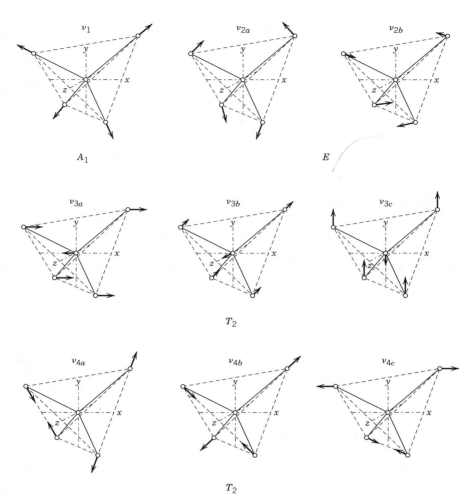

Figure 6.10 Normal modes of vibration of a tetrahedral XY_4 molecule. [Adapted with permission from G. Herzberg, *Molecular Spectra and Molecular Structure: II. Infrared and Raman Spectra of Polyatomic Molecules,* reprint 1991 with corrections, Krieger Publishers, Malabar, FL.]

The individual modes giving rise to these four fundamental frequencies of XY_4 are shown in Fig. 6.10.* The frequency ν_1 is a symmetric stretching motion, also described as a "breathing" mode. The doubly degenerate ν_2 frequency arises from two bending deformations, sometimes described as "skating on the sphere" of the four Y atoms. The frequency ν_3 arises from three degenerate antisymmetric stretching motions. The frequency ν_4 arises from three bending motions in which opposite pairs of Y atoms move in opposite phases. The composite motion of these three modes is a kind of "umbrella" deformation (cf. Appendix C).

*Appendix C also shows the forms of the normal vibrational modes for tetrahedral XY_4, as well as those for some other common structures. In these figures, doubly and triply degenerate modes are shown with a single drawing representing the linear combination of the individual degenerate modes.

6.2   Symmetry-Based Selection Rules and Their General Consequences

We have seen in Section 6.1 that the various normal modes of a molecule can be catalogued and analyzed in terms of their symmetry with respect to the overall molecular symmetry. The question now arises whether or not these vibrations can be observed in the infrared and Raman spectra. What we seek is a determination of the *spectroscopic activity* of the normal modes, also called a determination of the *spectroscopic selection rules*. These selection rules indicate which normal modes are active (allowed) or inactive (forbidden) in each kind of spectrum.

The spectroscopic activity of any normal mode depends upon quantum mechanical restrictions that can be analyzed in terms of the symmetry of the wave functions involved. For the moment we will confine our discussion to fundamentals, which for any normal mode involve a transition for which $\Delta \mathbf{v} = \pm 1$. The basic quantum mechanical selection rule for any such transition is that *the vibrational transition will have nonzero intensity in either the infrared or Raman spectrum if the appropriate transition moment is nonzero*. Normal modes that do not meet this criterion for one or the other kind of spectrum cannot be observed directly by that technique.

For infrared absorption to occur, the normal mode must have an oscillating molecular dipole moment with the same frequency as the oscillating electric field of the radiation. In terms of quantum mechanics, the transition moment for the fundamental of a normal mode ($\mathbf{v} = 0 \rightarrow \mathbf{v} = 1$) can be written

$$M(0, 1) = \int \psi_0 \mu \psi_1 \, d\tau \qquad (6.14)$$

where ψ_0 and ψ_1 are wave functions for the ground and excited vibrational states, and μ is the oscillating electric dipole moment vector as a function of the normal coordinate for the normal mode. The *normal coordinate*, usually given the symbol Q, is a single reference coordinate by which the progress of a normal mode can be followed.* Absorption will occur if $M \neq 0$. The vector μ is a resultant of its components μ_x, μ_y, and μ_z, so we may rewrite Eq. (6.14) as

$$M_x = \int \psi_0 \mu_x \psi_1 \, d\tau$$

$$M_y = \int \psi_0 \mu_y \psi_1 \, d\tau \qquad (6.15)$$

$$M_z = \int \psi_0 \mu_z \psi_1 \, d\tau$$

If any one of these components is nonzero, the entire transition moment will be nonzero.

---

*For a heteronuclear diatomic molecule, AB, the normal coordinate is expressed in terms of the displacements of the two atoms as $Q = \Delta r_A + \Delta r_B$, where $\Delta r_A / \Delta r_B = m_B / m_A$, the ratio of the masses of the atoms. The normal coordinate expressions are more complex for polyatomic molecules. See E. B. Wilson, Jr., J. C. Decius, and P. C. Cross, *Molecular Vibrations: The Theory of Infrared and Raman Vibrational Spectra*, McGraw-Hill, New York, 1955, for details.

We can analyze whether or not an integral of the form of Eqs. (6.14) and (6.15) will be nonvanishing on the basis of the symmetries of the wave functions and the components of μ. To do this we use a general result of quantum mechanics, presented here without proof, that *an integral of the product of two functions,* $\int f_A f_B \, d\tau$, *can be nonzero only if it is invariant under all operations of the molecule's point group. This can occur only if the direct product of* $f_A f_B$ *is or contains the totally symmetric representation of the point group.* Applying this to the present case, if the integral for the transition moment transforms as the totally symmetric representation, the vibrational transition will be infrared allowed. Now, the ground-state vibrational wave function ψ_0 is totally symmetric for all molecules (except free radicals) and the excited state wave function ψ_1 has the symmetry of the normal mode. As we have seen in Section 3.5, the direct product of the totally symmetric representation with any non-totally symmetric representation is the non-totally symmetric representation (e.g., in C_{2v}, $A_1 \times B_1 = B_1$). In the present case this means that by itself the product $\psi_0 \psi_1$ has the symmetry of ψ_1. However, the product of an irreducible representation with itself is or contains the totally symmetric representation (cf. Section 3.5). Thus, if any component of μ has the same symmetry as ψ_1, the product $\mu \psi_1$ will be totally symmetric and the integral will be nonvanishing. Since a dipole component can be represented by a vector, we can see that the symmetry species of μ_x, μ_y, and μ_z are the same as the unit vector transformations x, y, and z, as listed in the character table. Therefore, *a normal mode belonging to the same symmetry species as any of the unit vector transformations x, y, or z will be active in the infrared spectrum.* In other words, once we know the symmetries of the normal modes, all we need to do is look at the unit vector symmetries in the character table to see which modes are infrared active.

In the Raman experiment (Fig. 6.11), incident laser radiation with an oscillating electric field E impinges on a sample and induces an electric moment P in the sample. The induced moment arises because the molecule's electrons are attracted toward the positive pole of the field and the nuclei are attracted

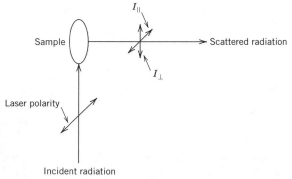

Figure 6.11 Typical experimental arrangement for Raman spectroscopy.

to the negative pole. The induced moment is related to the field strength of the incident radiation by

$$P = \alpha E \tag{6.16}$$

where the proportionality constant α is the *polarizability,* whose magnitude changes as the molecule oscillates. For a normal mode to be Raman active there must be a nonzero change in the polarizability with the normal coordinate at the equilibrium configuration; that is, $(\partial \alpha/\partial Q)_0 \neq 0$. The polarizability is best expressed as a tensor, and in this form Eq. (6.16) can be recast as

$$\begin{bmatrix} P_x \\ P_y \\ P_z \end{bmatrix} = \begin{bmatrix} \alpha_{xx} & \alpha_{xy} & \alpha_{xz} \\ \alpha_{yx} & \alpha_{yy} & \alpha_{yz} \\ \alpha_{zx} & \alpha_{zy} & \alpha_{zz} \end{bmatrix} \begin{bmatrix} E_x \\ E_y \\ E_z \end{bmatrix} \tag{6.17}$$

where $\alpha_{ij} = \alpha_{ji}$. If the change in any one of these components is nonzero $[(\partial \alpha_{ij}/\partial Q)_0 \neq 0]$, then the mode will be Raman active.

In terms of the polarizability tensor, the transition moment for Raman activity of a fundamental may be expressed as

$$P(0, 1) = \int \psi_0 \alpha E \psi_1 \, d\tau = E \int \psi_0 \alpha \psi_1 \, d\tau \tag{6.18}$$

An integral of this form can be written for every component α_{ij}, giving components P_{ij}. If there is any component for which $P_{ij} \neq 0$, then the entire moment will be nonvanishing $[P(0, 1) \neq 0]$, and the transition will be Raman active. As with the transition moment expression for infrared activity [Eq. (6.14) and (6.15)], the integral will be nonzero if it is totally symmetric. As before, ψ_0 is totally symmetric and ψ_1 has the symmetry of the normal mode. Thus, the integral will be nonzero if ψ_1 and any component α_{ij} have the same symmetry. The symmetries of the α_{ij} components are the same as the binary direct products of vectors, whose transformation properties are listed in the last column of the character table. From this it follows that *a normal mode will be Raman active if it belongs to the same symmetry species as one of the binary direct products of vectors listed in the character table for the molecule.*

In Section 6.1 we saw that SO_2 has three normal modes: $\nu_1(A_1)$, $\nu_2(A_1)$, $\nu_3(B_2)$. Looking at the character table for C_{2v}, we find the following unit vector and direct product transformation properties*:

C_{2v}		
A_1	z	x^2, y^2, z^2
A_2		xy
B_1	x	xz
B_2	y	yz

*The transformations of the rotational vectors R_x, R_y, and R_z have been omitted, because they have no relevance to determining infrared and Raman activity.

Both A_1 and B_2, the species of the three normal modes, have listings for unit vectors and direct products. Therefore, all three modes are active in both the infrared and Raman spectra. In other words, we should be able to observe the same three frequencies in both the infrared and Raman spectra.

Likewise, we found in Section 6.1 that a tetrahedral XY_4 molecule has four frequencies for its nine normal modes: $\nu_1(A_1)$, $\nu_2(E)$, $\nu_3(T_2)$, $\nu_4(T_2)$. Looking at the character table for T_d we find the following unit vector and direct product transformation properties:

T_d		
A_1		$x^2 + y^2 + z^2$
A_2		
E		$(2z^2 - x^2 - y^2, x^2 - y^2)$
T_1		
T_2	(x, y, z)	(xy, xz, yz)

All three unit vectors transform degenerately as T_2, so only normal modes with T_2 symmetry can be infrared active. For tetrahedral XY_4, this means that only ν_3 and ν_4 can be observed by infrared spectroscopy. In contrast, A_1, E, and T_2 all have direct product listings, which means that modes with these symmetries will be Raman active. Thus, we should be able to observe all four frequencies by Raman spectroscopy. In summary, we have the following activities:

A_1	E	$2T_2$
ν_1	ν_2	ν_3, ν_4
Raman	Raman	Infrared and Raman

Note that only the ν_3 and ν_4 modes are active in both the infrared and Raman spectra. As such, each is said to be *coincident* in both spectra. This means that we should expect to find a band for ν_3 at the same frequency in both the infrared and Raman spectra. The same is true for ν_4.*

Given the symmetry-based selection rules, examination of the character tables allows us to draw the following general conclusions:

1. *Infrared-active modes can be distributed among no more than three symmetry species,* since activity is associated with the transformation properties of the three unit vectors. *Raman-active modes can be distributed among as many as four symmetry species,* depending on the point group, because the direct products typically span two to four species.

---

*Although coincident bands have the same frequency in both infrared and Raman spectra, they generally do not have comparable intensities. This is a result of the very different phenomena giving rise to the two kinds of spectra: photon absorption for infrared spectroscopy and inelastic photon scattering for Raman spectroscopy. Thus, for example, a band for a given coincident mode might be very strong in the infrared spectrum but very weak in the Raman spectrum, or vice versa.

2. In centrosymmetric point groups (those that have inversion, *i*), unit vectors transform as *ungerade* species, and direct products transform as *gerade* species. Therefore, *infrared-active modes will be Raman inactive, and vice versa, for centrosymmetric molecules.* In other words, the infrared and Raman spectra of centrosymmetric molecules should have no fundamental frequencies in common. This requirement is known as the *rule of mutual exclusion.*

3. Unit vector and direct product transformations do not span all species in some groups. Thus, some irreducible representations are not associated with either a unit vector or direct product transformation. Nonetheless, the molecule may have a normal mode that transforms by one of these species. Thus, *it is possible to have some normal modes that cannot be observed as fundamentals in either the infrared or Raman spectra.* These spectroscopically inactive modes are often called *silent modes.*

4. The totally symmetric representation in every point group is associated with one or more direct product transformations. Therefore, *normal modes that are totally symmetric will always be Raman active.* Totally symmetric modes may or may not be infrared active, depending on the point group.

Normal modes that are totally symmetric can be identified experimentally in the Raman spectrum by measuring the *depolarization ratio, ρ.* As shown in Fig. 6.11, the scattered Raman radiation has a polarization that can be resolved into two intensity components, $I_\perp$ and $I_\parallel$, whose polarities are respectively perpendicular and parallel to that of the exciting laser radiation. The intensities of the separate components for each spectroscopic band can be observed by using a Polaroid analyzer (e.g., a plane polarizing camera lens filter). The spectrum is recorded with the analyzer in one orientation, say $I_\parallel$, and then recorded again in the other orientation, $I_\perp$. Using the recorded band intensities, the depolarization ratio can be calculated as

$$\rho = \frac{I_\perp}{I_\parallel} \tag{6.19}$$

The value of ρ depends upon the symmetry of the polarizability, α, as it oscillates with the particular normal mode. In liquid and gas samples, with which we are concerned here, the polarizability tensor for an individual molecule is randomly oriented. Therefore, throughout the sample all possible orientations are represented. In terms of Eq. (6.16), when the oscillating electric vector of the incident radiation, E, interacts with the sample, the resulting induced moment, P, is the result for the average of all orientations of α. For plane-polarized incident radiation, as produced by lasers, it can be shown from scattering theory that the measured depolarization ratio from such samples should have (a) a value in the range $0 < \rho < 3/4$ for any mode that is totally symmetric and (b) the fixed value $\rho = 3/4$ for any mode that is not totally symmetric. A band for which $0 < \rho < 3/4$ is said to be *polarized,* and a

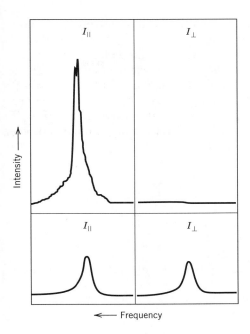

Figure 6.12   Raman band intensities $I_\parallel$ and $I_\perp$ of CCl_4 for ν_1 *(top)* at 458 cm^{-1} and ν_2 *(bottom)* at 218 cm^{-1}. The ν_1 band shows isotopic splitting due to ^{35}Cl and ^{37}Cl.

band for which $\rho = 3/4$ is said to be *depolarized*. In practice, for highly symmetric molecules one often finds $\rho \approx 0$ for polarized bands. This makes it very easy to assign such bands to totally symmetric modes, even without calculating the ratio. On the other hand, owing to experimental imperfections, one often finds that ρ is only approximately 3/4 for depolarized bands from nonsymmetric modes. Nonetheless, these deviations rarely cause confusion when making assignments. Figure 6.12 shows the intensities in both the $I_\parallel$ and $I_\perp$ orientations for the $\nu_1(A_1)$ and $\nu_2(E)$ Raman bands of CCl_4. As can be seen, in the $I_\perp$ orientation the $\nu_1(A_1)$ band virtually vanishes. Carefully obtained values show that $\rho = 0.005 \pm 0.002$ for $\nu_1(A_1)$ and $\rho = 0.75 \pm 0.02$ for $\nu_2(E)$.

6.3   *Spectroscopic Activities and Structures of Nonlinear Molecules*

We have seen in the preceding sections that the number and spectroscopic activity of normal modes depends upon the molecule's symmetry. On this basis we might expect that infrared and Raman spectroscopy could be used to distinguish between two or more possible structures that a particular molecule might have. Indeed, this approach has been taken for many simple molecules. Of course, as molecular complexity increases and symmetry decreases the ability to distinguish between various proposed structures diminishes, since the predicted numbers and activities of normal modes are likely to be nearly the same. Moreover, the ability to observe any predicted distinctions experimentally also declines. Nonetheless, within the limitations of relatively small molecules in gas and liquid samples, predictions from symmetry and group theory can be effective tools to interpret vibrational spectra and deduce structures.

Suppose we consider a compound whose molecular formula is XY_4. As we have seen, if the molecule is tetrahedral we should expect the $3n - 6 = 9$ normal modes to give rise to four fundamental frequencies, all of which are active in the Raman spectrum, and two of which are also active in the infrared spectrum. Furthermore, the Raman band for ν_1 (A_1) should be polarized. We can summarize these predictions as follows:

	T_d
Infrared	2 $(2T_2)$
Raman	4 $(A_1, E, 2T_2)$
Polarized	1 (A_1)
Coincidences	2 $(2T_2)$

This, then, sets the pattern of spectroscopic activities that characterizes a tetrahedral XY_4 molecule.

Now suppose we consider the possibility that XY_4 might be square planar; that is,

$$
\begin{array}{c}
Y \\
| \\
Y\!-\!X\!-\!Y \\
| \\
Y
\end{array}
$$

Counting nonshifted atoms from each operation and using the contributions per nonshifted atom in each case (the sum of A_{2u} and E characters), we obtain the following work sheet:

D_{4h}	E	$2C_4$	C_2	$2C_2'$	$2C_2''$	i	$2S_4$	σ_h	$2\sigma_v$	$2\sigma_d$		Σ	n_i		
N_i	5	1	1	3	1	1	1	5	3	1					
χ_i	3	1	-1	-1	-1	-3	-1	1	1	1					
Γ_{3n}	15	1	-1	-3	-1	-3	-1	5	3	1		Σ	n_i		
A_{1g}	15	2	-1	-6	-2	-3	-2	5	6	2		16	1		$x^2 + y^2,$
A_{2g}	15	2	-1	6	2	-3	-2	5	-6	-2		16	1	R_z	
B_{1g}	15	-2	-1	-6	2	-3	2	5	6	-2		16	1		$x^2 - y^2$
B_{2g}	15	-2	-1	6	-2	-3	2	5	-6	2		16	1		xy
E_g	30	0	2	0	0	-6	0	-10	0	0		16	1	(R_x, R_y)	(xz, yz)
A_{1u}	15	2	1	-6	-2	3	2	-5	-6	-2		0	0		
A_{2u}	15	2	1	6	2	3	2	-5	6	2		32	2	z	
B_{1u}	15	-2	1	-6	2	3	-2	-5	-6	2		0	0		
B_{2u}	15	-2	1	6	-2	3	-2	-5	6	-2		16	1		
E_u	30	0	2	0	0	6	0	10	0	0		48	3	(x, y)	

From this we have $\Gamma_{3n} = A_{1g} + A_{2g} + B_{1g} + B_{2g} + E_g + 2A_{2u} + B_{2u} + 3E_u$. Subtracting the species of the rotations and translations, as indicated above, leaves $\Gamma_{3n-6} = A_{1g} + B_{1g} + B_{2g} + A_{2u} + B_{2u} + 2E_u$, for a total of seven expected frequencies. From the unit vector and direct product transformations listed above we obtain the following activities:

A_{1g}	B_{1g}	B_{2g}	A_{2u}	B_{2u}	$2E_u$
ν_1	ν_2	ν_4	ν_3	ν_5	ν_6, ν_7
Raman (pol)	Raman	Raman	Infrared	—	Infrared

The forms of these normal modes are shown in Appendix C. Note that only ν_1 (A_{1g}), the symmetric stretching mode, preserves the complete symmetry of D_{4h}. Therefore it is the only mode expected to have a polarized band in the Raman spectrum (indicated "pol" in the table above). The two doubly degenerate frequencies, ν_6 and ν_7, each consist of a pair of normal modes whose individual motions are at right angles to one another in the plane of the molecule. The motions depicted for ν_6 and ν_7 in Appendix C are linear combinations of the pairs of modes in each case. Notice that one of the normal modes, $\nu_5(B_{2u})$, which might be described as a deformation of the molecular plane, is not active in either spectrum; that is, it is a silent mode. Therefore, instead of seven frequencies, we should expect to observe only six directly: three in the Raman spectrum and three in the infrared spectrum.

These predictions contrast significantly with those for a tetrahedral structure:

	T_d	D_{4h}
Infrared	2 ($2T_2$)	3 ($A_{2u}, 2E_u$)
Raman	4 ($A_1, E, 2T_2$)	3 (A_{1g}, B_{1g}, B_{2g})
Polarized	1 (A_1)	1 (A_{1g})
Coincidences	2 ($2T_2$)	None
Silent modes	None	1 (B_{2u})

First, there are three more frequencies than for a tetrahedral species. Second, as a centrosymmetric molecule, planar XY_4 is subject to the rule of mutual exclusion; that is, none of the Raman frequencies is active in the infrared spectrum and vice versa. These predictions suggest that it should be possible to distinguish between tetrahedral and square planar XY_4 structures on the basis of the infrared and Raman spectra. Although the numbers of bands and polarized Raman bands are not greatly different between the two models, the predicted absence of coincidences for the square planar structure probably would be a most telling distinction. Indeed, considerations of this sort were used as part of the early verification of the tetrahedral structure of CH_4.*

If one or more atoms in a compound are substituted by some other element, the symmetry often changes and with it the vibrational selection rules.

*See G. Herzberg, *Molecular Spectra and Molecular Structure. II. Infrared and Raman Spectra of Polyatomic Molecules,* Van Nostrand, Princeton, NJ, 1945, pp. 306–307.

For example, consider substituting one hydrogen atom in CH_4 with a deuterium atom to produce CH_3D. While the new species is essentially tetrahedral, the fact that the deuterium atom is approximately twice the mass of the hydrogen atom reduces the symmetry to C_{3v}. Consequently, the number of vibrational frequencies and their activities are those appropriate for this new point group. We could determine the new selection rules *de novo*, but CH_4 and CH_3D are closely related by their group–subgroup relationship $T_d \rightarrow C_{3v}$. Thus, we can think of the frequencies of CH_3D as arising from perturbations of the CH_4 frequencies, caused by the descent in symmetry. This suggests a correlation approach, making use of the correlation table in Appendix B that links the symmetry species of T_d and C_{3v}.

Taking this approach, we can construct the correlation diagram shown in Fig. 6.13. The nondegeneracy of the A_1 mode (ν_1) and the double degeneracy of the E modes (ν_2) are retained with the descent in symmetry, but the threefold degeneracy of the T_2 modes (ν_3 and ν_4) is lifted to become a nondegenerate A_1 mode and a pair of doubly degenerate E modes in the new point group C_{3v}. In both cases, deuterium substitution causes one of the three degenerate modes to acquire a unique frequency from the remaining two. Thus the four frequencies of CH_4 become six frequencies for CH_3D. The normal modes of these six frequencies (cf. XY_3Z in Appendix C) have A_1 and E symmetries, both of which in C_{3v} are associated with unit vector and direct product transformations (cf. character table in Appendix A). Thus, all six frequencies of CH_3D are active in both infrared and Raman spectra. Of these, the three with A_1 symmetry should be polarized in the Raman spectrum.

We can take the same approach with CH_2D_2, using the correlation from T_d to C_{2v} (Appendix B). The C_{2v} group does not allow degeneracy, so all degenerate modes are split into distinct modes with unique frequencies on descent from T_d (Fig. 6.14). Thus all nine normal modes of CH_2D_2 have corresponding individual frequencies. From the transformation properties listed in the C_{2v} character table, we conclude that eight are both infrared and Raman active, but the A_2 mode is exclusively Raman active.

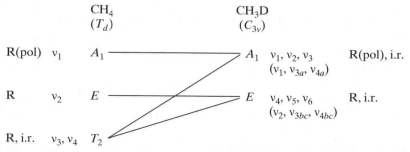

Figure 6.13 Correlation diagram for the normal modes of CH_4 and CH_3D. CH_3D frequencies are indicated by their appropriate numbers for C_{3v} and the corresponding numbers for CH_4 (in parentheses).

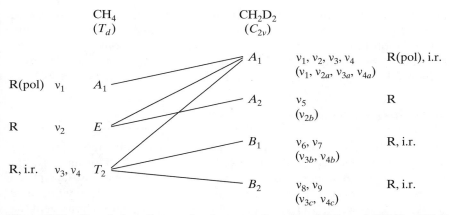

Figure 6.14 Correlation diagram for the normal modes of CH_4 and CH_2D_2. CH_2D_2 frequencies are indicated by their appropriate numbers for C_{2v} and the corresponding numbers for CH_4 (in parentheses).

We can summarize these results for CH_4, CH_3D, and CH_2D_2 as follows:

	CH_4 T_d	CH_3D C_{3v}	CH_2D_2 C_{2v}
Infrared	$2\ (2T_2)$	$6\ (3A_1, 3E)$	$8\ (4A_1, 2B_1, 2B_2)$
Raman	$4\ (A_1, E, 2T_2)$	$6\ (3A_1, 3E)$	$9\ (4A_1, A_2, 2B_1, 2B_2)$
Polarized	$1\ (A_1)$	$3\ (3A_1)$	$4\ (4A_1)$
Coincidences	$2\ (2T_2)$	$6\ (3A_1, 3E)$	$8\ (4A_1, 2B_1, 2B_2)$

These predictions agree well with the observed frequencies for CH_4, CH_3D, and CH_2D_2 (Table 6.1).

Note that we could not use our results for CH_3D to correlate directly to CH_2D_2. Instead we began again with CH_4. This was necessary because there is no group–subgroup relationship between C_{3v} and C_{2v}. By the same restriction, we could not deduce the number of frequencies and activities for square

Table 6.1 Relationships Among Vibrational Frequencies (cm^{-1}) of CH_4, CH_3D, and CH_2D_2.[a]

$CH_4\ (T_d)$	$\nu_1(A_1)$	$\nu_2(E)$		$\nu_3(T_2)$			$\nu_4(T_2)$		
	2914	1526		3020			1306		
$CH_3D\ (C_{3v})$	$\nu_2(A_1)$	$\nu_5(E)$		$\nu_1(A_1)$	$\nu_4(E)$		$\nu_3(A_1)$	$\nu_6(E)$	
	2205	1477		2982	3030		1306	1156	
$CH_2D_2\ (C_{2v})$	$\nu_2(A_1)$	$\nu_3(A_1)$	$\nu_5(A_2)$	$\nu_1(A_1)$	$\nu_6(B_1)$	$\nu_8(B_2)$	$\nu_4(A_1)$	$\nu_7(B_1)$	$\nu_9(B_2)$
	2139	1450	1286	2974	3030	2255	1034	1090	1235

[a]Data from K. Nakamoto, *Infrared Spectra of Inorganic and Coordination Compounds*, 2nd ed., John Wiley & Sons, New York, 1970, p.113.

planar XY_4 by correlation with the tetrahedral case, because T_d and D_{4h} do not have a group–subgroup relationship. In all such cases, where a group–subgroup relationship does not exist, a direct correlation among chemically related structures is not possible. However, if two chemical structures share a common subgroup, a correlation through that subgroup may be possible.

6.4 Linear Molecules

The procedure we have seen for nonlinear molecules can be applied to linear cases. The reducible representation Γ_{3n} is constructed in the same way, but only five nongenuine modes (three translations and two rotations, R_x and R_y) are subtracted to obtain the representation for the $3n - 5$ normal vibrational modes, Γ_{3n-5}. However, to reduce this representation we must circumvent the problem of the infinite order of the molecule's point group (either $D_{\infty h}$ or $C_{\infty v}$), since Eq. (3.1) is meaningless when $h = \infty$. As discussed in Section 3.4, it is convenient to set up the problem in a finite subgroup (usually either D_{2h} or C_{2v}) and carry out the reduction in that group. After subtracting the translations and rotations, the results in the finite group can be correlated to the actual infinite-order group, using Table 3.8 ($C_{\infty v} \leftrightarrow C_{2v}$) or Table 3.9 ($D_{\infty h} \leftrightarrow D_{2h}$). The activities of the normal modes can be determined, as before, from the unit vector and direct product listings in the character table for $D_{\infty h}$ or $C_{\infty v}$, as appropriate.

As an example of this approach, consider C_3O_2, carbon suboxide:

$$\ddot{O}=C=C=C=\ddot{O}$$

This is a centrosymmetric linear molecule, so the point group is $D_{\infty h}$. To avoid the problem of the group's infinite order, we will set up the representation Γ_{3n} in D_{2h}, taking the molecular axis as z. By counting nonshifted atoms and multiplying by the contributions per nonshifted atom for each operation, we obtain the following results:

D_{2h}	E	$C_2(z)$	$C_2(y)$	$C_2(x)$	i	$\sigma(xy)$	$\sigma(xz)$	$\sigma(yz)$
N_i	5	5	1	1	1	1	5	5
χ_i	3	−1	−1	−1	−3	1	1	1
Γ_{3n}	15	−5	−1	−1	−3	1	5	5

This reduces as $\Gamma_{3n} = 2A_g + 2B_{2g} + 2B_{3g} + 3B_{1u} + 3B_{2u} + 3B_{3u}$, which has the required dimension of 15. By inspecting the unit vector and direct product listings in the D_{2h} character table we determine that $\Gamma_{trans} = B_{1u} + B_{2u} + B_{3u}$ and $\Gamma_{rot} = B_{2g} + B_{3g}$. Note that Γ_{rot} is composed of only the two species corresponding to R_x and R_y, not R_z (B_{1g}). Thus, $\Gamma_{3n-5} = 2A_g + B_{2g} + B_{3g} + 2B_{1u} + 2B_{2u} + 2B_{3u}$, which has the required $3n - 5 = 10$ dimension. Using Table 3.9 we can construct the correlation diagram shown in Fig. 6.15, which shows that in $D_{\infty h}$ $\Gamma_{3n-5} = 2\Sigma_g^+ + \Pi_g + 2\Sigma_u^+ + 2\Pi_u$. From the $D_{\infty h}$ character table we see that Σ_g^+ and Π_g are associated with the direct product transfor-

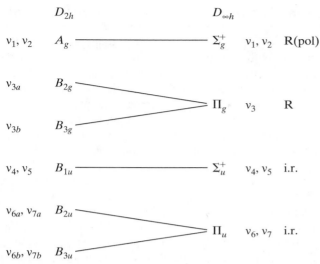

Figure 6.15   Correlation from the working group D_{2h} to the true group $D_{\infty h}$ for the normal modes of C_3O_2.

mations z^2 and (xz, yz), respectively, indicating that the normal modes with these symmetries are Raman active. The two modes with Σ_g^+ symmetry, the totally symmetric representation, will be polarized. Likewise, we see that Σ_u^+ and Π_u are associated with the unit vector transformations z and (x, y), respectively, indicating that the normal modes with these symmetries are infrared active. Note that C_3O_2, being a centrosymmetric molecule, is subject to the rule of mutual exclusion, so the infrared-active normal modes are not Raman active, and vice versa. In summary, our predictions for C_3O_2 are as follows:

	$D_{\infty h}$
Infrared	$4\ (2\Sigma_u^+ + 2\Pi_u)$
Raman	$3\ (2\Sigma_g^+ + \Pi_g)$
Polarized	$2\ (2\Sigma_g^+)$
Coincidences	None

The infrared and Raman spectra of C_3O_2 are consistent with these expectations. However, demonstrating that the spectra were uniquely consistent with a linear structure proved to be a less-than-straightforward task in this case. As a result, the linear geometry of C_3O_2 was for many years in doubt, since the spectra are complicated and could be plausibly interpreted on the basis of alternative structures. Indeed, the controversy over the structure of C_3O_2 provides a case study of the difficulties that can be encountered in using vibrational spectra for structure elucidation.*

*See F. A. Miller, D. H. Lemmon, and R. E. Witkowski, *Spectrochim. Acta,* **1965,** *21,* 1709; and W. H. Smith and G. E. Leroi, *J. Chem. Phys.,* **1966,** *45,* 1765, as well as earlier references cited in these papers.

6.5   *Overtones, Combinations, and Other Complications*

We have seen that it is possible to predict the number and spectroscopic activities of the fundamental transitions of the normal modes of a polyatomic molecule. When applying these predictions to actual spectra a number of complications can arise. On the one hand, weak intensities or instrumental limitations can result in fewer observable frequencies than predicted. But more often there are more peaks in the spectrum than one would predict for the structure. Several factors can give rise to additional peaks in the infrared and Raman spectra, the most common being the presences of *overtone bands* and *combination bands*.

Although the fundamental selection rule for a harmonic oscillator allows only transitions for which $\Delta \mathbf{v} = \pm 1$, anharmonicity in the oscillations of real molecules gives rise to weak spectroscopic bands from transitions for which $\Delta \mathbf{v} = \pm 2, \pm 3, \ldots, \pm n$. For a normal mode ν_i, such transitions represent the first, second, and succeeding overtones of the fundamental, customarily designated $2\nu_i, 3\nu_i, \ldots, n\nu_i$. The probabilities of these transitions, and hence their band intensities, fall off rapidly with higher $\Delta \mathbf{v}$, so generally only first and (to a lesser extent) second overtones are encountered in routine work. The observed frequencies of overtones are nearly the appropriate whole number multiplies of the frequency of the corresponding fundamental. However, since the spacings between successive levels of an anharmonic oscillator are progressively smaller (cf. Fig. 6.3), overtone frequencies are almost always slightly less than the whole-number multiple values.

It is also possible for excitations of two separate normal modes to couple, giving rise to *combination bands* of the type $\nu_k + \nu_i$. Less commonly, combinations with and among overtones, having the general form $n\nu_k + m\nu_i$, can be observed. There can also be subtractive combinations, called *difference bands*, such as $\nu_k - \nu_i$, or more generally $n\nu_k - m\nu_i$. As with overtones, intensities of both summation and difference combinations are typically much less than those of fundamentals. This is especially true for combinations involving overtones or differences. In most cases, owing to anharmonicity, the observed frequencies tend to be slightly lower than the numerical sum or difference of the frequencies of the combined fundamentals.

Spectroscopic selection rules for overtones and combinations follow the same considerations as those we have applied to fundamentals. For infrared activity, an expression for the transition moment M, in the form of Eq. (6.15), must be nonzero. For Raman activity, an expression for the transition moment P, in the form of Eq. (6.18), must be nonzero. Thus, an overtone or combination will be infrared allowed if it belongs to the same irreducible representation as one or more components of the electric dipole, equivalent to the symmetry species of the unit vectors listed in the next-to-last column of the appropriate character table. Likewise, an overtone or combination will be Raman allowed if it belongs to the same irreducible representation as one or more components of the polarizability tensor, equivalent the symmetry species of the binary direct products listed in the last column of the appropriate character table. Although these basic selection rules allow spectroscopic activity for many overtones and combinations, generally only a few have sufficient intensity to be observed. They tend to be more prevalent in the infrared than the Raman spectrum. In light of this, spectroscopic selection rules are not used to predict the numbers of such bands in the spectra, but rather to justify an assignment of a suspect weak band. The question becomes "Is this overtone or combination allowed in this spectrum?" With these kinds of bands, more than with fundamentals, the selection rules tell what can be, rather than what probably will be.

The symmetries of overtones and combinations can be determined by taking the direct products of the irreducible representations of the fundamentals involved. From the properties of direct products discussed in Section 3.5, we can anticipate that the resulting representation will be irreducible if one or both of the component fundamentals of the overtone or combination are nondegenerate. If both components are degenerate, the direct product will be a reducible representation, requiring decomposition into its component irreducible representations. We can combine our knowledge of the general properties of direct products with the symmetry-based infrared and Raman selection rules developed in Section 6.2 to make some general predictions about the activities of overtones and combinations. As we develop these generalizations, we will illustrate with specific examples from the infrared and Raman spectra of CH_4, taken from data listed in Herzberg's book.* For reference, the fundamental frequencies of CH_4 are listed in Table 6.1.

Recall that the direct product of any nondegenerate irreducible representation with itself is the totally symmetric representation. From this we may conclude that the first overtone of any nondegenerate normal mode will belong to the totally symmetric representation. Since the totally symmetric representation is always the symmetry species of at least one component of the polarizability tensor, we may conclude that *all first overtones of nondegenerate normal modes will be Raman allowed.* Furthermore, inspection of the character tables reveals that *first overtones will be infrared forbidden, except for molecules belonging to the point groups C_1, C_s, C_n, and C_{nv}.* Only these small groups have at least one of the unit vectors transforming as the totally symmetric representation. In the case of CH_4, the only nondegenerate fundamental is $\nu_1 = 2914 \text{ cm}^{-1}$. By what we have just seen, we might expect an overtone $2\nu_1$ in the Raman spectrum near $2 \times 2914 \text{ cm}^{-1} = 5828 \text{ cm}^{-1}$. However, none has been reported, probably owing to extremely weak intensity. This is an illustration of the point previously made that the selection rules for overtones and combinations are merely permissive or prohibitive, and not predictive.

The direct product of any *degenerate* representation with itself contains the totally symmetric representation along with other irreducible representa-

*G. Herzberg, *Molecular Spectra and Molecular Structure. II. Infrared and Raman Spectra of Polyatomic Molecules,* Van Nostrand, Princeton, NJ, 1945, p. 308.

tions. Therefore, we can extend the previous generalization to all fundamentals; i.e., *all first overtones are Raman allowed*. Such overtones may be infrared allowed as well, without restriction to certain point groups, depending on the other symmetry species that may compose the reducible representation. For example, consider the possible activities of $2\nu_3$ and $2\nu_4$ of CH_4. Both fundamentals belong to T_2, and as such are active in both infrared and Raman spectra. The direct product of T_2 with itself is

T_d	E	$8C_3$	$3C_2$	$6S_4$	$6\sigma_d$
T_2	3	0	-1	-1	1
T_2	3	0	-1	-1	1
$\Gamma(T_2T_2)$	9	0	1	1	1

This reduces as $\Gamma(T_2T_2) = A_1 + E + T_1 + T_2$. As expected, this contains the totally symmetric representation, A_1, but it also contains E and T_2, which are likewise symmetry species of polarizability tensor components. Thus, the overtones $2\nu_3$ and $2\nu_4$ are Raman allowed by virtue of the three components $A_1 + E + T_2$. The T_2 species is the only one in T_d that is associated with infrared activity, so these overtones are also infrared allowed. The symmetry species T_1, which is also a component of $\Gamma(T_2T_2)$, is not associated with either infrared or Raman activity and can be ignored for the present purposes. Herzberg lists $2\nu_3$ as 6006 cm^{-1} (cf. 2×3020 cm^{-1} = 6040 cm^{-1}) and $2\nu_4$ as 2600 cm^{-1} (cf. 2×1306 cm^{-1} = 2612 cm^{-1}) from infrared data.

The activities of combinations can be determined in similar manner. Consider the combination $\nu_2 + \nu_4$ of CH_4. The species of the two fundamentals are E and T_2, respectively, and their direct product is

T_d	E	$8C_3$	$3C_2$	$6S_4$	$6\sigma_d$
E	2	-1	2	0	0
T_2	3	0	-1	-1	1
$\Gamma(ET_2)$	6	0	-2	0	0

This reduces as $\Gamma(ET_2) = T_1 + T_2$. This combination contains T_2, so it is both infrared and Raman allowed. Again, the T_1 component can be ignored. An infrared band at 2823 cm^{-1} has been assigned to this combination (cf. 1526 cm^{-1} + 1306 cm^{-1} = 2832 cm^{-1}).

The activities of difference bands are determined in the same way as the positive combinations. The symmetries of both positive and negative combinations are the same. Thus, if a combination for $\nu_k + \nu_i$ is allowed or forbidden, then a combination for $\nu_k - \nu_i$ is likewise allowed or forbidden. For example, combinations $\nu_3 \pm \nu_4$ for CH_4 have the symmetry $\Gamma(T_2T_2) = A_1 + E + T_1 + T_2$ and are allowed in both infrared and Raman spectra. An infrared band at 4313 cm^{-1} has been assigned to $\nu_3 + \nu_4$ (cf. 3020 cm^{-1} +

$1306 \text{ cm}^{-1} = 4326 \text{ cm}^{-1}$), and another at approximately 1720 cm^{-1} has been assigned to $\nu_4 - \nu_3$ (cf. $3020 \text{ cm}^{-1} - 1306 \text{ cm}^{-1} = 1714 \text{ cm}^{-1}$).

The assignment of overtone and combination bands is usually uncertain, particularly when degenerate modes are involved. In such cases, anharmonicity may cause splitting into sub-bands that correspond to the individual symmetry species comprising the direct product. Even without this complication, assignment is often ambiguous and must be regarded as tentative. Often several allowed overtones and combinations may have potentially similar frequencies or may be obscured by stronger fundamentals. Of course, none is expected to have much intensity, which makes such bands difficult to discern above the spectroscopic noise.

Even with these complications, overtones may give useful information not readily available by other means. Recall that for a planar XY_4 species there is one normal mode, $\nu_5(B_{2u})$, which is not active in either infrared or Raman spectra (cf. Section 6.3). However, the first overtone of this mode should be totally symmetric (A_{1g}) and allowed in the Raman spectrum. In the case of XeF_4, a weak feature at 442 cm^{-1} in the Raman spectrum has been assigned to $2\nu_5$, implying that ν_5 is approximately 221 cm^{-1}.* This value is reasonable, given the other observed frequencies. This fortuitous observation is virtually the only way to estimate the value of ν_5 from the spectra.

Although generally weak, overtones and combinations can sometimes have surprisingly strong intensities when they fall near a fundamental with the same symmetry. In these cases, the overtone or combination "borrows" intensity from the fundamental. The two bands mix and split, losing their individual identities. One feature moves to higher frequency and the other to lower, with both having comparable intensities. The phenomenon is known as *Fermi resonance*. Fermi resonance is a necessary consequence of symmetry. If two different states with the same symmetry were to have equal energies, this would imply a degeneracy higher than that of the states' symmetry species. For example, if a totally symmetric fundamental were to have the same frequency as a first overtone of a nondegenerate fundamental (also totally symmetric) there would exist a double degeneracy for a nondegenerate symmetry species (the totally symmetric representation). This is an oxymoron. Moreover, it is forbidden by the symmetry of the system. Therefore, *accidental degeneracy among states with the same symmetry is forbidden*. By mixing and separating the two states, Fermi resonance avoids the symmetry contradiction that would otherwise occur.

A well-known case of Fermi resonance occurs in the Raman spectrum of CCl_4 (Fig. 6.16). The fundamental $\nu_3(T_2)$ has a frequency of 776 cm^{-1}. The combination $\nu_1 + \nu_4$ has the same symmetry ($A_1T_2 = T_2$) and on the basis of normal coordinate analysis has an expected frequency of $459 + 314 \text{ cm}^{-1} = 773 \text{ cm}^{-1}$, which is virtually the same as ν_3, given the breadth of the bands. As a result, ν_3 and $\nu_1 + \nu_4$ mix to form a Fermi resonant doublet. The Fermi

*H. H. Claassen, C. L. Chernick, and J. G. Malm, *J. Am. Chem. Soc.,* **1963,** *85, 1927.*

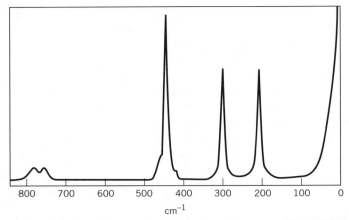

Figure 6.16   Medium resolution Raman spectrum of liquid CCl_4 showing the Fermi doublet of ν_3 and $\nu_1 + \nu_4$ with maxima at 762 cm^{-1} and 790 cm^{-1}. Neither band can be assigned exclusively to one component.

doublet, shown in Fig. 6.16, has maxima at approximately 762 cm^{-1} and 790 cm^{-1}. Neither of these bands can be assigned exclusively to ν_3 or $\nu_1 + \nu_4$. The position of the fundamental ν_3 can be estimated from the trough between the two bands. The same Fermi doublet is also infrared active by virtue of its T_2 symmetry and is easily observed (Fig. 6.17).

   In addition to overtones, combinations, and Fermi resonance, "extra" peaks may appear in the spectrum from *isotopic splitting* if one of the elements in the compound exists as two or more isotopes with relatively high

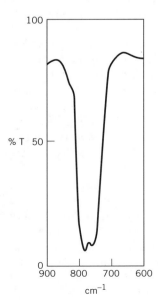

Figure 6.17   Infrared spectrum of the Fermi doublet of CCl_4.

abundances. In such cases, some spectroscopic bands may appear as multiplets. Although it is customary to refer to this fine structure as splitting, in reality it is merely the superimposition of frequencies from the same or similar modes of all the possible isotopically substituted molecular species. The effect tends to be most noticeable when relatively light atoms are involved, since the masses among the isotopes are significantly different. We have seen the effect of deuterium substitution on the vibrational frequencies of CH_4 (Section 6.3), where $m_D/m_H = 2$. In a naturally occurring sample of methane the isotopic abundance of ^1D is only 0.015%, so band splitting from the presence of CH_3D, CH_2D_2, CHD_3, and CD_4 is undetectable. However, if the compound contains elements such as boron (20% ^{10}B, 80% ^{11}B) or chlorine (75.77% ^{35}Cl, 24.23% ^{37}Cl), some bands may show fine structure from the mix of isotopically substituted species in the sample. Bands arising from normal modes in which the isotopically substituted atoms have highest amplitude (e.g., stretching modes) are most likely to show splitting.

The band for ν_1 in the Raman spectrum of CCl_4 at high resolution (Fig. 6.18) shows components from all possible isotopically substituted species: $C\ ^{35}Cl_4$, $C\ ^{37}Cl\ ^{35}Cl_3$, $C\ ^{37}Cl_2\ ^{35}Cl_2$, $C\ ^{37}Cl_3\ ^{35}Cl$, $C\ ^{37}Cl_4$. The actual symmetries of the various components are T_d, C_{3v}, and C_{2v}, depending on whether they have four, three, or two atoms of one of the isotopes. However, the mass differences between ^{35}Cl and ^{37}Cl are relatively small ($m_{37}/m_{35} = 1.03$), so all molecules are virtually tetrahedral. The observed peaks arise from essentially the same symmetric stretching motion from each isotopically substituted molecular species. Other bands in the spectrum are too broad to show comparable fine structure.

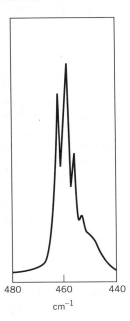

480   460   440
cm^{-1}

Figure 6.18 High-resolution Raman spectrum of the symmetric stretching band (ν_1) of CCl_4.

Problems

6.1   Determine the number of frequencies, their symmetries, and the infrared and Raman activities of the normal modes for the following molecules. Indicate the number of polarized Raman bands and the number of frequencies that should be coincident between the two spectra. Representations of the normal modes for these structures can be found in Appendix C. (a) NH_3, (b) $FeCl_6^{3-}$, (c) H_2CO, (d) PF_5, (e) C_2H_6 (staggered configuration), (f) H_2O_2.

6.2   Determine the number of frequencies, their symmetries, and the infrared and Raman activities of the normal modes for the following molecules. Indicate the number of polarized Raman bands and the number of frequencies that should be coincident between the two spectra. (a) SeF_5^-, (b) AsF_4^-, (c) BeF_3^-, (d) OSF_4, (e) *trans*-FNNF, (f) *cis*-FNNF, (g) $S_2O_3^{2-}$, (h) B_2H_6.

6.3   H. D. Rix [*J. Chem. Phys.*, **1954**, *22*, 429] interpreted incomplete vibrational data for C_3O_2 on the basis of a C_{2h} structure. Determine the number of frequencies, symmetries, infrared and Raman activities, and the number of polarized Raman bands expected for this structure. Compare these results to those predicted for linear C_3O_2 (cf. Section 6.4).

6.4   There are several other structures that one might propose for XY_4 molecules, besides tetrahedral and square planar. Determine the number of frequencies, their symmetries, the infrared and Raman activities, number of polarized Raman bands, and coincidences for the following alternative structures. [*Hint:* You may find it more expedient to use a correlation from either tetrahedral XY_4 or square planar XY_4, as appropriate.]

(a) A distorted tetrahedron, as predicted by VSEPR theory for four bond pairs and one lone pair about the central X atom
(b) A distorted tetrahedron resulting from slightly squashing a perfect tetrahedron along one of its C_2 axes
(c) A square pyramid in which the X atom is at the apex
(d) Planar XY_4 with two *trans*-related bonds longer than the other two

6.5   Using a correlation approach, show the relationships among the normal modes and their infrared and Raman activities for the substituted octahedral structures I, II, III, and IV in Fig. 3.1. Summarize your results in a table listing infrared active frequencies, Raman active frequencies, polarized Raman bands, and coincidences.

6.6   Determine the number of frequencies, their symmetries, and the infrared and Raman activities of the normal modes for the following linear molecules. Indicate the number of polarized Raman bands and the number of frequencies that should be coincident between the two spectra. (a) CO_2, (b) OCN^-, (c) $H-C\equiv C-H$, (d) $Cl-C\equiv C-H$, (e) $H-C\equiv C-C\equiv C-H$.

6.7   The nitrate ion in aqueous solution shows a tendency to associate with a variety of metal cations (e.g., Ag^+, Cu^{2+}, Zn^{2+}, Hg^{2+}, Ca^{2+}, Al^{3+}, Ce^{3+}, Th^{4+}). Depending on concentration and cation, the association may involve direct ion pairing or pairing through one or more solvated water molecules. In dilute solutions, where such association is minimal and NO_3^- has nearly its idealized D_{3h} symmetry, four frequencies can be observed: 1400 cm^{-1} (infrared, Raman), 1050 cm^{-1} (Raman, polarized), 830 cm^{-1} (infrared), 720 cm^{-1} (infrared, Raman).

[See D. E. Irish, A. R. Davis, and R. A. Plane, *J. Chem. Phys.* **1969**, *50*, 2262; D. E. Irish and G. E. Walrafen, *J. Chem. Phys.* **1967**, *46*, 378; R. E. Hester and R. A. Plane, *J. Chem. Phys.* **1964**, *40*, 411.]

(a) Assign the four frequencies of the "free" nitrate ion by frequency number (ν_1, ν_2, ν_3, ν_4) and symmetry species. The numbering in this case is systematic by symmetry species, giving priority to nondegenerate modes.

(b) Whether or not solvent water is involved, anion–cation association reduces the effective symmetry of NO_3^-, causing certain of the spectroscopic bands to split into two components. Two geometries of cation–anion association are plausible: (1) association along the C_3 axis of the NO_3^- ion and (2) association in the NO_3^- ion plane to one of the oxygen atoms. Predict the changes in the infrared and Raman spectra that would occur with each of these modes of association.

(c) In 7.2 M $Ca(NO_3)_2$ solution, where direct cation–anion association has been postulated, six Raman frequencies have been observed: 1450 cm^{-1}, 1358 cm^{-1}, 1052 cm^{-1}, 823 cm^{-1}, 743 cm^{-1}, 717 cm^{-1}. The band at 1052 cm^{-1} is strongly polarized, and the bands at 1358 cm^{-1} and 743 cm^{-1} appear to be weakly polarized. From these data, which of the two modes of association appears to occur in concentrated calcium nitrate solutions? [See D. E. Irish and G. E. Walrafen, *J. Chem. Phys.* **1967**, *46*, 378.]

(d) The laser Raman spectra of 0.3–2.3 M $Al(NO_3)_3$ solutions show an overtone at approximately 1660 cm^{-1} of the infrared-active fundamental at 830 cm^{-1} ("free" ion). Explain why the overtone can be observed in the Raman spectrum, but its corresponding fundamental cannot. Why is this overtone not observed in the infrared spectrum? [See W. L. Grossman and G. Chottard, *Spectrochim. Acta* **1970**, *26*, 2379.]

6.8 Justify or refute the following generalizations:

(a) All even-number overtones ($2\nu_i$, $4\nu_i$,...) are Raman allowed.

(b) All even-number overtones ($2\nu_i$, $4\nu_i$,...) are infrared forbidden.

(c) If a normal mode is infrared active, its odd-number overtones ($3\nu_i$, $5\nu_i$, ...) will be infrared allowed.

(d) All combinations between a totally symmetric fundamental and any other fundamental ($\nu_s \pm \nu_i$) will be allowed or forbidden in the same manner as the nonsymmetric fundamental (ν_i).

(e) All combinations between a totally symmetric fundamental and the first overtone of any other fundamental ($\nu_s \pm 2\nu_i$) will be Raman allowed.

6.9 The following frequencies and their assignments have been reported for isotopically pure $^{100}MoF_6$ in the gas phase [R. S. McDowell, R. J. Sherman, L. B. Asprey, and R. C. Kennedy, *J. Chem. Phys.* **1975**, *62*, 3974.]:

Raman (cm^{-1})	Assignment	Infrared (cm^{-1})	Assignment
741 ± 0.8 (pol)	ν_1	1479.4 ± 0.5	$\nu_1 + \nu_3$
652.0 ± 0.5	ν_2	1386.4 ± 0.5	$\nu_2 + \nu_3$
531 ± 3	$2\nu_4$	913.1 ± 0.5	$\nu_2 + \nu_4$
380 ± 3	$\nu_4 + \nu_6$	739.3 ± 0.5	ν_3
317 ± 1	ν_5	262.7 ± 0.5	ν_4
233 ± 2 (pol)	$2\nu_6$		

(a) From the Raman bands assigned as $\nu_4 + \nu_6$ and $2\nu_6$, the frequency of the fundamental ν_6 can be estimated to be 117 cm^{-1}. Why is no band reported at this frequency in either the infrared or Raman spectrum?

(b) The Raman band at 741.8 cm^{-1} is nearly the same frequency as the infrared band at 739.3 cm^{-1}. How can we be certain that these bands are not from the same normal mode?

(c) The data above show three Raman bands and three infrared bands assigned to various overtones and combinations. Show that each of these assignments is plausible on the bases of frequency, symmetry, and expected spectroscopic activity.

(d) None of the overtones or combinations observed in the Raman spectrum is observed in the infrared spectrum. On the basis of your results in part (c), is this an expected result?

(e) Estimates of the frequency of ν_6 come from the Raman data only. Is any combination of the type $\nu_i + \nu_6$ ($i = 1, 2, 3, 4, 5$) allowed in the infrared? If so, approximately what frequency would it have? Account for the absence of any such combination in the reported infrared spectrum.

CHAPTER 7

Transition Metal Complexes

It is not our intention in this chapter to duplicate the kinds of presentations of transition metal chemistry that can be found in most standard advanced inorganic chemistry texts. In particular, we will gloss over most quantitative aspects of the subject and ignore almost completely the descriptive chemistry of coordination compounds. Rather, our focus will be on the physical consequences that follow from the symmetry of transition metal complexes. Nonetheless, a certain amount of duplication with other treatments will be unavoidable but should ensure that our examination of the symmetry aspects of transition metal chemistry is fully appreciated. In particular, it will be necessary to have had some prior exposure to Russell–Saunders coupling terms for various equivalent d^n configurations. Consequently, we will undertake a brief review of the fundamental concepts and terminology before addressing the symmetry-induced effects. More detailed discussions of Russell–Saunders terms can be found in most standard physical chemistry texts.*

7.1 Crystal Field Theory

In 1929 Hans Bethe published his classic paper on the splitting of terms in crystals, which laid the foundation for *crystal field theory*.[†] The work, in German, is long and difficult, requiring a thorough understanding of group theory and quantum mechanics. Little wonder, then, that few English-speaking scientists of the time, with the notable exception of theoreticians such as J. H. Van Vleck,[‡] adopted this approach for interpreting the magnetic and spectroscopic properties of transition metal complexes. Instead, the valence bond (VB) approach, championed by Linus Pauling and others,[§] was the

---

*For example, see any of the following: P. Atkins, *Physical Chemistry*, 5th ed., W. H. Freeman, New York, 1994, pp. 451–456; R. A. Alberty and R. J. Silbey, *Physical Chemistry*, John Wiley & Sons, New York, 1992, pp. 370–378; G. M. Barrow, *Physical Chemistry*, 6th ed., McGraw-Hill, New York, 1996, pp. 495–502.

†H. Bethe, *Ann. Physik*, **1929**, *3* [5], 135. An English translation of this paper, under the title *Splitting of Terms in Crystals*, is available from Books on Demand, a Division of University Microfilms, International, Ann Arbor, Michigan.

‡J. H. Van Vleck, *The Theory of Electric and Magnetic Susceptibilities*, Oxford University Press, Oxford, England, 1932; *Phys. Rev.*, **1932**, *41*, 208; *J. Chem. Phys.*, **1935**, *3*, 803 and 807.

§L. Pauling, *J. Am. Chem. Soc.*, **1931**, *53*, 1367; J. C. Slater, *Phys. Rev.* **1931**, *38*, 1109. Also see L. Pauling, *The Nature of the Chemical Bond*, 3rd ed., Cornell University Press, Ithaca, NY, 1960.

dominant theory until the 1950s. This was probably because VB theory was more accessible to most chemists and was more generally applicable to a wide variety of compounds, not just transition metal complexes. Eventually, the need to have theoretical tools to interpret the electronic spectra of complexes prompted chemists to look back to the work of Bethe and Van Vleck. Lead by L. E. Orgel,* many began to adopt the Crystal Field Theory (CFT) approach, and subsequently the *Ligand Field Theory* (LFT) and Molecular Orbital (MO) Theory approaches, as well.

CFT, originally developed to account for the magnetic and spectroscopic properties of crystals, looks at the effects on the electronic state of an atom in a nonhomogeneous field created by electrostatic (ionic) interactions with neighboring groups. This model may have some validity in the case of ionic solids, but, when applied to a transition metal surrounded by several ligands, the "electrostatic-only" starting assumption represents an admittedly extreme approximation in most cases. Nonetheless, the qualitative results of CFT agree remarkably well with ligand field theory, which employs empirically determined corrections to account for metal–ligand orbital interactions. As we shall see, it also agrees with results of the molecular orbital theory approach, which can account for virtually any degree of metal–ligand orbital overlap. As Van Vleck first pointed out, the various approaches, while superficially different, yield comparable results. Even though we may regard the starting assumptions of molecular orbital theory as fundamentally "better," the directness of CFT makes it a useful shorthand for routine discussions of transition metal complex chemistry.

Transition metal ions are characterized by the presence of an incompletely filled d subshell. Therefore, CFT looks at the relative energies of the d orbitals on a central metal ion surrounded by a certain number of nucleophilic ligands, arranged in a particular geometry. Changes in d orbital energies brought about by the ligand environment and the new electron configurations made possible by it alter the overall electronic energy state of the system. These kinds of changes occur for complexes of virtually any geometry, but octahedral six coordination is the most common among transition metal ions. Therefore, we will consider the case of ML_6 O_h first and in most detail.

An isolated metal ion belongs to the infinite-order rotation–inversion point group R_3,[†] consisting of all possible operations whose elements pass through a common point. With such high symmetry, all five d orbitals (and for that matter the several orbitals comprising any subshell, all of which share a given pair of n and l values) can be degenerate. But if we surround the ion with six ligands in an octahedral configuration, the symmetry will descend to O_h, in which no higher than threefold degeneracy is possible. Note that the highest dimension of any irreducible representation in O_h is $d_i = 3$; for ex-

---

*L. E. Orgel, *J. Chem. Soc.* **1952**, 4756; *J. Chem. Phys.* **1955**, 1004 and 1819.

[†]This group and its totally rotational subgroup are identified by several other notations, including K_h and K; $R_h(3)$ and $R(3)$; $R(3)$ and $R^+(3)$; and $\theta(3)$ and $\theta^+(3)$. We will only have occasion to refer to the full group, and thus we have adopted the simple R_3 notation.

ample, T_{1u}, T_{2g}. Thus, the degeneracy among the five d orbitals must be lifted. If we associate the various d orbitals with their corresponding direct products of the same notation, we can see from the O_h character table (cf. Appendix A) that d_{xz}, d_{yz}, and d_{xy} transform degenerately as T_{2g}, and $d_{2z^2-x^2-y^2}$ and $d_{x^2-y^2}$ transform degenerately as E_g.* Thus, the d orbitals in an O_h ML_6 complex must be split into two sets, one labeled t_{2g} and the other labeled e_g, with different energies.

We can verify that the d orbitals in an O_h field must divide into two degenerate sets by determining which orbitals are related to each other by operations of the group. The effect of a symmetry operation on an orbital is merely a change in coordinates without a change in energy. Thus, if some operation converts two or more orbitals into each other, they must be energetically equal. Conversely, any orbitals that are not related to each other by any symmetry operation cannot be degenerate (except, rarely, as an accidental degeneracy).

For the purpose of determining the degeneracies in O_h we only need to look at one symmetry element, a threefold axis about which C_3 and C_3^2 are performed. Figure 7.1 shows such an axis in relation to the x, y, z coordinates of an octahedral system. Before determining the effects of C_3 and C_3^2 on the d orbitals, it will be useful to recognize their effects on the x, y, z coordinates themselves. Assuming clockwise rotations, we can see that the following transformations will be effected:

	C_3	C_3^2
$x \rightarrow$	z	y
$y \rightarrow$	x	z
$z \rightarrow$	y	x

Thus, x, y, and z are transformed into one another and must be degenerate in O_h. Indeed, as the character table shows, they transform as T_{1u}, a triply de-

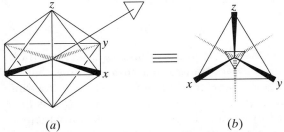

(a)   (b)

Figure 7.1 Two views of the same C_3 axis of an octahedron: (a) the axis emerging from a triangular face of the octahedron; and (b) looking down the axis, between the x, y, and z axes of the coordinate system.

*The notation $d_{2z^2-x^2-y^2}$ is the full label in O_h for the orbital more commonly identified as d_{z^2}. The more complete notation will be useful in the discussion that follows.

generate symmetry species. We may note at this point that the three p orbitals (p_x, p_y, p_z) transform as the vectors x, y, and z. Therefore, the interconversion of these vectors by C_3 and C_3^2 implies that p_x, p_y, and p_z orbitals in the same subshell on a central atom in O_h must still be degenerate, as they are for a free atom.

We can use the interchanges among the x, y, and z coordinates to deduce the transformations of the d orbitals by the C_3 and C_3^2 operations about our chosen axis. All we need to do is make the appropriate changes in the subscript designations of the various d orbitals. By this approach, we see that C_3 and C_3^2 effect the following transformations:

	C_3	C_3^2
$d_{xy} \rightarrow$	d_{zx}	d_{yz}
$d_{yz} \rightarrow$	d_{xy}	d_{zx}
$d_{zx} \rightarrow$	d_{yz}	d_{xy}
$d_{2z^2-x^2-y^2} \rightarrow$	$d_{2y^2-z^2-x^2}$	$d_{2x^2-y^2-z^2}$
$d_{x^2-y^2} \rightarrow$	$d_{z^2-x^2}$	$d_{y^2-z^2}$

From the first three lines of these results we readily see that d_{xy}, d_{yz}, and d_{zx} interchange among themselves and must be degenerate. In keeping with this, the corresponding direct products are listed in the O_h character table as a degenerate set transforming by T_{2g}. The degeneracy between $d_{2z^2 - x^2 - y^2}$ and $d_{x^2 - y^2}$ (last two lines) is a little less obvious from these results, because the transformed orbitals do not correspond to the conventional d orbitals with which we are most familiar. However, we can define these new orbitals as the following linear combinations of the conventional wave functions:

$$d_{2y^2-z^2-x^2} = -(1/2)d_{2z^2-x^2-y^2} - (3/2)d_{x^2-y^2} \qquad (7.1a)$$

$$d_{2x^2-y^2-z^2} = -(1/2)d_{2z^2-x^2-y^2} + (3/2)d_{x^2-y^2} \qquad (7.1b)$$

$$d_{z^2-x^2} = +(1/2)d_{2z^2-x^2-y^2} - (1/2)d_{x^2-y^2} \qquad (7.1c)$$

$$d_{y^2-z^2} = -(1/2)d_{2z^2-x^2-y^2} - (1/2)d_{x^2-y^2} \qquad (7.1d)$$

From these relationships we see that, while $d_{z^2-x^2-y^2}$ and $d_{x^2-y^2}$ are not transformed directly into each other, they are transformed into functions that are linear combinations of each other. These combinations must have the same energy as the original orbitals comprising them, so the starting orbitals must have identical energies to each other, too. Note that the transformations of the t_{2g} orbitals do not involve the e_g orbitals and vice versa. Indeed, there is no operation of O_h that relates any of the orbitals in one set to those in the other. This indicates that the two sets are not degenerate with each other and therefore should have distinct energies.

Although we have demonstrated that the d orbitals in an octahedral ML$_6$ complex divide into a triply degenerate set and a doubly degenerate set, we have not yet shown that the symmetries of these sets are t_{2g} and e_g, as indicated by the O_h character table. To verify this, we must take the five wave functions as a basis for a reducible representation in O_h and then decompose this representation into its component irreducible representations. The reducible representation can be constructed by using relationships that give the characters, $\chi[R]$, for operations in spherical symmetry (the group R_3) as a function of the angular momentum quantum number, j, of the wave function or state under consideration. These relationships can be used with O_h or any other point group, because all point groups are subgroups of R_3. As we have seen in Section 3.3, the character for a group operation that is retained in a subgroup is the same in the lower-order group. Thus, the following five relationships can be used as needed with any point group*:

$$\chi(E) = 2j + 1 \tag{7.2}$$

$$\chi[C(\theta)] = \frac{\sin(j + 1/2)\theta}{\sin \theta/2} \tag{7.3}$$

$$\chi(i) = \pm (2j + 1) \tag{7.4}$$

$$\chi[S(\theta)] = \pm \frac{\sin(j + 1/2)(\theta + \pi)}{\sin(\theta + \pi)/2} \tag{7.5}$$

$$\chi(\sigma) = \pm \sin(j + 1/2)\pi \tag{7.6}$$

In these equations, E, C, i, S, and σ indicate the operations, where θ is the rotation angle for C and S, and $\pi = 180°$. The quantum number j can be replaced by l when considering an orbital, or it can be replaced by L when considering the total orbital angular momentum of a Russell–Saunders term (cf. Section 7.4). Likewise, j can be replaced by s when considering electron spin, or it can be replaced by S when considering the spin state of a term. The variable sign ($\pm$) in Eqs. (7.4)–(7.6) is taken as $+1$ for *gerade* states and -1 for *ungerade* states. In the case of one-electron wave functions, orbitals with even-valued l (s, d, g, etc.) are *gerade* (no sign change with inversion), so the positive value is used. Orbitals with odd-valued l (p, f, h, etc.) are *ungerade* (sign changes with inversion), so the negative value is used.

Applying Eqs. (7.2)–(7.6) to the case of d orbitals ($l=2$) in an octahedral field, we obtain the following reducible representation:

O_h	E	$8C_3$	$6C_2$	$6C_4$	$3C_2$	i	$6S_4$	$8S_6$	$3\sigma_h$	$6\sigma_d$
Γ_d	5	-1	1	-1	1	5	-1	-1	1	1

This reduces, as expected, to $\Gamma_d = E_g + T_{2g}$. As we have shown by our considerations of the effects of C_3 and C_3^2, the $d_{2z^2-x^2-y^2}$ and $d_{x^2-y^2}$ orbitals are

*R. L. DeKock, A. J. Kromminga, and T. S. Zwier, *J. Chem. Educ.* **1979**, *56*, 510.

doubly degenerate, and so must constitute the e_g set. Likewise, the d_{xy}, d_{yz}, and d_{zx} orbitals were shown to be triply degenerate, and so must constitute the t_{2g} set.

It now remains for us to determine the relative energy order of the d orbitals. If a transition metal ion were placed in a spherical field equivalent to the charges on six ligands, the energies of all five d orbitals would rise together (degenerately) as a result of the repulsions between the negative charges on the ligands and the negative charges of the electrons in the metal orbitals. Now imagine localizing the ligand charges equidistant from the metal ion along the axes of a Cartesian coordinate system, an octahedral arrangement. Rearranging the charges in this manner should not cause any net change in the energy of the system. However, under the influence of the directional field produced by the octahedral environment, the metal d orbitals of the e_g set will experience greater repulsions than those of the t_{2g} set. This occurs because the lobes of the orbitals in the e_g set point directly at the ligands, while the lobes of the orbitals in the t_{2g} set point between ligands (Fig. 7.2). Relative to the energy of the hypothetical spherical field, the e_g set will rise in energy and the t_{2g} set will fall in energy (Fig. 7.3), creating an energy separation of Δ_o or $10Dq$ between the two sets of d orbitals. If the system is to maintain the same overall energy, the energy increase of the e_g orbitals and the energy decrease of the t_{2g} orbitals must be balanced relative to the energy of the hypothetical spherical field (sometimes called the barycenter).

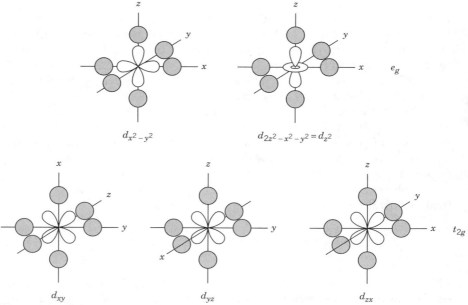

Figure 7.2   Orientations of the d orbitals relative to six octahedrally arranged ligands. (Note that axis orientations vary, so as to project orbitals in the plane of the page.)

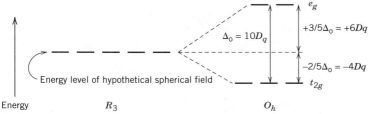

Figure 7.3   Splitting of d orbitals in an octahedral field.

Accordingly, the energy of each of the two orbitals of the e_g set rises by $+3/5\Delta_o = +6Dq$ while the energy of each of the three t_{2g} orbitals falls by $-2/5\Delta_o = -4Dq$. This results in no net energy change for the system; that is

$$\Delta E = E(e_g) + E(t_{2g}) = (2)(+3/5\Delta_o) + (3)(-2/5\Delta_o)$$

$$= (2)(+6Dq) + (3)(-4Dq)$$

$$= 0 \tag{7.7}$$

The magnitude of Δ_o depends upon both the metal ion and the attaching ligands. It increases for similar transition metal ions in successive periods (i.e., first row < second row < third row), and it increases as the charge on the metal ion increases (i.e., $M^{2+} < M^{3+}$). For the same metal ion, Δ_o increases for common ligands according to the spectrochemical series*

$$I^- < Br^- < S^{2-} < SCN^- < Cl^- < NO_3^- < F^- < OH^- < ox < H_2O <$$
$$NCS^- < CH_3CN < NH_3 < en < dipy < phen < NO_2^- < CN^- < CO$$

In the context of CFT, the spectrochemical series represents an empirical result that cannot be rationalized in terms of simple point charges. This is most evident in the case of CO, a neutral ligand that nonetheless produces the largest Δ_o splitting. The spectrochemical series can, however, be rationalized in terms of models that acknowledge orbital interactions between metal and ligands.

The ground-state electronic configurations of various transition metal ions in octahedral complexes can be determined by filling the d orbitals in the two levels according to the usual Aufbau process. Electrons are added to individual orbitals of the lower t_{2g} level and then to the upper e_g level, in keeping with Hund's rule of maximum multiplicity and the Pauli exclusion principle. For the configurations d^1, d^2, d^3, d^8, d^9, and d^{10} there is no ambiguity to the assignments (Fig. 7.4). However, for configurations d^4 through d^7 both *high spin* and *low spin* configurations are possible, as shown in Fig. 7.4. The relative magnitudes of Δ_o and the *mean pairing energy, P*, determine which spin state results in these cases. The mean pairing energy results from coulombic repulsions between electrons in the same orbital, as well as from the loss of *exchange energy* produced by distributing electrons across multiply degenerate orbitals.

---

*Abbreviations in this list: ox = oxalate, en = ethylenediamine, *dipy* = dipyradine, *phen* = o-phenanthroline.

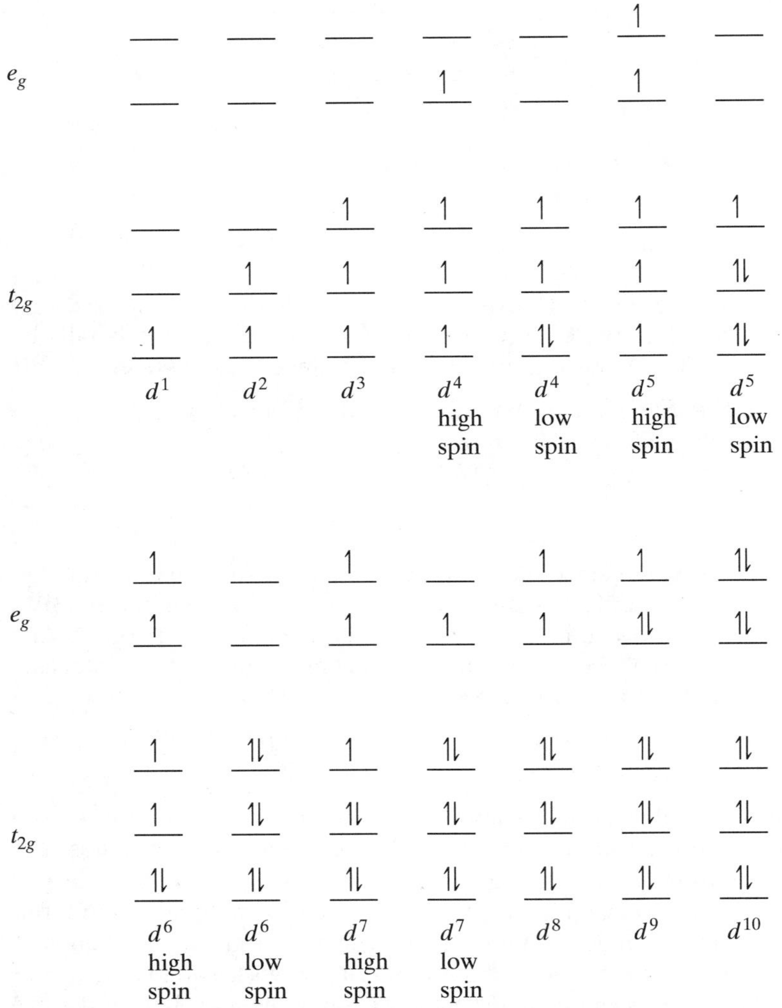

Figure 7.4 Configurations for d^1–d^{10} metal ions in an octahedral field.

The coulombic contribution to the pairing energy tends to fall off in the order $3d > 4d > 5d$, as the orbitals become larger and the electron interactions are lessened. A high spin configuration avoids pairing by spreading the electrons across both the t_{2g} and e_g levels, while a low spin configuration avoids occupying the higher energy e_g level by pairing electrons in the t_{2g} level.

In free atoms, the energy difference between subshells tends to be greater than the pairing energy. Consequently, it is usually energetically more favorable to pair electrons in the same subshell, rather than promote them to the next higher energy subshell. In contrast, the Δ_o energy gap in octahedral complexes of transition metals is relatively small and is comparable to typical pairing energies. In d^{4-7} O_h cases a weak crystal field (small Δ_o) favors the high spin configuration, and a strong crystal field (large Δ_o) favors the low spin configuration. For a given first-row transition metal ion, the magnitude of the field depends largely on the nature of the ligand—that is, where it falls in the spectrochemical series. However, second- and third-row transition metals tend to have larger Δ_o and smaller P values, which favor low spin configurations. In these heavier transition elements the more expansive $4d$ and $5d$ orbitals achieve more effective interactions with ligand orbitals, resulting in larger energy separations among them. Moreover, when two electrons occupy the same $4d$ or $5d$ orbital, their interelectronic repulsions are less than they would be in a more compact $3d$ orbital, thereby lowering the pairing energy.

We can apply similar CFT considerations to tetrahedral ML_4 complexes. As with the octahedral case, the operations of C_3 and C_3^2 make the orbitals d_{xy}, d_{yz}, and d_{zx} triply degenerate and the orbitals $d_{2z^2-x^2-y^2}$ and $d_{x^2-y^2}$ doubly degenerate. Applying Eqs. (7.2)–(7.6) shows that the two sets transform as T_2 and E, in keeping with the direct product listings in the T_d character table. Thus the d_{xy}, d_{yz}, and d_{zx} orbitals are labeled t_2, and the $d_{2z^2-x^2-y^2}$ and $d_{x^2-y^2}$ orbitals are labeled e. Unlike the octahedral case, the relative energies of the two levels are reversed. Although no d orbitals point directly at ligands, the t_2 orbitals are closer to ligands than are the e orbitals. This can be seen by comparing the orientations of the $d_{x^2-y^2}$ orbital (e set) and d_{xy} orbital (t_2 set) relative to the four ligands, as shown in Fig. 7.5. The difference results in an energy split between the two levels of Δ_t or $10Dq'$.* As Fig. 7.6 shows, the e level is lower by $-3\Delta_t/5 = -6Dq'$, and the t_2 level is higher by $+2\Delta_t/5 = +4Dq'$ relative to the barycenter defined by the hypothetical spherical field. Similar to the octahedral case, we might expect both high and low spin configurations for $d^3 - d^6$. However, Δ_t is much smaller than Δ_o. For a given ligand at the same M–L distances, it can be shown that $\Delta_t = 4\Delta_o/9$, which is much smaller than the pairing energy, P, in ordinary complexes. Thus, with extremely rare exceptions, only high spin configurations are observed.

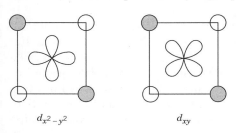

$d_{x^2-y^2}$ d_{xy}

Figure 7.5 Orientations of $d_{x^2-y^2}$ and d_{xy} orbitals relative to the four ligands forming a tetrahedral field, looking down the z axis. The plane of the projection passes through the orbitals. Shaded circles represent ligands that lie above the plane of the projection, and unshaded circles represent ligands that lie below it.

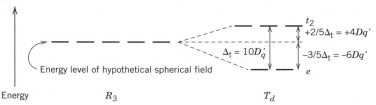

Energy R_3 T_d

Figure 7.6 Splitting of d orbitals in a tetrahedral field.

*The Δ_t energy gap is often called $10Dq$, like the octahedral case. We shall use $10Dq'$ to avoid confusion.

7.2   Jahn–Teller Distortion and Other Crystal Fields

We can deduce the CFT splitting of d orbitals in virtually any ligand field by (a) noting the direct product listings in the appropriate character table to determine the ways in which the d orbital degeneracies are lifted and (b) carrying out an analysis of the metal-ligand interelectronic repulsions produced by the complex's geometry. In a number of cases it is useful to begin with either the octahedral or tetrahedral results and consider the effects brought about by distorting the perfect geometry to bring about the new configuration. The results for the perfect and distorted geometries can then be correlated through descent in symmetry, using the appropriate correlation tables. Real situations in which such an approach might be taken include distortions produced by ligand substitution or by intermolecular associations.

Aside from these obvious cases, there is a more fundamental cause of distortion, called the *Jahn–Teller theorem*, which operates even with isolated complexes composed of only one kind of ligand.* The Jahn–Teller theorem requires that *for any nonlinear molecular system in a degenerate electronic state a distortion will occur so as to lower the symmetry and remove the degeneracy.* The theorem does not predict the exact nature of the distortion. However, *if the system is centrosymmetric, inversion symmetry will be preserved.*

A Jahn–Teller distortion results in partial or complete lifting of the degeneracies among some orbitals. In so doing, electrons may occupy lower-energy orbitals, resulting in a lower overall energy state for the system. One consequence of this is that the "perfect" geometries really cannot exist as stable species for certain electronic configurations, since the distorted molecule is the energetically preferred structure. In light of this, describing certain complexes as octahedral, tetrahedral, or square planar is often really an approximation of their true structure.

We will examine electronic states and their associated terms in more detail in Section 7.4. For the present purposes, we can identify octahedral ground-state configurations subject to the Jahn–Teller effect by considering the cases shown in Fig. 7.4. A degenerate electronic state results whenever the electrons in either the t_{2g} or e_g levels can be distributed in two or more ways among degenerate orbitals. For example, the d^1 ground-state configuration can have the single electron in any one of the three t_{2g} orbitals, so the electronic state is triply degenerate. In similar manner, we can see that with equal probability any one of the three t_{2g} orbitals could be vacant in the ground state for d^2 (t_{2g}^2), so this too is a triply degenerate state. Likewise, d^3 (t_{2g}^3) is nondegenerate, d^4 high spin ($t_{2g}^3 e_g^1$) is doubly degenerate, d^4 low spin (t_{2g}^4) is triply degenerate, and so forth.[†] By similar considerations for all the

*H. A. Jahn and E. Teller, *Proc. R. Soc.* **1937**, *A161*, 220; H. A. Jahn, *Proc. R. Soc.* **1938**, *A164*, 117.
[†]We are only concerned with the ground state here. In most cases (e.g., d^2) other states arise from a given d^n configuration. See Section 7.4.

configurations shown in Fig. 7.4, it becomes apparent that the majority represent degenerate ground states and must be distorted. Indeed, only the configurations d^3, d^5 high spin, d^6 low spin, d^8, and d^{10} will be immune to Jahn–Teller distortions in their ground states.

CFT considerations predict that distortions will be more pronounced for the doubly degenerate configurations $t_{2g}^3 e_g^1$, $t_{2g}^6 e_g^1$, and $t_{2g}^6 e_g^3$, which have an imbalance in the filling of the e_g level. Lesser distortions result from triply degenerate states, which have an imbalance in the distribution among t_{2g} orbitals (viz., t_{2g}^1, t_{2g}^2, t_{2g}^4, t_{2g}^5, $t_{2g}^4 e_g^2$, $t_{2g}^5 e_g^2$). The difference can be understood by considering shielding effects and the orientations of the t_{2g} and e_g orbitals. For example, consider the d^9 case, which is doubly degenerate as a result of the equivalence of the two specific configurations $t_{2g}^6 (d_{x^2-y^2})^2 (d_{z^2})^1$ and $t_{2g}^6 (d_{x^2-y^2})^1 (d_{z^2})^2$. In the first of these configurations, the pair of electrons in the $d_{x^2-y^2}$ orbital would more effectively shield ligands in the xy plane from the metal ion's charge than the single electron in the d_{z^2} would shield ligands along the z axis. If this were to occur, the ligands along z would be more strongly attracted to the central metal ion and their M–L bond lengths would be shortened relative to those in the xy plane. In the second configuration, the reverse condition and effect would exist. In either case, the difference in shielding would create an inequality among the M–L distances, producing a distortion from perfect octahedral geometry. Shielding effects are less pronounced for triply degenerate configurations, because the orbitals' lobes are oriented between the ligands. Thus, the resulting distortions are not as severe.

Although the exact nature of the resulting distortion cannot be predicted from the Jahn–Teller theorem, the foregoing analysis of the d^9 case suggests that a *tetragonal distortion* might result. A tetragonal distortion to an octahedron results from any change in geometry that preserves a C_4 axis. This occurs whenever two *trans*-related ligands are differentiated from the remaining four. For example, a tetragonal distortion would occur if the M–L bonds of two ligands lying along the z axis were either stretched or compressed equally while maintaining equivalence among the four remaining ligands in the xy plane (Fig. 7.7).* By either process, the symmetry would descend from O_h to D_{4h}. The descent in symmetry causes a partial lifting of the degeneracies among the d orbitals in the octahedral field. From the correlation table that links the groups O_h and D_{4h} (Appendix B) we see that the two e_g orbitals of the octahedral field become nondegenerate as a_{1g} and b_{1g} in the tetragonal field. From the direct product listings in the D_{4h} character table (Appendix A) we see that the a_{1g} orbital is $d_{2z^2-x^2-y^2}$ (d_{z^2}), and the b_{1g} orbital is $d_{x^2-y^2}$ in D_{4h}. Similarly, the correlation table shows that the degeneracy among the t_{2g} orbitals in O_h is partially lifted to become b_{2g} and e_g in the D_{4h} tetragonal field. As the D_{4h} character table indicates, the b_{2g} orbital is d_{xy} and the e_g orbitals are d_{xz} and d_{yz}.

*This is only one of several possible tetragonal distortions. For example, the two M–L bonds along z could be changed nonequivalently, in which case the symmetry would descend to C_{4v}. This, however, is not a centrosymmetric group, so this type of distortion would not result from the Jahn–Teller effect.

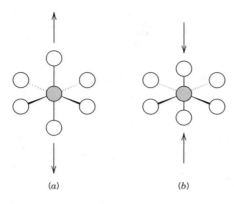

Figure 7.7 Two examples of a tetragonal distortion on an octahedral complex: (*a*) a stretch along z, and (*b*) a compression along z.

The relative energy ordering of the orbitals depends on the direction and magnitude of the tetragonal distortion. A distortion in which the two M–L bonds along z are progressively stretched is an interesting case to consider, because at its limit the two ligands would be removed, resulting in a square planar ML_4 complex. Moving the two ligands away from the central metal ion lowers the repulsions between ligand electrons and the metal electrons in d orbitals that have substantial electron distribution along z. Thus the energies of the d_{xz}, d_{yz}, and d_{z^2} orbitals are lowered. If we assume that the stretch along z is accompanied by a counterbalancing contraction in the xy plane, so as to maintain the overall energy of the system, then the orbitals with substantial electron distribution in the xy plane will experience increased repulsions. Thus, the d_{xy} and $d_{x^2-y^2}$ orbitals rise in energy. As a result, the degeneracies among the t_{2g} and e_g levels of the octahedral field are lifted in the manner shown in Fig. 7.8. The upper e_g orbitals of the perfect octahedron

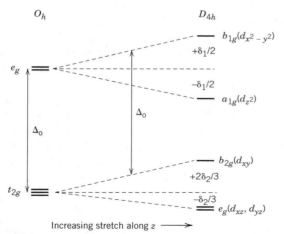

Figure 7.8 Crystal field effects of a tetragonal distortion on an octahedral ML_6 complex, deformed by stretching the two M–L bonds along z.

split equally by an amount δ_1, with the $d_{x^2-y^2}$ orbital (b_{1g} in D_{4h}) rising by $+\delta_1/2$ and the d_{z^2} orbital (a_{1g} in D_{4h}) falling by $-\delta_1/2$. The lower t_{2g} orbitals of the perfect octahedron split by an amount δ_2, with the d_{xy} orbital (b_{2g} in D_{4h}) rising by $+2\delta_2/3$ and the degenerate d_{xz} and d_{yz} orbitals (e_g in D_{4h}) falling by $-\delta_2/3$. Both the δ_1 and δ_2 splittings, which are very small compared to Δ_o, maintain the barycenters defined by the e_g and t_{2g} levels of the undistorted octahedron. The energy gap δ_1 is larger than that of δ_2, since the $d_{x^2-y^2}$ and d_{z^2} orbitals are directed at ligands. Note that the distortion has the same effect on the energies of both the $d_{x^2-y^2}$ and d_{xy} orbitals; that is, $\delta_1/2 = 2\delta_2/3$. As a result, their energies rise in parallel, maintaining a separation equal to the Δ_o of the undistorted octahedral field.

If we carry out the opposite tetragonal distortion (compression along z), the octahedral degeneracies will be lifted in the same manner, as required by symmetry, but the ordering of the orbitals across both the δ_1 and δ_2 gaps will be reversed. The energy of the $d_{x^2-y^2}$ orbital (b_{1g}) will fall by $-\delta_1/2$, and the energy of the d_{z^2} orbital (a_{1g}) will rise by $+\delta_1/2$. Likewise, the energy of the d_{xy} orbital (b_{2g}) will fall by $-2\delta_2/3$, and the energy of the d_{xz} and d_{yz} orbitals (e_g) will rise by $+\delta_2/3$. In this case, the energy of the d_{xy} (b_{2g}) and $d_{x^2-y^2}$ (b_{1g}) orbitals will fall equally with increasing compression along z (i.e., $-\delta_1/2 = -2\delta_2/3$), maintaining a separation equal to Δ_o.

If we imagine continuing the stretching of M–L bonds along z, the orbital splittings will become progressively greater, producing successively larger values of δ_1 and δ_2. Eventually the two ligands will be removed, resulting in a square planar ML_4 complex. At some point before this extreme the a_{1g} (d_{z^2}) level may cross and fall below the b_{2g} (d_{xy}) level, resulting in the splitting scheme shown in Fig. 7.9, the orbital energy level scheme for a square planar complex.* Most square planar complexes are d^8 and less often d^9. In virtually

---

*The ordering of the lower four d orbitals probably varies among square planar complexes and has been the subject of much debate. See A. B. P. Lever, *Inorganic Electron Spectroscopy*, 2nd ed., Elsevier, Amsterdam, 1984, p. 537*ff.* and references therein.

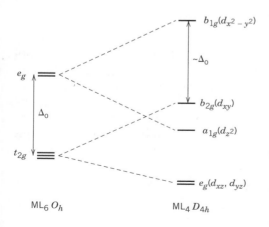

Figure 7.9 Crystal field splitting of d orbitals for a square planar ML_4 complex and its relationship to the splitting for an octahedral ML_6 complex.

all d^8 cases a low spin configuration is observed, leaving the upper b_{1g} ($d_{x^2-y^2}$) level vacant in the ground state. This is expected, since square planar geometry in first-row transition metal ions is usually forced by strong field ligands. Recalling that the energy gap between the b_{2g} (d_{xy}) and b_{1g} ($d_{x^2-y^2}$) levels is equivalent to Δ_o, we would expect strong field ligands to produce a large Δ_o value, which would favor a low-spin configuration. For example, Ni^{2+} ion tends to form square planar, diamagnetic complexes with strong-field ligands (e.g., $[Ni(CN)_4]^{2-}$), but tends to form tetrahedral, paramagnetic complexes with the weaker-field ligands (e.g., $[Ni(Cl)_4]^{2-}$). With second and third row transition metal ions the Δ_o energies are inherently larger, and square planar geometry can occur even with relatively weak field ligands (e.g., square planar $[PtCl_4]^{2-}$).

7.3 Molecular Orbital Approach to Bonding in Complexes

The quantitative predictions of CFT, which are based on a purely electrostatic model, require empirical corrections in order to give satisfactory agreement with experimental results. With these corrections the model is known as Modified Crystal Field Theory, or more commonly *Ligand Field Theory* (LFT). The need for corrections to CFT arises from metal–ligand orbital overlap, which implies a certain amount of covalence in the M–L interactions. One manifestation of this is the observation from absorption spectra that there is less repulsion between d electrons in a complex ion than in the free gaseous ion. Covalent interaction with ligands allows metal electrons to be delocalized onto the ligands, lessening repulsions. In effect, taking a CFT view, the d orbitals have been "expanded" by the presence of the ligands. This is the so-called *nephelauxetic effect* (Greek, *nephelē* = cloud + *auxēsis* = growth; hence, "cloud-expanding"), which depends upon both the metal ion and ligand. For a given metal ion, the ability of ligands to induce this cloud expanding increases according to a *nephelauxetic series**:

$$F^- < H_2O < NH_3 < en < ox < SCN^- < Cl^- < CN^- < Br^- < I^-$$

By using empirically determined constants for both ligands and the central metal ion, it is possible to reconcile the ligand field model of a complex with quantitative spectroscopic results. Discussion of these techniques is beyond the intent of this chapter and can be found in some advanced inorganic chemistry texts.[†] For our purposes we would simply note that the need to modify CFT to account for the nephelauxetic effect suggests that a molecular orbital approach might be useful. Such an MO model could be adjusted for various degrees of M–L orbital overlap, representing a range from polar covalent

---

*Note that the ordering of ligands in the nephelauxetic series is not the same as the spectrochemical series.
[†]For example, see D. F. Shriver, P. Atkins, and C. H. Langford, *Inorganic Chemistry*, 2nd ed., W. H. Freeman, New York, 1994, pp. 585–595.

bonding to nearly ionic interactions. Furthermore, an MO approach might allow us to understand the relationship between orbital overlap and the energy separations among d orbitals in fields of various geometries.

We will only illustrate the molecular orbital approach for octahedral ML_6 and tetrahedral ML_4 complexes, but of course the methodology can be applied to any complex coordination or geometry. We will first examine the MO model for ML_6 complexes that have only sigma metal–ligand interactions, such as $M(NH_3)_6^{n+}$ complexes of first-row transition metal ions with $+2$ or $+3$ charges. Unlike the CFT and LFT models, we will include consideration of interactions with metal ion s and p orbitals. Once we have constructed our MO model, we will be interested to see how it accounts for what in the CFT and LFT approaches is the Δ_o separation between the d orbitals and of course the nephelauxetic effect.

For the case of six sigma-bonding ligands, we take six vectors pointing toward the center of a Cartesian coordinate system as our basis for a reducible representation of SALCs, Γ_σ (Fig. 7.10). By ascribing a positive unit contribution for each vector that remains nonshifted by any operation of a class, we obtain the following representation:

O_h	E	$8C_3$	$6C_2$	$6C_4$	$3C_2(=C_4^2)$	i	$6S_4$	$8S_6$	$3\sigma_h$	$6\sigma_d$
Γ_σ	6	0	0	2	2	0	0	0	4	2

This reduces as $\Gamma_\sigma = A_{1g} + E_g + T_{1u}$. Thus, we can define six SALCs with three different symmetries, which can form bonding and antibonding combinations with like symmetry AOs on the central metal ion.

To identify the symmetries of metal ion AOs we recall that an s orbital transforms as the totally symmetric representation, p orbitals transform as the three unit vectors, and d orbitals transform as their matching direct products. Thus, in O_h we have

$$s = a_{1g}$$

$$p_x, p_y, p_z = t_{1u}$$

$$d_{x^2-y^2}, d_{z^2} = e_g$$

$$d_{xy}, d_{xz}, d_{yz} = t_{2g}$$

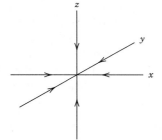

Figure 7.10 Vector basis for a representation of six sigma-bonding SALCs for an octahedral ML_6 complex.

The symmetries of the d orbitals are, of course, the same as noted in our considerations of CFT. Comparing these symmetries with those of the SALCs we conclude that the s, p_x, p_y, p_z, $d_{x^2-y^2}$, and d_{z^2} orbitals have the proper symmetries to form bonding and antibonding combinations with matching symmetry SALCs. The three t_{2g} orbitals (d_{xy}, d_{xz}, d_{yz}), however, have no matching SALCs and must remain nonbonding. This is a consequence of the orientation of these orbitals relative to the ligands. We define the positions of the ligands as lying along the Cartesian axes, so the lobes of the t_{2g} orbitals lie between the axes, precluding effective overlap with ligand sigma orbitals.

Recognizing the matches with AOs, we can write the six SALCs in the following forms, in which the six ligand sigma orbitals (σ) are identified by their positions on the Cartesian coordinates.* The a_{1g} SALC, which matches with the metal ns orbital, is

$$\Sigma_a = \frac{1}{\sqrt{6}}(\sigma_x + \sigma_{-x} + \sigma_y + \sigma_{-y} + \sigma_z + \sigma_{-z}) \tag{7.8}$$

The two e_g SALCs, which match with $(n-1)d_{z^2}$ and $(n-1)d_{x^2-y^2}$ orbitals, are

$$\Sigma_{z^2} = \frac{1}{2\sqrt{3}}(2\sigma_z + 2\sigma_{-z} - \sigma_x - \sigma_{-x} - \sigma_y - \sigma_{-y}) \tag{7.9a}$$

$$\Sigma_{x^2-y^2} = \frac{1}{2}(\sigma_x + \sigma_{-x} - \sigma_y - \sigma_{-y}) \tag{7.9b}$$

The three t_{1u} SALCs, which match with the np_z, np_x, and np_y orbitals, respectively, are

$$\Sigma_z = \frac{1}{\sqrt{2}}(\sigma_z - \sigma_{-z}) \tag{7.10a}$$

$$\Sigma_x = \frac{1}{\sqrt{2}}(\sigma_x - \sigma_{-x}) \tag{7.10b}$$

$$\Sigma_y = \frac{1}{\sqrt{2}}(\sigma_y - \sigma_{-y}) \tag{7.10c}$$

These SALCs and their matching AOs are shown in Fig. 7.11.

Figure 7.12 shows the resulting qualitative MO scheme for an octahedral complex with only sigma interactions between the metal ion and ligands. This scheme should be regarded as only an approximation for real complexes, and the order and nature of the MOs may differ in individual cases. Nonetheless, this scheme is sufficient for our purposes. Note that the 12 electrons provided by the ligands alone are sufficient to fill the lowest three levels of MOs (a_{1g}, t_{1u}, and e_g). Any electrons provided by the metal ion will result in an equivalent filling of the t_{2g} level and if necessary the e_g level. Thus, the electron filling above the six MOs in the lowest three levels is identical to the presumed filling

---

*The mathematical expressions for the SALCs can be determined either by a pictorial analysis or by applying the techniques of projection operators, as described in detail in Section 5.1.

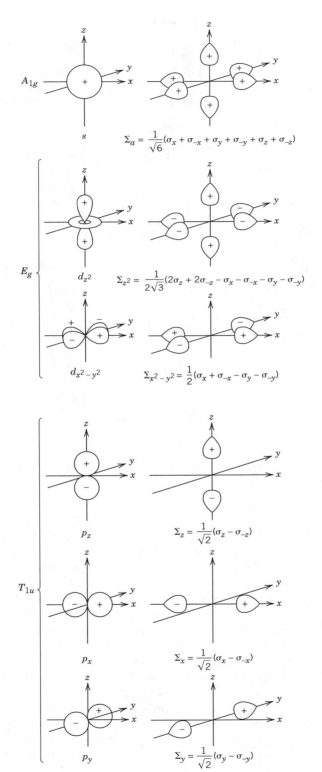

$$A_{1g} \qquad s \qquad \Sigma_a = \frac{1}{\sqrt{6}}(\sigma_x + \sigma_{-x} + \sigma_y + \sigma_{-y} + \sigma_z + \sigma_{-z})$$

$$E_g \qquad d_{z^2} \qquad \Sigma_{z^2} = \frac{1}{2\sqrt{3}}(2\sigma_z + 2\sigma_{-z} - \sigma_x - \sigma_{-x} - \sigma_y - \sigma_{-y})$$

$$d_{x^2-y^2} \qquad \Sigma_{x^2-y^2} = \frac{1}{2}(\sigma_x + \sigma_{-x} - \sigma_y - \sigma_{-y})$$

$$T_{1u} \qquad p_z \qquad \Sigma_z = \frac{1}{\sqrt{2}}(\sigma_z - \sigma_{-z})$$

$$p_x \qquad \Sigma_x = \frac{1}{\sqrt{2}}(\sigma_x - \sigma_{-x})$$

$$p_y \qquad \Sigma_y = \frac{1}{\sqrt{2}}(\sigma_y - \sigma_{-y})$$

Figure 7.11 Ligand SALCs and their matching symmetry metal ion AOs.

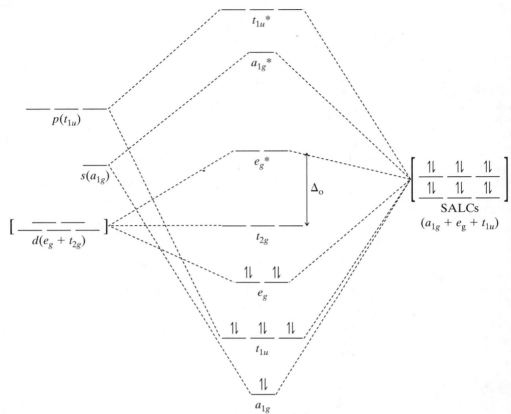

Figure 7.12 Molecular orbital diagram for an octahedral complex with only sigma bonding between metal ion and ligands.

of d orbitals in the CFT model. As with the CFT model, both high- and low-spin ground states are possible for d^4 through d^7 metal ion configurations. In the MO scheme Δ_o or $10Dq$ is defined as the energy separation between the t_{2g} and e_g^* levels. The lower t_{2g} orbitals are nonbonding and can be taken as essentially the d_{xy}, d_{xz}, and d_{yz} orbitals of the metal ion, which is not materially different from the CFT view. However, the upper e_g^* orbitals are now seen as antibonding molecular orbitals. These are combinations of the d_{z^2} and $d_{x^2-y^2}$ metal ion orbitals with the matching ligand SALCs of Eqs. (7.9a) and (7.9b), taken with a negative sign. Although antibonding, the e_g^* MOs when occupied involve sharing of electron density between the metal ion and the ligands.

We can make allowances for varying degrees of covalent interaction between the metal ion and ligands by adjusting the MO scheme in Fig. 7.12. Our primary interest is in interactions with the metal ion d orbitals. Without admitting the possibility of pi-bonding, no adjustment of the scheme can change the localized character of the t_{2g} orbitals. However, electrons occupying the

e_g* MOs will have more or less delocalization onto the ligands depending upon the relative energies of the metal ion d orbitals and the ligand sigma orbitals. If metal d orbitals lie higher in energy than ligand sigma orbitals (Fig. 7.13a), the e_g* MOs will lie closer in energy to the metal d orbitals and have more metal ion character than ligand character. In this case, e_g* electron density will be more localized on the metal. If the disparity in levels is extreme, this becomes an ionic model in which the e_g* MOs are essentially metal d orbitals, like the CFT approach. Thus, the CFT model is a special case in the MO approach. As the energies of the metal ion d orbitals and the ligand sigma orbitals become more comparable the degree of electron sharing (covalence) will become greater. More of the e_g* electron density will be delocalized toward the ligands (Fig. 7.13b). If the ligand sigma orbitals were to lie significantly higher than the metal ion d orbitals, e_g* electron density would be predominantly localized on the ligands (Fig. 7.13c). We note that the weakest ligands in the nephelauxetic series (F^-, H_2O, and NH_3) have low-energy atomic or molecular orbitals relative to transition metal ion d orbitals. This is more in keeping with the situation portrayed in Fig. 7.13a. Thus, for complexes with these ligands, both t_{2g} and e_g* electron density is essentially localized in metal d orbitals, not unlike the assumptions of the CFT model.*

In principle, an MO approach such as we have just described gives a fundamentally more complete view of the bonding. We can appreciate that the im-

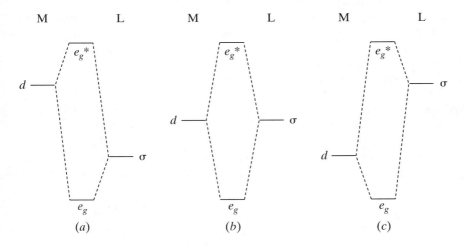

Increasing delocalization of e$_g$*electron density toward ligands $\longrightarrow$

Figure 7.13   Relationship between the relative energies of metal ion d orbitals and ligand sigma orbitals to the delocalization of e_g* electrons toward either metal ion or ligands.

---

*As we shall see shortly, F^- has occupied $2p$ orbitals with appropriate symmetry for pi-bonding with t_{2g} metal ion d orbitals. This does not essentially alter our analysis of why F^- is a weak nephelauxetic effect ligand.

plied covalence of the occupied bonding and antibonding MOs could account for the discrepancies between CFT predictions about d orbital repulsions and the experimental results. However, adjusting the scheme to reflect the quantitative realities of the complex is far from trivial. If our interest is more qualitative—say, to understand the magnetic properties or visible spectra—the MO model does not really add any information that is not already available from CFT. To understand these properties we only need to concern ourselves with the electronic configuration in the t_{2g} and e_g orbitals, which is precisely the concern of the CFT model. CFT just ignores the lower filled MO levels, which have little relevance to these problems. In the region defining Δ_o the two approaches are equivalent. The difference lies in how we view the nature of the orbitals: metal ion d orbitals in the CFT approach versus whole complex molecular orbitals in the MO approach. As chemists, we should be glad that this essential equivalence exists, because it allows us to use the inherently simpler CFT model for routine work, without the need to resort to the complications of the more "correct" and quantitatively capable MO approach.

This kind of equivalence between the MO and CFT models persists even if the ligands are capable of pi interactions. However, in these cases the MO approach has the advantage of allowing us to understand the ability of some strong-field ligands, especially those that are neutral, to produce large Δ_o splittings and (where possible) low-spin configurations. Given that the limited sigma-only case presents computational difficulties, we can anticipate that the situation will be even more complicated if we include pi-interactions. Nonetheless, it is useful to consider at least qualitatively the implications of pi bonding between the metal ion and ligands to see how they relate to the simplifications of the CFT model.

To include pi-bonding in our MO scheme for octahedral ML_6 complexes we use the 12 vectors shown in Fig. 7.14 as a basis for a representation of SALCs. These vectors might indicate occupied p orbitals (other than those engaged in sigma bonding), such as the np_x and np_y orbitals on halide ligands in complexes like CrX_6^{3-} ($X = F^-$, Cl^-). These are classified as donor ligands, because they have electrons to contribute to the pi system of the complex. Alternately, the vectors might indicate other unoccupied pi symmetry

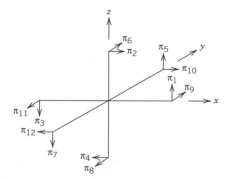

Figure 7.14   Vector basis for a representation of octahedral π-SALCs.

AOs or MOs on the ligands, such the empty π^* antibonding MOs of CO and CN^- in complexes like $Cr(CO)_6$ and $[Fe(CN)_6]^{4-}$. Since such ligands receive electron density from the pi system, they are classified as acceptor ligands.

To determine the characters of our reducible representation of pi symmetry SALCs, we ascribe a +1 contribution for each nonshifted vector, a −1 contribution for each vector transformed into the negative of itself, and a zero contribution for all vectors moved off their original positions by any symmetry operation. This gives the following reducible representation:

O_h	E	$8C_3$	$6C_2$	$6C_4$	$3C_2(=C_4^2)$	i	$6S_4$	$8S_6$	$3\sigma_h$	$6\sigma_d$
Γ_π	12	0	0	0	−4	0	0	0	0	0

This reduces as $\Gamma_\pi = T_{1g} + T_{2g} + T_{1u} + T_{2u}$.

Recalling the symmetries we identified for the AOs on the central metal ion in the sigma-only case, we recognize that now we can form pi-bonding and antibonding combinations between the t_{2g} orbitals (d_{xy}, d_{xz}, d_{yz}) and the π-SALCs of the same symmetry. This will change the character of the t_{2g} level, which we previously had identified as nonbonding in the sigma-only MO scheme (Fig. 7.12). In addition, we have three T_{1u} SALCs, which on the basis of their symmetry match with the three np orbitals on the metal ion could form π-MOs. However, we have already used these metal ion np AOs to form bonding and antibonding σ-MOs with the T_{1u} σ-SALCs (Figs. 7.11 and 7.12). The sigma interactions are likely to result in more effective overlaps, so we will assume that the np orbitals have only minimally effective interactions with the T_{1u} π-SALCs. The T_{1u} π-SALCs, then, will be virtually nonbonding in most cases, although they might be weakly bonding in certain complexes. The remaining six π-SALCs with T_{1g} and T_{2u} symmetry have no matching AOs on the metal ion. Therefore, these SALCs must be strictly nonbonding.

We are naturally most interested in the T_{2g} symmetry π-SALCs, which form bonding and antibonding combinations with the t_{2g} metal ion d orbitals. Using the notation in Fig. 7.14, these SALCs have the following forms:

$$\Pi_{xz} = \frac{1}{2}(\pi_1 + \pi_2 + \pi_3 + \pi_4) \qquad (7.11a)$$

$$\Pi_{yz} = \frac{1}{2}(\pi_5 + \pi_6 + \pi_7 + \pi_8) \qquad (7.11b)$$

$$\Pi_{xy} = \frac{1}{2}(\pi_9 + \pi_{10} + \pi_{11} + \pi_{12}) \qquad (7.11c)$$

Π_{xz}, Π_{yz}, and Π_{xy} form bonding and antibonding combinations with d_{xz}, d_{yz}, and d_{xy} AOs, respectively. Figure 7.15 shows the form of the LCAOs for the bonding combinations.

The other potentially bonding π-SALCs, with T_{1u} symmetry, have the following forms:

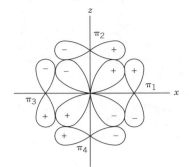

Figure 7.15   Bonding π-LCAO formed by d_{xz} with the Π_{xz} SALC. Similar combinations occur for d_{yz} with Π_{yz} and d_{xy} with Π_{xy}.

$$\Pi_z = \frac{1}{2}(\pi_1 - \pi_3 + \pi_5 - \pi_7) \tag{7.12a}$$

$$\Pi_x = \frac{1}{2}(\pi_2 - \pi_4 + \pi_{10} - \pi_{12}) \tag{7.12b}$$

$$\Pi_y = \frac{1}{2}(\pi_6 - \pi_8 + \pi_9 - \pi_{11}) \tag{7.12c}$$

where Π_z, Π_x, and Π_y have the proper combinations of ligand AOs to form interactions with metal ion np_z, np_x, and np_y AOs, respectively. As noted above, these SALCs are less likely to have as effective overlaps with the metal np AOs as the σ-SALCs of the same symmetry [Eqs. (7.10)]. This can be seen by comparing the LCAOs for pi-bonding and sigma-bonding formed with the same metal np orbital (Fig. 7.16).*

If we consider expanding our sigma-only MO scheme to include the pi-bonding, pi-antibonding, and nonbonding interactions we have identified, we immediately recognize that the task is fraught with difficulties and uncertainties. The identities of the central metal ion and the ligand, the relative energies of the orbitals on each, the nature and effectiveness of their sigma and

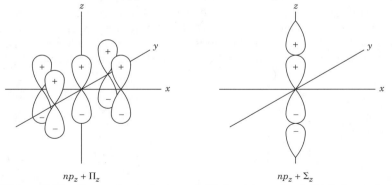

$np_z + \Pi_z$ $np_z + \Sigma_z$

Figure 7.16   Comparison of a t_{1u} π-LCAO with a t_{1u} σ-LCAO formed with the same metal np_z orbital. The sigma combination results in more effective overlap. Similar LCAOs are formed with np_x and np_y.

*The expressions for the nonbonding T_{1g} and T_{2u} SALCs should be apparent from the symmetry relationships indicated by their Mulliken symbols (or the characters of the irreducible representations themselves), and verification is left to you as an exercise.

pi orbital interactions, and even the electron filling in ligand orbitals will all have effects in determining the energies and bonding characters of the molecular orbitals. Thus, it is not possible to construct a detailed MO scheme that will have general applicability to a range of octahedral complexes. The best we can hope for is a simplified scheme that identifies interacting orbitals by symmetry type, approximates their bonding type, and arranges MOs of the same type in a plausible relative energy order. Figure 7.17 shows such a scheme. Note that this scheme makes no attempt to distinguish between the

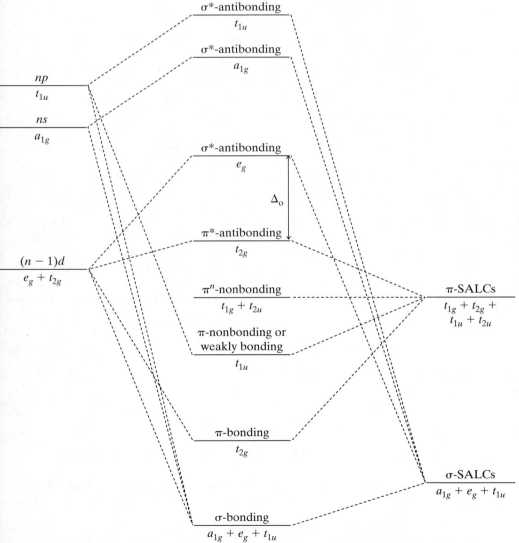

Figure 7.17   Simplified qualitative MO scheme for an octahedral ML_6 complex with pi-bonding.

energies of same-type orbitals with different symmetries. In spite of its limitations, this scheme provides us with a starting point for understanding the relationships between the CFT and MO approaches to pi-bonded complexes.

To illustrate the filling of electrons into the scheme of Fig. 7.17, let us consider the complex CrF_6^{3-}. The Cr^{3+} ion has a d^3 configuration, and therefore it supplies three electrons. Assuming that the $2s$ electrons are nonbonding, each F^- ion supplies six electrons, making a total of 36 electrons from ligands. Thus, we should fill our scheme with 39 electrons. Thirty-six electrons are sufficient to fill all levels through the nonbonding t_{1g} and t_{2u} MOs. The remaining three electrons occupy individual t_{2g} π^*-MOs, resulting in a configuration $(t_{2g}^*)^3$, equivalent to the CFT model's configuration t_{2g}^3. On the basis of this MO scheme, Δ_o is defined as the energy gap between the pi-antibonding t_{2g}^* level and the sigma-antibonding e_g^* level. The energies of the t_{2g}^* and e_g^* levels will be sensitive to differences in the effectiveness of metal–ligand pi and sigma interactions, respectively. Hence, the interplay between sigma- and pi-bonding strength affects the magnitude of Δ_o. Likewise, the relative abilities of a ligand to engage in these modes of bonding are important factors in determining its position in the spectrochemical series.*

Tetrahedral ML_4 complexes may also involve both sigma and pi metal–ligand bonding. To set up the problem we assume that each of the ligands possesses one or more sigma orbitals directed at the central metal ion and pairs of pi orbitals perpendicular to the M–L bond axis. Let us assume that the ligands are monatomic ions, such as halide ions, which could use ns and np_z orbitals for sigma interactions and np_x and np_y orbitals for pi interactions with the metal ion $(n-1)d$, ns, and np orbitals. For simplicity we will assume that the ligand ns orbitals are essentially nonbonding and that only the np orbitals have significant overlap with the metal ion orbitals. Before proceeding to the determination of symmetries of SALCs, it will be useful to recognize the symmetries of the AOs on the central metal atom. From the T_d character table we have

$$s = a_1$$

$$p_x, p_y, p_z = t_2$$

$$d_{x^2-y^2}, d_{z^2} = e$$

$$d_{xy}, d_{xz}, d_{yz} = t_2$$

Once again, the symmetries of the d orbitals are the same as we noted in the CFT approach.

The vector basis for a reducible representation of σ-SALCs is identical to that we considered in the case of methane (Fig. 4.15). Thus the resulting

---

*The relationships between ligand sigma donor, pi donor, and pi acceptor abilities and the magnitude of Δ_o are more fully discussed in many advanced inorganic chemistry texts. For example, see B. E. Douglas, D. H. McDaniel, and J. J. Alexander, *Concepts and Models of Inorganic Chemistry*, 3rd ed., John Wiley & Sons, New York, 1994, pp. 471–472.

representation and its decomposition are the same; that is, $\Gamma_\sigma = A_1 + T_2$. In the case of methane, the SALCs were formed as various combinations of hydrogen $1s$ wave functions [Eqs. (4.21a)–(4.21d)]. For ML_4 we obtain a similar set of expressions using p_z orbitals. The A_1 σ-SALC has appropriate symmetry to form sigma combinations with metal ns orbitals, although the effectiveness of the overlap may be limited. The T_2 σ-SALCs have appropriate symmetry to form sigma combinations with np_z, np_y, and np_x orbitals on the metal ion. However, the d_{xz}, d_{yz}, and d_{xy} orbitals also have T_2 symmetry and can likewise form combinations with these SALCs. From this we can anticipate that there may be some degree of d–p mixing in the t_2 σ-MOs. In constructing our MO scheme we will assume, for simplicity, that the t_2 σ-MOs are formed principally with the metal np orbitals, although d–p mixing may be appreciable in specific complexes.

The vector basis for a representation of π-SALCs is shown in Fig. 7.18. A pair of mutually perpendicular vectors is located at each ligand, oriented at right angles to the M–L bond axis, for a total of eight vectors. All the operations of T_d, except identity and the threefold rotations, move the vectors off their positions, resulting in zero characters. In the case of C_3 or C_3^2 about any one of the M–L bond axes, the positions of the two vectors on the ligand become intermixed. This is similar to what we saw for x and y components of a general vector in Section 2.4 [cf. Eq. (2.16)]. Assuming a clockwise rotation, the positions of the two vectors after a C_3 rotation are described by the expression

$$\begin{bmatrix} -1/2 & -\sqrt{3}/2 \\ \sqrt{3}/2 & -1/2 \end{bmatrix} \begin{bmatrix} x \\ y \end{bmatrix} = \begin{bmatrix} x' \\ y' \end{bmatrix} \tag{7.13}$$

From the operator matrix we obtain the character -1. With this result the reducible representation for π-SALCs is

T_d	E	$8C_3$	$3C_2$	$6S_4$	$6\sigma_d$
Γ_π	8	-1	0	0	0

This reduces as $\Gamma_\pi = E + T_1 + T_2$. The T_1 SALCs have no match in metal atom AOs and will be nonbonding. The E SALCs will form pi combinations

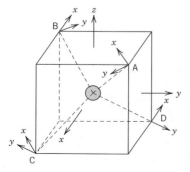

Figure 7.18 Vector basis for a representation of π-SALCs for a tetrahedral ML_4 complex. The x and y vector orientations on the ligands are defined by taking the M–L axes as individual z axes. The x, y, and z axes of the tetrahedron are defined in the conventional manner, relative to the symmetry elements of T_d.

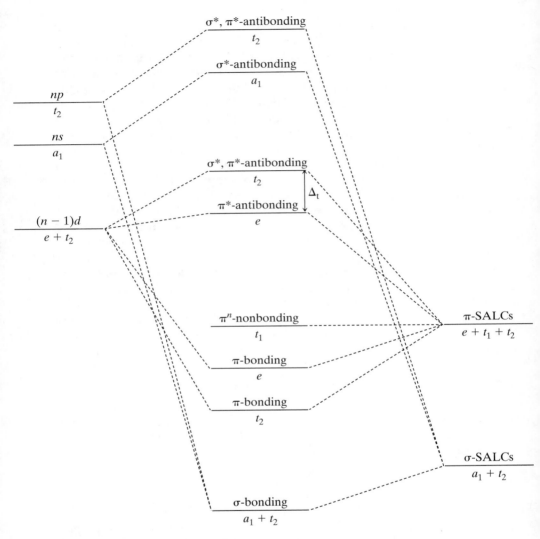

Figure 7.19 Simplified qualitative MO scheme for a tetrahedral ML$_4$ complex.

with the $d_{x^2-y^2}$ and d_{z^2} orbitals on the metal atom. The T_2 π-SALCs, like the σ-SALCs of the same symmetry, can potentially form combinations with both t_2 $(n-1)d$ and np orbitals on the metal atom. Once again, the MOs that are formed may involve some degree of d–p mixing. Since we have assumed that the t_2 σ-MOs mainly use the np orbitals, we will assume in similar manner that the t_2 π-MOs are formed principally with the metal $(n-1)d$ orbitals; that is, d_{xy}, d_{xz}, d_{yz}. Nonetheless, the distinction between t_2 σ-MOs and t_2 π-MOs is not as clean as we might like. None of the metal t_{2g} orbitals is directed at ligands (the ideal orientation in sigma-bonding), nor is any one oriented at right angles to the bond axis (the ideal orientation in pi-bonding). Therefore, each type of MO has some of the character of the other type in this case. We shall assume that the bonding t_{2g} MOs are essentially either sigma or pi, and that the mixing is more pronounced in the antibonding MOs.

The preceding considerations enable us to construct the simplified qualitative MO scheme shown in Fig. 7.19. As with the similar scheme for the octahedral case (Fig. 7.17), no attempt has been made to distinguish energies between MOs of the same bonding type, and the ordering of levels is only meant to be suggestive of a plausible arrangement. The nature and ordering of MOs will depend upon the peculiarities of the complex in question.

Regardless of the ordering of lower lying levels, a scheme such as Fig. 7.19 allows us to see the equivalence of the MO approach with the CFT model. Suppose we fill the scheme with the appropriate number of electrons for a complex such as $NiCl_4^{2-}$. The four Cl^- ligands supply six electrons each, for a total of 24. Since Ni^{2+} is a d^8 ion, the total number of electrons is 32. Twenty-four electrons will fill all lower levels through the t_1 nonbonding level in our scheme. The remaining eight electrons will fill the antibonding e and t_2 levels, giving a configuration $(e*)^4(t_2*)^4$, which is paramagnetic owing to two unpaired electrons in the upper t_2* orbitals. This is equivalent to the CFT configuration $e^4t_2^4$. Also similar to the CFT model, Δ_t is defined in the MO approach as the energy separation between the antibonding $e*$ and t_2* MOs. Thus, like the octahedral case, the essential parameters of the CFT model are similarly defined in the MO model.

7.4 Terms of Free Ions with d^n Configurations

Our usual notation for electronic configurations simply indicates the number of electrons that occupy particular sets of degenerate orbitals. Thus, we usually do not presume to know which specific orbitals the electrons are occupying at any time, except when a degenerate set of orbitals is half-filled or fully filled. Furthermore, except for a fully filled subshell, we rarely presume to know which of the two possible spin states individual electrons have. For example, we may know that two electrons in a d subshell have parallel spins (the ground-state configuration of d^2), but we cannot know which orbitals they may occupy at any time nor whether the orientations of their spins are both $m_s = +1/2$ or $m_s = -1/2$. Of course, this does not prevent us from identifying and cataloging all the possible pairs of m_l and m_s values each of the electrons in the configuration might have, consistent with the Pauli exclusion principle. When we do this, each conceivable set of individual m_l and m_s values constitutes a *microstate* of the configuration. Some of these microstates may be allowable arrangements in the ground state, and others may be allowable arrangements in some higher-energy excited state.

For example, in the case of a single electron in a degenerate set of five d orbitals (nd^1), the electron can have any of the values $m_l = +2, +1, 0, -1, -2$ and either of the values $m_s = \pm 1/2$. Thus, there are 10 ways of arranging the electron with a particular spin and orbital assignment, making a total of 10 microstates. In this very simple case all microstates are part of the ground-state configuration. With additional electrons the number of microstates rises dramatically, and some will be associated with different energy states of the configuration.

In general, for any allowed number of electrons in a set of degenerate orbitals (called *equivalent electrons**), the number of possible microstates is given by

$$D_t = \frac{(2N_o)!}{(2N_o - N_e)!N_e!} \tag{7.14}$$

where D_t is the number of possible microstates, called the *total degeneracy* of the configuration; N_o is the number of degenerate orbitals in the set or sub-shell; and N_e is the number of electrons in the configuration. Equation (7.14) predicts the following numbers of microstates for nd^{1-10} configurations of a free transition metal ion[†]:

Configuration	d^1	d^2	d^3	d^4	d^5	d^6	d^7	d^8	d^9	d^{10}
Microstates	10	45	120	210	252	210	120	45	10	1

In any microstate both the individual orbital magnetic moments (related to m_l) and spin magnetic moments (related to m_s) will interact with one another, resulting in an energy state or *term* for the configuration. Except for a fully filled configuration, no one microstate uniquely gives rise to a particular energy for the configuration. Instead, a number of microstates generally contribute to a single term. Thus the terms are usually degenerate according to the number of microstates giving rise to them. In Section 7.5 we will consider the ways in which these degeneracies are partially lifted in ligand fields of various symmetries.

The ways in which individual m_l and m_s values interact are not easily evaluated for a real atom or ion. In fact, the notion that we can assign individual m_l and m_s values to the electrons and assess their interactions on that basis is really an extreme extension of the one-electron wave mechanical model. However, in the absence of a better model (at least one that is practical), this assumption forms a reasonably good first approximation for assessing the origins of the term energies in many cases. When it is appropriate to invoke this assumption, the *Russell–Saunders coupling scheme* usually gives an adequate approximation of observed behavior. Relevant to our con-

*For a free atom or ion, equivalent electrons occupy the same subshell and therefore have the same pair of n and l values—for example, $3d^2$. Nonequivalent electrons with the same l value differ in their n values—for example, $3d^1 4d^1$.

[†]In the case of d^{10} and all other fully filled configurations Eq. (7.14) has 0! in the denominator. Recall that by convention 0! = 1, and therefore the equation remains determinate with a value of $D_t = 1$. This result is consistent with the easily demonstrated fact that there is only one way to arrange $2N_o$ electrons (as N_o pairs) in N_o degenerate orbitals.

cerns in this chapter, the Russell–Saunders coupling scheme can be applied successfully to first and second row transition metals, but it is less successful with the third row transition elements. It is hopelessly inadequate with *f*-block transition elements (i.e., lanthanides and actinides). A detailed description of the process by which the Russell–Saunders terms can be identified from the microstates of a configuration would be a needless digression for our purposes. Therefore we will simply outline the concepts involved so as to define the terminology and make the subsequent discussion more understandable.

In the Russell–Saunders coupling scheme the various terms that can exist for a particular configuration are indicated by a *term symbol* of the form

$$^{2S+1}L$$

L and S are quantum numbers that relate to the overall orbital and spin angular momenta for the system of electrons. Values of L may be 0, 1, 2,..., and values of S may be 0, 1/2, 1, 3/2,.... Both are analogous to the quantum numbers l and s for single electrons. The number $2S + 1$, which appears as the left superscript in the term symbol, is called the *multiplicity* of the term. A third quantum number, J, is often added to term symbols as a right subscript. J relates to the total angular momentum arising from spin-orbital coupling. Its allowed values are $L + S$, $L + S - 1$, $L + S + 2$, ..., $|L - S|$. For a given L value, the various values of J represent closely spaced energy sublevels of the term energy. The multiplicity $2S + 1$ equals the number of J values (and hence the number of sublevels) for the particular L value when $L > S$.* For a given L value the energy differences between the sublevels of various J values are small and can be ignored for our purposes. Thus, in keeping with common practice, we will omit the J values from our term symbols. However, the multiplicity indicated by the superscript $2S + 1$ is retained, because it relates directly to the spin state (and hence the number of unpaired electrons) of the term. Accordingly, it is often called the *spin multiplicity* and used without reference to J.

Let us consider in more detail the quantum numbers that define the term symbols. L is the *overall orbital angular momentum quantum number* for the configuration and defines an energy state. It is related to the *resultant orbital angular momentum* and to the *resultant orbital magnetic moment* of the system, obtained by vectorial addition of the vectors related to the l quantum numbers of the individual electrons. In keeping with this, L is sometimes called the *resultant orbital quantum number*. The orbital angular momentum for individual electrons has a magnitude of $[l(l + 1)^{1/2}](h/2\pi)$, and the resultant orbital angular momentum has a magnitude of $[L(L + 1)]^{1/2}(h/2\pi)$. There is a variety of ways in which the individual l values can add vectorially,

---

*The multiplicity cannot be equated to the number of J values when $L = 0$ or $L < S$. In such cases the number of J values is $2L + 1$. For example, if $L = 0$ only the single value $J = S$ is possible.

so a variety of L values can result for a given configuration of electrons. This is true even when all electrons have the same individual l values. Figure 7.20 shows the L values that result from the possible ways in which l values can be combined for the configurations p^2 ($l = 1$) and d^2 ($l = 2$).*

In the term symbol notation the values of L are given capital letter designations, which correspond to the familiar lowercase atomic orbital notations (s, p, d, f, etc.):

L value:	0	1	2	3	4	5	...
State:	S	P	D	F	G	H	...

After $L = 2$ the notation proceeds alphabetically (with the omission of J, to avoid confusion with the total angular momentum quantum number, J). Thus, we see that two p electrons give rise to the terms S, P, and D, and two d electrons give rise to the terms S, P, D, F, and G, as indicated in Fig. 7.20.

For a given term the magnitude of the resultant orbital angular momentum is fixed as $[L(L+1)]^{1/2}(h/2\pi)$. However, the vector for the momentum can have a number of allowed orientations in space relative to an applied magnetic field, which defines the z direction of the system. The various allowed orientations are associated with the *overall orbital magnetic quantum*

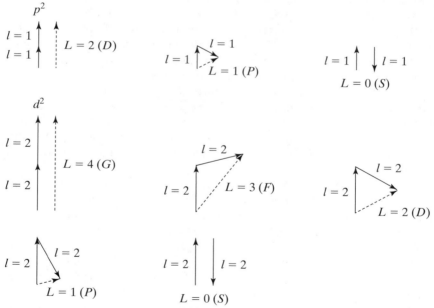

Figure 7.20 Vector addition of individual l vectors to give L for configurations p^2 and d^2.

*As noted, the component and resultant vectors have magnitudes of $[l(l + 1)]^{1/2}(h/2\pi)$ and $[L(L + 1)]^{1/2}(h/2\pi)$, respectively. In Figure 7.20 vectors have been drawn as if their magnitudes were $l(h/2\pi)$ and $L(h/2\pi)$, permitting the values of L to be obtained directly from the drawings.

number, M_L, which can take on the $2L + 1$ values $M_L = L, L - 1, ..., 1 - L,$ $- L$. Given this number of possible orientations, $2L + 1$ may be regarded as the *orbital multiplicity* or *orbital degeneracy* of the term. As shown in Fig. 7.21, a D term ($L = 2$, $2L+1 = 5$) has five possible orientations corresponding to $M_L = +2, +1, 0, -1, -2$. Each of these orientations has a projection on z whose magnitude is $M_L(h/2\pi)$. In the Russell–Saunders coupling scheme values of M_L can be obtained as the sum of the m_l values of the individual electrons; that is, $M_L = \Sigma m_l$. Thus it is possible to assign an M_L value for each and every microstate of a configuration. Since M_L represents the possible orientations of the orbital angular momentum vector, it follows that a given L value must arise from a complete set of microstates with the $2L + 1$ values $M_L = L, L - 1, ..., 1 - L, - L$, which identify these orientations.

The *overall spin quantum number*, S, defines the spin state of the term, and $2S + 1$ defines the spin multiplicity. If the overall configuration associated with the term has no unpaired electrons, then $S = 0$, and the multiplicity is $2S + 1 = 1$, called a *singlet* state. One unpaired electron gives a *doublet* state ($S = \frac{1}{2}$, $2S + 1 = 2$), two unpaired electrons give a *triplet* state ($S = 1$, $2S + 1 = 3$), three unpaired electrons give a *quartet* state ($S = \frac{3}{2}$, $2S + 1 = 4$), and so forth. The physical meaning of S is related to the *resultant spin angular momentum* and to the *resultant spin magnetic moment* of the system. Like L, S can be obtained by vectorial addition of the spin angular momentum vectors related to the s quantum numbers of the individual electrons. In terms of S, the magnitude of the resultant spin angular momentum is $[S(S + 1)]^{1/2}(h/2\pi)$. S is related to an *overall spin magnetic quantum number*, M_S,

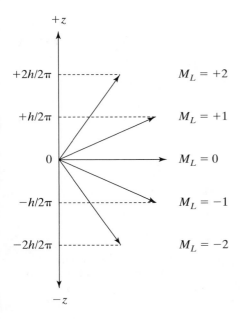

Figure 7.21 Possible orientations of the resultant orbital angular momentum vector for a D term ($L = 2$). The magnitude of the vector is $\sqrt{6}(h/2\pi)$ and its projections on the z axis have magnitudes of $M_L(h/2\pi)$.

whose allowed values are $M_S = S, S + 1, ..., 1 - S, - S$. These $2S + 1$ values indicate the allowed orientations of the spin angular momentum vector relative to an applied magnetic field, which defines the z direction of the system. From this we can see that the spin multiplicity, given as $2S + 1$, represents the *spin degeneracy* of a particular spin state.

For a given spin state the magnitude of the spin angular momentum is fixed as $[S(S + 1)]^{1/2}(h/2\pi)$, but its projections on the z axis in the allowed orientations are given by $M_S(h/2\pi)$. Figure 7.22 shows the three allowed orientations for the spin state $S = 1$. In the Russell–Saunders scheme we assume that M_S is the sum of m_s values of the individual electrons; that is, $M_S = \Sigma m_s$, where $m_s = \pm \frac{1}{2}$. In this way each microstate can be assigned a value of M_S. Since M_S represents the possible orientations of the spin angular momentum vector, it follows that a given S value must arise from a complete set of microstates with the $2S + 1$ values $M_S = S, S - 1, ..., 1 - S, - S$.

From the preceding relationships it follows that a term having particular values of both L and S must arise from the set of microstates that has the necessary $2L + 1$ values of M_L and also the necessary $2S + 1$ values of M_S. This means that one can identify all the allowed terms of a configuration by systematically arranging all microstates in such a way as to be able to cull the sets of M_L and M_S values with the appropriate ranges that define the various terms. This is a straightforward but tedious process, especially for configurations with large numbers of microstates. We shall not concern ourselves with the mechanics of this task here, except to note that a variety of techniques have been developed to carry out the labor.*

We have seen by vector addition that the configuration d^2 gives rise to the terms $S, P, D, F,$ and G (cf. Fig. 7.20). With two electrons the only possible spin states are $S = 0$ (paired) and $S = 1$ (unpaired). Thus, the spin multiplicities of the terms can only be singlets and triplets. If the two electrons are

*Two good methods have been described by K. E. Hyde, *J. Chem. Educ.* **1975**, *52*, 87 and by E. R. Tuttle, *Am. J. Phys.* **1967**, *35*, 26.

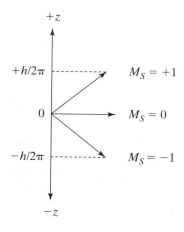

Figure 7.22   Possible orientations of the resultant spin angular momentum vector for a triplet term ($S = 1, 2S + 1 = 3$). The magnitude of the vector is $\sqrt{2}(h/2\pi)$, and its projections on the z axis have magnitudes of $M_S(h/2\pi)$.

Table 7.1   Terms for Free-Ion d^n Configurations

d^n	Free-Ion Terms[a]	Total Degeneracy
d^0, d^{10}	^1S	1
d^1, d^9	^2D	10
d^2, d^8	$^1S, {}^1D, {}^1G, {}^3P, {}^3F$	45
d^3, d^7	$^2P, {}^2D(2), {}^2F, {}^2G, {}^2H, {}^4P, {}^4F$	120
d^4, d^6	$^1S(2), {}^1D(2), {}^1F, {}^1G(2), {}^1I, {}^3P(2), {}^3D, {}^3F(2), {}^3G, {}^3H, {}^5D$	210
d^5	$^2S, {}^2P, {}^2D(3), {}^2F(2), {}^2G(2), {}^2H, {}^2I, {}^4P, {}^4D, {}^4F, {}^4G, {}^6S$	252

[a]Terms for configurations d^n and d^{10-n} are the same. A number in parentheses indicates the number of times a term occurs, if more than once.

in different subshells (e.g., $3d^14d^1$), all terms will occur as both singlets and triplets. But if we stipulate that the two electrons are equivalent, meaning a configuration nd^2 within the same subshell, the Pauli exclusion principle will limit the possible combinations of m_l and m_s. By any of the systematic methods for determining terms from microstates it can be shown that the allowed terms for the configuration nd^2 are $^1S, {}^3P, {}^1D, {}^3F, {}^1G$. By Eq. (7.14) we know that the terms for two equivalent d electrons arise from 45 microstates. Therefore the sum of the degeneracies of all these terms must equal this number. *The degeneracy of each term, equivalent to the number of microstates giving rise to it, is the product of its spin degeneracy times its orbital degeneracy; that is, (2S + 1)(2L + 1).* Thus, for the set of terms for nd^2 we have $(1)(1) + (3)(3) + (1)(5) + (3)(7) + (1)(9) = 45$. Table 7.1 lists the Russell–Saunders terms for all d^n configurations of equivalent electrons. Note that in each case the total degeneracy of terms is equal to the number D_t given by Eq. (7.14), as must be the case. The ground-state term can be identified by applying Hund's rules, but in general the actual energies of the terms, and hence their relative ordering, must be determined from analysis of spectroscopic data.

7.5 *Splitting of Terms*

The orbital term symbols for free atoms and ions are identical to the symbols for the appropriate symmetry species in the spherical group R_3. The irreducible representations of R_3 include all possible degeneracies, so there are no inherent symmetry restrictions on possible orbital degeneracies. Thus, for free-ion terms we can have fivefold degenerate D terms, sevenfold degenerate F terms, ninefold degenerate G terms, and so on. However, when a transition metal ion is subjected to a ligand field, the new point group usually places restrictions on the maximum orbital term degeneracies. In O_h and T_d, for example, the highest dimension irreducible representations are threefold degenerate. Consequently, for octahedral and tetrahedral complexes, free-ion terms with orbital degeneracies greater than three $(D, F, G,...)$ must split into new terms, each of which can have no higher than threefold degeneracy. In effect, the higher orbital multiplicity terms [viz., $(2L + 1) > 3$] split as a result of the descent in symmetry from R_3 to the finite point group of the complex. In the ligand field all the term symbols, including those that are not split, are redefined and newly designated with the appropriate Mulliken symbols of their corresponding irreducible representations in the finite point group of the complex.

From a physical standpoint, lifting the degeneracy among the d orbitals can destroy the equivalence among microstates that give rise to a particular free-ion term. Orbital assignments that were energetically equivalent in the free ion may now be quite distinct in the environment of the complex. These differences result in new collections of equivalent microstates, each of which gives rise to a distinct *ligand-field term*. However, the total number of microstates for the configuration, as represented by D_t, remains the same.

For example, for the free-ion configuration d^1, placing the electron in any one of the five d orbitals with either spin orientation is energetically equivalent. These 10 microstates give rise to a ^2D term. In an octahedral field, the single electron may have either the configuration t_{2g}^1 or e_g^1, corresponding to the ground state and excited state, respectively. In the ground state, the electron can be in any of the three t_{2g} orbitals with either spin orientation ($m_s = \pm\frac{1}{2}$). This makes six equivalent microstates. Since there are three equivalent orbital assignments, the overall orbital degeneracy (orbital multiplicity) is three. Likewise, since there are only two overall spin orientations ($M_s = \pm\frac{1}{2}$), the spin degeneracy (spin multiplicity) is two. As we shall see shortly, the resulting term is $^2T_{2g}$, in which the Mulliken symbol for the orbital term is appropriately threefold degenerate. In the excited state configuration e_g^1 there are two possible orbital assignments, each with two possible spin orientations, making a total of four microstates. The associated term is ^2E_g, in which the Mulliken symbol for the orbital term is twofold degenerate. Note that the total degeneracy of each ligand-field term, equivalent to the number of microstates giving rise to it, is the product of its spin degeneracy times its orbital degeneracy. Thus, for $^2T_{2g}$ we have $(2)(3) = 6$ and for ^2E_g we have $(2)(2) = 4$. Moreover, the sum of total degeneracies of the ligand-field terms, $6 + 4 = 10$, is equivalent to D_t for the configuration d^1.

The fate of any free-ion term in the point group of a complex can be determined by applying Eqs. (7.2)–(7.6). Although it is possible to apply these equations to both the spin and orbital terms (S and L states), the field does not interact directly on the electron spin in a chemical environment such as a complex ion. This means that the new ligand-field terms will retain the original spin multiplicities of the free-ion terms from which they originate.* Thus,

---

*However, if spin–orbital coupling (L–S coupling) and hence the J states are to be considered, the symmetry effects on both the spin and orbital functions must be evaluated and combined. For our purposes, the J states are unimportant, so we will ignore this aspect of the problem. See R. L. DeKock, A. J. Kromminga, and T. S. Zwier, *J. Chem. Educ.* **1979**, *56*, 510.

we only apply these equations to the L state of a free-ion term to determine the identities of the terms that result from splitting in the ligand field. As we proceed to this task, one important point needs to be made specifically regarding the use of Eqs. (7.4)–(7.6). These expressions have variable sign ($\pm$), depending on whether the function is *gerade* or *ungerade*. We will be concerned solely with terms arising from configurations of d electrons, which are inherently *gerade*. Therefore we will choose the positive expression in all cases.* Nonetheless, in noncentrosymmetric point groups (e.g., T_d, D_{3h}) the resulting Mulliken symbol for the new state will not have a g subscript notation, which would be inappropriate in such groups.

Let us now consider the possible splittings of S, P, D, and F terms arising from d^n configurations in an octahedral field. An S state, for which $L = 0$, is nondegenerate. As with an s orbital, it has no angular dependence and no orientation in space. Consequently, without resorting to Eqs. (7.2)–(7.6), we can conclude that in any point group an S term will not be split and will bear the Mulliken symbol for the totally symmetric representation. In O_h this is A_{1g}. For a P term, for which $L = 1$, Eqs. (7.2)–(7.6) give the following representation in O_h:

O_h	E	$8C_3$	$6C_2$	$6C_4$	$3C_2$	i	$6S_4$	$8S_6$	$3\sigma_h$	$6\sigma_d$
Γ_P	3	0	-1	1	-1	3	1	0	-1	-1

Inspection of the character table shows that this is T_{1g}. Thus, in O_h a P term is not split, but becomes a triply degenerate T_{1g} term.† A D term, for which $L = 2$, has a fivefold orbital degeneracy, as do d orbitals. Thus, in O_h, which allows no higher than threefold degeneracy, the term must be split. Indeed, applying Eqs. (7.2)–(7.6) yields a representation identical to that which we generated for the d orbitals themselves (cf. Section 7.1), and the term is therefore split into a doubly degenerate E_g term and a triply degenerate T_{2g} term. Likewise, an F term ($L = 3$) is sevenfold degenerate and must split in an O_h field. The reducible representation is

O_h	E	$8C_3$	$6C_2$	$6C_4$	$3C_2$	i	$6S_4$	$8S_6$	$3\sigma_h$	$6\sigma_d$
Γ_F	7	1	-1	-1	-1	7	-1	1	-1	-1

This reduces as $\Gamma_F = A_{2g} + T_{1g} + T_{2g}$. Thus an F state will split in an O_h field into a nondegenerate A_{2g} state and two triply degenerate states, T_{1g} and T_{2g}.

---

*By contrast, configurations of p or f electrons are inherently *ungerade*, requiring use of the negative sign in Eqs. (7.4)–(7–6).

†Recall that the threefold degenerate p orbitals transform as T_{1u} in O_h, but as we now see a P state transforms as T_{1g}. The transformations are different because the p orbitals are inherently *ungerade*, but the P state arising from a d configuration is inherently *gerade*. Thus, when applying Eqs. (7.4)–(7.6), the negative sign is used with the p orbitals and the positive sign is used with the P state.

Table 7.2 Splitting of Free-Ion Terms of d^n Configurations in an Octahedral Field

Free-Ion Term	Terms in O_h
S	A_{1g}
P	T_{1g}
D	$E_g + T_{2g}$
F	$A_{2g} + T_{1g} + T_{2g}$
G	$A_{1g} + E_g + T_{1g} + T_{2g}$
H	$E_g + 2T_{1g} + T_{2g}$
I	$A_{1g} + A_{2g} + E_g + T_{1g} + 2T_{2g}$

The splittings of other states (G, H, I, etc.) can be determined in similar manner and are shown in Table 7.2 for free-ion terms corresponding to $L = 0 - 6$. The splittings of free-ion terms and the Mulliken symbols for the ligand-field terms in other point groups can be obtained in similar manner by using Eqs. (7.2)–(7.6), but it is usually more efficient to use the correlation tables (Appendix B) with the results of Table 7.2. For example, inspection of the correlation table for O_h and T_d shows that the splittings are identical in both groups, except for the omission of the subscript g for the tetrahedral states. Correlations with other groups (e.g., D_{4h}, D_3, D_{2d}) are not as trivial, but are equally straightforward.

We now know from group theory how various free-ion terms will split in an octahedral field. It remains for us to determine how the ligand-field terms for a certain configuration are ordered by energy and how their energies will change with changes in the strength of the ligand field. Group theory alone, of course, cannot provide quantitative answers. However, it is possible to address the problem at least qualitatively with a *correlation diagram*, which shows how the energies of terms change as a function of the ligand field strength, measured as Δ_o. To construct the correlation diagram, we look at two extremes: a weak field, just strong enough to lift the R_3 free-ion term degeneracies, and a hypothetical extremely strong field. On the left of this diagram we show the energies of the free-ion terms and the Mulliken symbols for the terms into which they are split in a weak octahedral field. On the right, at the limit of an extremely large Δ_o separation between the t_{2g} and e_g orbitals, we show the energies of the possible electronic configurations for the ground state and all excited states. At this limit we will assume that the interactions between electrons in separate orbitals are negligible. We can then identify the terms that will emerge from each of these configurations in a slightly less strong field, where electronic interactions begin to be felt. The job of constructing the diagram amounts to determining the correlations between terms in the weak field and the terms in the strong field. We will carry out this task for the case of a d^2 configuration in an octahedral field.

Bethe developed a general approach, called the *method of descending symmetry*, that can be used to construct the complete correlation diagram for any configuration. Carrying out this procedure for the d^2 case is a straightforward but somewhat involved process.* For purposes of illustrating the underlying principles, we can take a somewhat less systematic approach to the d^2 case. Our strategy will concentrate primarily on identifying the triplet states arising from the allowed d^2 configurations in the strong-field case and correlating them with the appropriate terms for the weak-field case. We focus on these terms because they have the same spin multiplicity as the ground-state term for d^2. As we shall see (Section 7.6), terms with the same spin multiplicity as the ground state are of primary importance to understanding the visible spectra of transition metal complexes. Moreover, taking this approach in this case will also reveal the correlations for all the terms, both singlets and triplets.

The free-ion terms and the ligand-field terms into which they are split in a weak octahedral field, taken from Table 7.2, are listed below in order of increasing energy of the free-ion terms (left to right).

Free-ion terms	^3F	^1D	^3P	^1G	^1S
Octahedral terms (weak field)	$^3A_{2g}$	^1E_g	$^3T_{1g}$	$^1A_{1g}$	$^1A_{1g}$
	$^3T_{1g}$	$^1T_{2g}$		^1E_g	
	$^3T_{2g}$			$^1T_{1g}$	
				$^1T_{2g}$	

The free-ion terms are listed in order of increasing energy running up the left side of the correlation diagram we seek to construct (cf. Fig. 7.23). As we attempt to correlate these terms with the terms from the extremely strong field case, we will observe this principle: *Only terms of the same spin state are linked in both weak and strong fields.* Thus, a singlet state in the weak field does not correlate with a triplet state in the very strong field, and vice versa. Moreover, a term does not change its orbital identity as a result of the field strength. Thus, for example, a T_{1g} term in the weak field remains a T_{1g} term in the stronger field.

When we examine the 11 terms listed above for d^2, we note that both the ^3F and ^3P free-ion terms give rise to $^3T_{1g}$ states, each of which is a distinct state that must correlate uniquely to a triplet state from the extremely strong field. To avoid confusion, we will label the terms $^3T_{1g}\,(F)$ and $^3T_{1g}\,(P)$, indicating their origins in the free-ion terms. Now the question may arise as to whether or not these two terms might reverse their relative energy order at some field strength, implying that their correlation lines on our diagram might cross. The answer to this question lies in a general observation called

*For a detailed development of the complete correlation for the d^2 case by the method of descending symmetry, see F. A. Cotton, *Chemical Applications of Group Theory*, 3rd ed., John Wiley & Sons, New York, 1990, pp. 270–273. For extension of the method to determine the strong-field terms of other d^n configurations, see D. W. Smith, *J. Chem. Educ.* **1996**, *73*, 504–507.

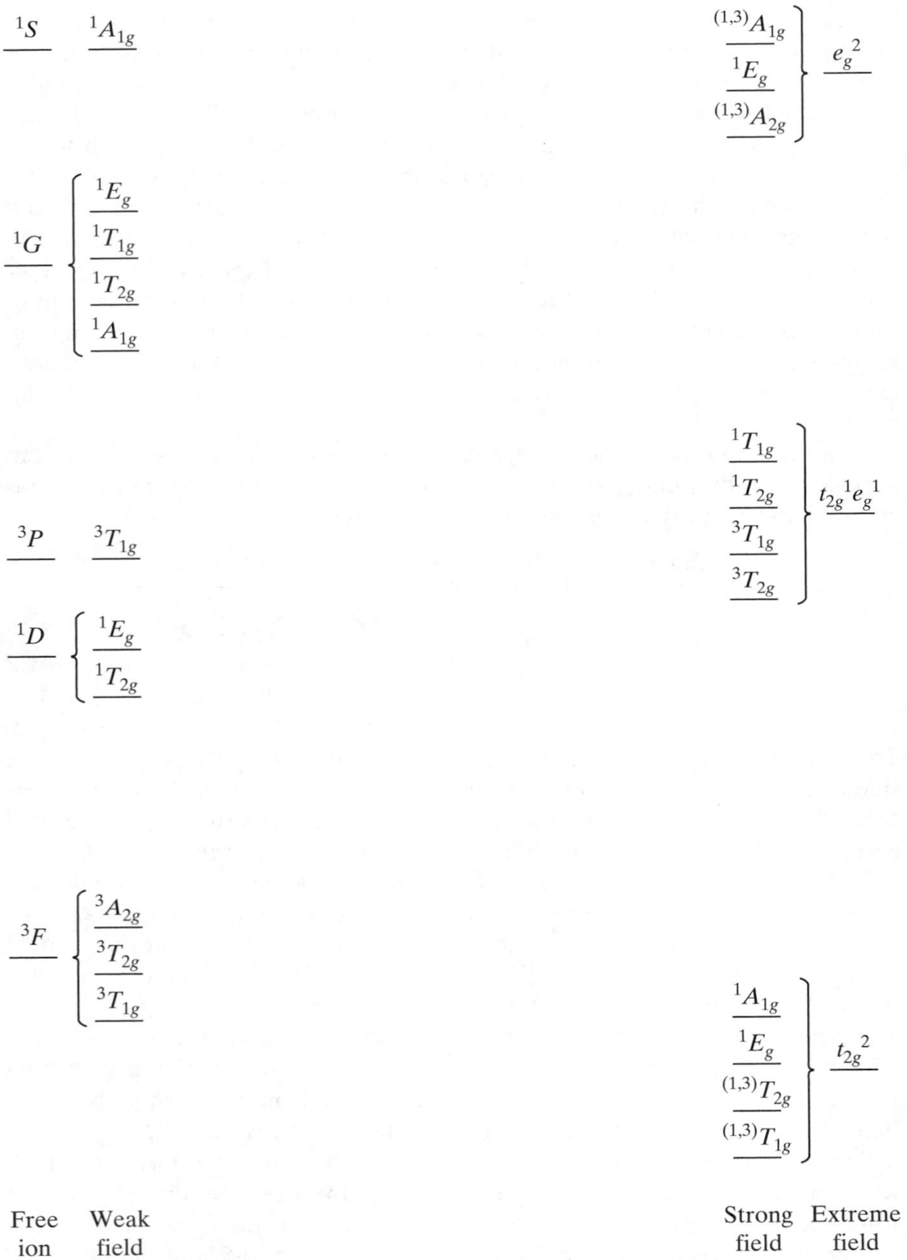

$\underline{^1S}$ $\underline{^1A_{1g}}$

$\left.\begin{array}{l}\underline{^{(1,3)}A_{1g}}\\[6pt]\underline{^1E_g}\\[6pt]\underline{^{(1,3)}A_{2g}}\end{array}\right\}e_g^{\,2}$

$\underline{^1G}\left\{\begin{array}{l}\underline{^1E_g}\\[4pt]\underline{^1T_{1g}}\\[4pt]\underline{^1T_{2g}}\\[4pt]\underline{^1A_{1g}}\end{array}\right.$

$\left.\begin{array}{l}\underline{^1T_{1g}}\\[6pt]\underline{^1T_{2g}}\\[6pt]\underline{^3T_{1g}}\\[6pt]\underline{^3T_{2g}}\end{array}\right\}t_{2g}^{\,1}e_g^{\,1}$

$\underline{^3P}$ $\underline{^3T_{1g}}$

$\underline{^1D}\left\{\begin{array}{l}\underline{^1E_g}\\[4pt]\underline{^1T_{2g}}\end{array}\right.$

$\underline{^3F}\left\{\begin{array}{l}\underline{^3A_{2g}}\\[4pt]\underline{^3T_{2g}}\\[4pt]\underline{^3T_{1g}}\end{array}\right.$

$\left.\begin{array}{l}\underline{^1A_{1g}}\\[6pt]\underline{^1E_g}\\[6pt]\underline{^{(1,3)}T_{2g}}\\[6pt]\underline{^{(1,3)}T_{1g}}\end{array}\right\}t_{2g}^{\,2}$

| Free ion | Weak field | | Strong field | Extreme field |

Figure 7.23 Setup for preparing a correlation diagram for a d^2 ion in an octahedral environment. The spin multiplicities of strong field terms indicated with superscript (1,3) are yet to be determined.

the *noncrossing rule*, which has firm roots in quantum mechanics: *States of the same symmetry and same multiplicity do not cross, but rather repel one another, thereby increasing their relative energy separation beyond a certain minimum as field strength increases.*

We now consider the states in the extremely strong field, represented on the right side of our diagram. At this hypothetical extreme we can have the following three configurations:

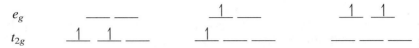

In the absence of interelectronic interactions, we can assume that the ground state is t_{2g}^2. The configuration $t_{2g}^1 e_g^1$ puts one electron in an orbital that lies Δ_o higher, so the energy of this state lies higher by Δ_o. Likewise, the configuration e_g^2 promotes both electrons by this amount, so the energy of this state is higher than the ground-state configuration by $2\Delta_o$.

If we now relax the field a little, so that the electrons just begin to interact, each of the strong-field configurations will give rise to a number of energy states, depending upon how electrons with specific spins occupy specific orbitals. Each of these new terms is uniquely associated with a collection of microstates. For example, by Eq. (7.14) the configuration t_{2g}^2 has a total degeneracy of 15. This means there are 15 ways of arranging the two electrons by individual spins and orbital assignments within the t_{2g} orbitals. Some of these microstates will have the electrons with the same spin and some will have the electrons with opposite spins. Therefore, both singlet and triplet terms arise from this configuration. Likewise, the configuration e_g^2, which has two electrons occupying two degenerate orbitals, has a total degeneracy of 6. Thus, there are six ways of arranging the two electrons in the two e_g orbitals, each resulting in a microstate. Again, both singlet and triplet states will arise from this configuration. The configuration $t_{2g}^1 e_g^1$ involves two nonequivalent electrons. There are clearly six ways of arranging one electron in the three t_{2g} orbitals (three possible orbital assignments with two possible spin orientations), and there are four ways of arranging one electron in the two e_g orbitals (two possible orbital assignments with two possible spin orientations). The two electrons are in separate degenerate sets of orbitals, so the possible microstates are not restricted by the Pauli exclusion principle. This results in 24 ways of arranging both electrons in the configuration $t_{2g}^1 e_g^1$; that is, $6 \times 4 = 24$ microstates. Here, too, we can have both singlet and triplet states arising from the configuration. Altogether these three configurations account for 45 microstates, equal to the total degeneracy for a d^2 configuration. As expected, subjecting the general configuration to an octahedral field, even a very strong one, does not change the overall number of microstates.

We need to know the term symbols arising from these 45 microstates in the slightly relaxed strong-field case. We can determine the Mulliken symbols for the orbital part of the term symbols by taking the direct products of the irreducible representations of the individual electrons in the three configurations. For the ground-state configuration t_{2g}^2 we take the direct product $t_{2g} \times t_{2g}$, which yields the following representation:

O_h	E	$8C_3$	$6C_2$	$6C_4$	$3C_2$	i	$6S_4$	$8S_6$	$3\sigma_h$	$6\sigma_d$
$\Gamma(t_{2g}^2)$	9	0	1	1	1	9	1	0	1	1

This reduces as $\Gamma(t_{2g}^2) = A_{1g} + E_g + T_{1g} + T_{2g}$. These, then, are the orbital terms arising from this configuration in a strong but not extreme field. But we also need to know the spin multiplicities of these terms if we are to make the correlations with the weak-field terms. We can begin to sort out the spin multiplicities by recalling that the total degeneracy of a configuration is equal to the sum of the products of the spin degeneracies times the orbital degeneracies over all terms. Here, $D_t = 15$, so we may write for the series of terms $A_{1g} + E_g + T_{1g} + T_{2g}$

$$(a)(1) + (b)(2) + (c)(3) + (d)(3) = 15$$

where the unknown coefficients are either 1 or 3. This is satisfied if either T_{1g} or T_{2g} is a triplet and the other two terms are singlets, or if both A_{1g} and E_g are triplets and the other two terms are singlets. In other words, at this point any one of the following assignments is possible:

$$^1A_{1g} + {}^1E_g + {}^1T_{1g} + {}^3T_{2g}$$

$$^1A_{1g} + {}^1E_g + {}^3T_{1g} + {}^1T_{2g}$$

$$^3A_{1g} + {}^3E_g + {}^1T_{1g} + {}^1T_{2g}$$

We can narrow the choices slightly by considering the possible orbital arrangements of two unpaired electrons (a triplet state) for the configuration t_{2g}^2:

t_{2g} ⇅ ↑ ↑ __   ↑ __ ↑   __ ↑ ↑

There are three orbital arrangements for two unpaired electrons, so the triplet state must have a triply degenerate orbital term (either $^3T_{1g}$ or $^3T_{2g}$ in this case). This means we have either $^1A_{1g} + {}^1E_g + {}^3T_{1g} + {}^1T_{2g}$ or $^1A_{1g} + {}^1E_g + {}^1T_{1g} + {}^3T_{2g}$; that is, we can rule out the third listed choice. Both remaining assignments satisfy the total degeneracy of the configuration, so we will need to gather more information before deciding which of these is correct.

In similar manner, to determine the orbital terms for the configuration e_g^2 we take the direct product $e_g \times e_g$ and obtain the representation

O_h	E	$8C_3$	$6C_2$	$6C_4$	$3C_2$	i	$6S_4$	$8S_6$	$3\sigma_h$	$6\sigma_d$
$\Gamma(e_g^2)$	4	1	0	0	4	4	0	1	4	0

From the reduction of $\Gamma(e_g{}^2)$ we obtain the terms $A_{1g} + A_{2g} + E_g$. As previously noted, the total degeneracy for this configuration is 6, so we may write

$$(a)(1) + (b)(1) + (c)(2) = 6$$

Clearly $c \neq 3$, so either $a = 3$ or $b = 3$ and the other two coefficients are 1. Therefore we have either $^3A_{1g} + {}^1A_{2g} + {}^1E_g$ or $^1A_{1g} + {}^3A_{2g} + {}^1E_g$. We can come to the same conclusion by looking at how we could arrange two unpaired electrons in two degenerate orbitals:

$$e_g \quad \underline{\uparrow} \quad \underline{\uparrow}$$

There is only one choice, so the orbital term for the triplet state must be non-degenerate (either A_{1g} or A_{2g} in this case). Again, we will need to obtain other information before deciding which possible assignment is correct.

Finally, the terms from the configuration $t_{2g}{}^1 e_g{}^1$ are obtained from the direct product $t_{2g} \times e_g$, which yields the representation

O_h	E	$8C_3$	$6C_2$	$6C_4$	$3C_2$	i	$6S_4$	$8S_6$	$3\sigma_h$	$6\sigma_d$
$\Gamma(t_{2g}\, e_g)$	6	0	0	0	-2	6	0	0	-2	0

On reduction this gives the terms $T_{1g} + T_{2g}$. As we have noted, the electrons in this configuration are unrestricted by the Pauli exclusion principle. Therefore they may have the same or opposite spins with all possible orbital assignments. This means that both terms occur as both singlets and triplets. As we have seen, the total degeneracy for $e_g{}^1 t_{2g}{}^1$ is $D_t = 24$, which is uniquely satisfied by the assignment $^1T_{1g} + {}^1T_{2g} + {}^3T_{1g} + {}^3T_{2g}$.

Although we have not unambiguously decided all the spin multiplicities, we can proceed to make the correlations. This will actually help us make the spin assignments. At this point in the process our correlation diagram looks like Fig. 7.23, where either the known or possible spin multiplicities have been indicated for each term on the strong-field side.* We begin by examining the split terms from the ^3F free-ion term on the left of Fig. 7.23. The $^3A_{2g}$ term is unique among the weak-field terms, and it must correlate with the unique A_{2g} term from $e_g{}^2$ on the strong-field side. This term, then, must also be a triplet. It now becomes evident that our choice of spin multiplicities for the terms from $e_g{}^2$ must be $^1A_{1g} + {}^3A_{2g} + {}^1E_g$. We turn now to the $^3T_{2g}$ term arising from ^3F. This is the only triplet T_{2g} term on the weak-field side (the two other T_{2g} terms are singlets), so there can be only one such term on the strong-field side. We have already identified a $^3T_{2g}$ term from $t_{2g}{}^1 e_g{}^1$ on the strong-field side, which must correlate with the $^3T_{2g}$ term from ^3F. This means that the T_{2g} term from $t_{2g}{}^2$ on the strong field side must be a singlet, and from this it follows that the spin multiplicities for terms from this configuration must be $^1A_{1g} + {}^1E_g + {}^3T_{1g} + {}^1T_{2g}$. We now can make the correlation for the

---

*You may find it useful to fill in the correlations and spin multiplicities on Figure 7.23 as they are explained in the rest of this paragraph.

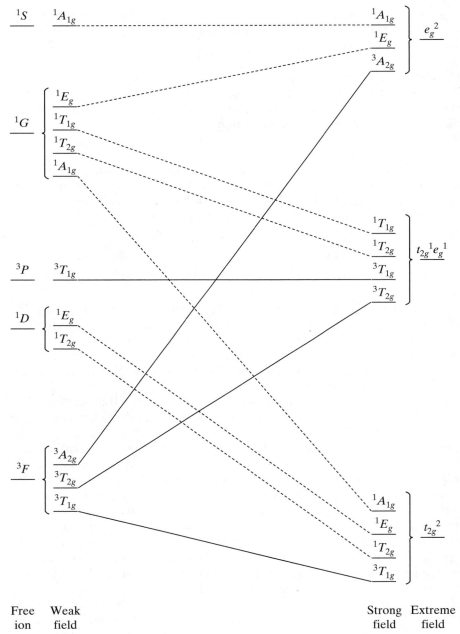

Free   Weak   Strong   Extreme
ion   field   field   field

Figure 7.24   Complete correlation diagram for a d^2 ion in an octahedral environment.

remaining two triplet terms on the weak-field side; namely, $^3T_{1g}$ (F) and $^3T_{1g}$ (P). From the noncrossing rule we conclude that $^3T_{1g}$ (F) connects with $^3T_{1g}$ from t_{2g}^2, and $^3T_{1g}$ (P) connects with $^3T_{1g}$ from $t_{2g}^1e_g^1$. The remaining correlations between singlet states on both sides can now be made by applying the noncrossing rule. The resulting correlation diagram is shown in Fig. 7.24.

In principle, we could construct the correlation diagram for any d^n configuration in an octahedral or other field by taking an approach similar to what we have shown here for the d^2 case. However, as the number of microstates and terms increases with the number of electrons, the labor of constructing the correlation diagrams "from scratch" becomes considerably more onerous. Fortunately, some general relationships between configurations, terms, and ligand environments minimize the needed effort. To begin, we noted in Table 7.1 that the free-ion terms for a configuration d^n and a configuration d^{10-n} are the same. Since each free-ion term splits into a specific collection of terms in any ligand field, it follows that the splitting of terms for a configuration d^n is identical to that for the configuration d^{10-n}. However, this does not mean that the correlation diagrams are identical. To the contrary, *the order of splitting of a given d^n term will show the reverse pattern of that of a d^{10-n} configuration in the same ligand field.* Thus, the same ligand-field term that becomes more stable (moves to lower energy) for a d^n ion will become less stable (moves to higher energy) for a d^{10-n} ion as Δ_o changes.

Consider the simplest pair of such configurations, d^1 and d^9, in an octahedral field. Both configurations give rise to a ^2D free-ion term, which is split into ^2E_g and $^2T_{2g}$ terms in an octahedral field (cf. Table 7.2). Not coincidentally this splitting is identical to the splitting of d orbitals in an octahedral field. In the d^1 case, the two terms in the octahedral field arise from the following configurations:

where the configuration t_{2g}^1 is the ground state and e_g^1 is the excited state (higher energy). As previously noted, there are three possible orbital assignments for the t_{2g}^1 configuration, giving rise to the $^2T_{2g}$ term, and there are two possible orbital assignments for the e_g^1 configuration, giving rise to the ^2E_g term. We can predict that as the Δ_o gap between the t_{2g} and e_g orbitals increases with increasing field strength the $^2T_{2g}$ term will become more stable and the ^2E_g term will become less stable. Thus, the separation between the two states will increase. In fact the separation is numerically equal to Δ_o, the magnitude of the field. Relative to the energy of the ^2D free-ion term, the $^2T_{2g}$ term will be stabilized by $-(2/5)\Delta_o$ and the ^2E_g term will be destabilized by $+(3/5)\Delta_o$. From these considerations we readily obtain the correlation diagram shown at the top of Fig. 7.25. Now the same terms will arise from the

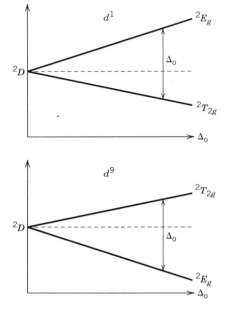

Figure 7.25   Correlation diagrams for d^1 (*top*) and d^9 (*bottom*) ions in an octahedral field, illustrating the effect of hole formalism.

^2D free-ion term of a d^9 configurations, but they now correspond to the following two configurations in the octahedral field:

e_g •• •○   •• ••

t_{2g} •• •• ••   •• •• •○

$\quad\quad\quad t_{2g}^6 e_g^3 \quad\quad\quad\quad t_{2g}^5 e_g^4$

We have introduced a hole symbol (○) here to mark the absence of an electron in an orbital. Using the hole as a marker, it is easy to see that the ground state $t_{2g}^6 e_g^3$ configuration consists of two equivalent orbital assignments and therefore must correspond to the ^2E_g term. Likewise, the three possible orbital assignments for the hole in the configuration $t_{2g}^5 e_g^4$ verify that it gives rise to the $^2T_{2g}$ term. Thus, the energy ordering of the terms for d^9 is the reverse of the d^1 case. For the d^9 (O_h) case the ^2E_g term is stabilized by $-(3/5)\Delta_o$ and the $^2T_{2g}$ term is destabilized by $+(2/5)\Delta_o$. From this we obtain the correlation diagram shown at the bottom of Fig. 7.25.

The relationship between d^n and d^{10-n} term splittings in the same-symmetry ligand field is sometimes called the *hole formalism*. The name comes from seeing d^n as a configuration of n electrons and d^{10-n} as a configuration of n positive holes (equivalent to positrons), as illustrated with the configurations shown for d^1 and d^9 above. A configuration of n electrons will interact with a ligand field in the same way as a configuration of n positrons, except that repulsions in the former case become attractions in the latter case. For example, for a d^1 ion in an octahedral field a transition from the ground state to the excited state involves promoting the electron by Δ_o. The same kind of transition for a d^9 ion involves demoting the hole by Δ_o. As we have seen, the consequence of this hole formalism is reversal in the order of ligand-field terms arising from the same free-ion term. This principle can be applied to more complicated cases. *The correlation diagram for d^{10-n} can be obtained by reversing the order of the sets of terms for the various $t_{2g}e_g$ configurations on the strong field side of the d^n diagram, relabeling for the appropriate d^{10-n} configurations, and redrawing the connecting lines, paying attention to the noncrossing rule.* Applying this technique, we can use the d^2 octahedral correlation diagram (Fig. 7.24) to obtain the d^8 diagram shown in Fig. 7.26. Note that in the d^2 diagram the sets of terms for the three $t_{2g}e_g$ configurations on the strong-field side are ordered

$$[^3A_{2g}, \, ^1E_g, \, ^1A_{1g}] < [^3T_{2g}, \, ^3T_{1g}, \, ^1T_{2g}, \, ^1T_{1g}] < [^3T_{1g}, \, ^1T_{2g}, \, ^1E_g, \, ^1A_{1g}]$$

but for the d^8 diagram they are ordered

$$[^3T_{1g}, \, ^1T_{2g}, \, ^1E_g, \, ^1A_{1g}] < [^3T_{2g}, \, ^3T_{1g}, \, ^1T_{2g}, \, ^1T_{1g}] < [^3A_{2g}, \, ^1E_g, \, ^1A_{1g}]^*$$

Comparing the two diagrams, we also note that the connections between certain weak-field terms and their strong-field counterparts have changed as a result of the noncrossing rule.

A generalization similar to the hole formalism allows us to relate the correlation diagrams for octahedral cases to those for tetrahedral cases. As we have noted, the ligand-field term symbols for the states in a tetrahedral field arising from any d^n free-ion term are the same as those in an octahedral field, except the labels for the tetrahedral terms omit the subscript g notation. However, the energies of the new terms in the tetrahedral field have an inverted order. To understand this, recall that the splitting of d orbitals into e and t_2 levels in a tetrahedral field is the inverse of the splitting into t_{2g} and e_g levels in an octahedral field. As this suggests, the tetrahedral and octahedral fields have similar but opposite effects on the d orbitals. The same is true for the terms arising from d^n configurations. Thus, *the correlation diagram for $d^n(T_d)$ can be obtained by reversing the order of the sets of terms for the various $t_{2g}e_g$ configurations on the strong-field side of the $d^n(O_h)$ diagram, relabeling for the appropriate d^n tetrahedral configurations, omitting the subscript g notations from all terms, and redrawing the connecting lines, paying attention to the noncrossing rule.* This is essentially the same process we have seen for configurations related by hole formalism. Therefore, the correlation diagram for $d^n(T_d)$ is qualitatively the same as that for $d^{10-n}(O_h)$, except for minor changes in labels of configurations and term symbols. For example, the $d^2(O_h)$ correlation shown in Fig. 7.24 is essentially the same as the correlation for $d^8(T_d)$, and the $d^8(O_h)$ correlation shown in Fig. 7.26 is essentially the same as the correlation for $d^2(T_d)$.

---

*The ordering of terms within each set is arbitrarily chosen for convenience in drawing the diagrams and therefore has no significance regarding relative energies within the set.

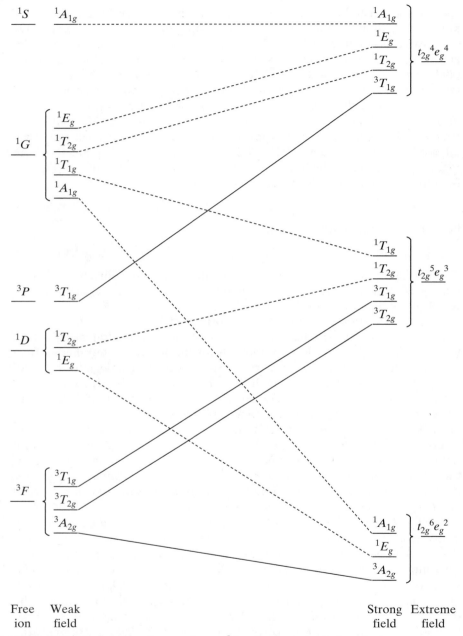

Figure 7.26 Correlation diagram for a d^8 ion in an octahedral environment.

The diagrams we have just constructed only show the splitting of terms and suggest very qualitatively the senses of energy changes among the new terms as the field strength varies. In actual practice, most chemists refer to a more detailed set of semiempirical diagrams for octahedral complexes, originally developed by Yukito Tanabe and Satoru Sugano in 1954.* A complete set of these diagrams for octahedral complexes of metal ions with the configurations d^2 through d^8 is shown in Appendix D.† The unique parameters and plotting methods of the Tanabe and Sugano diagrams can be understood by referring to a specific example, such as the d^7 case shown in Fig. 7.27. Like our qualitative correlation diagrams, these diagrams are plots of term energy versus field strength. However, the energies of all states are plotted relative to the energy of the ground-state term; that is, the ground-state energy forms the abscissa of the plot. Moreover, the term energies and field strengths are expressed as the variables E/B and Δ/B, respectively, where B is the *Racah parameter*. The Racah parameter is a measure of the interelectronic repulsion and is used to measure the energy difference between states of the same spin multiplicity. For example, in the d^7 case the difference between the ^4F and ^4P free-ion terms is $15B$, which for Co^{2+} is approximately 14,500 cm^{-1}. By us-

*Y. Tanabe and S. Sugano, *J. Phys. Soc. Japan*, **1954**, *9*, 753 and 766.
†Diagrams for d^1 and d^9 are not needed, since the diagrams shown in Figure 7.25 are complete for those simple cases.

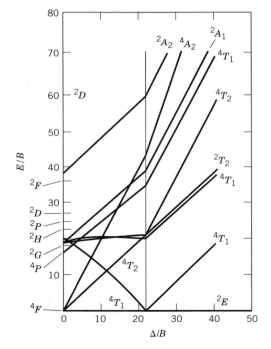

Figure 7.27 Tanabe and Sugano diagram for d^7 octahedral complexes. [Adapted with permission from Y. Tanabe and S. Sugano, *J. Phys. Soc. Japan* **1954,** *9*, 766.]

ing the appropriate values of the Racah parameter, the Tanabe and Sugano diagrams can be used with a variety of metal ions and complexes. The d^7 diagram, like all such diagrams for configurations that may be either high spin or low spin, has a perpendicular line near the middle marking the change in spin state. To the left of the line (low field strength, high spin), the ground state is $^4T_1$, emerging from the free-ion ^4F term.* To the right of the line (high field strength, low spin) the ground state is ^2E, and therefore it becomes the abscissa beyond the spin-state crossover point. There is no discontinuity in this. The high-spin ^2E ground state is a continuation of the line for the low-spin excited-state ^2E term, which emerges from the ^2G free-ion term. Note that the line for the former high-spin ground-state $^4T_1$ term ascends as an excited state on the low-spin (right) side of the diagram. With many lines emerging from certain free-ion terms, it sometimes can be difficult to trace back the free-ion origin of some of the octahedral terms, particularly on the diagrams for high-spin/low-spin configurations. Keep in mind that the spin multiplicities of the split terms must match those of the free-ion terms. Failure to recognize this has caused some texts to erroneously render the d^6 diagram (cf. Appendix D) with the high-spin ground-state $^1A_1$ term emerging from the ^3D free-ion term, rather than the correct ^1I term.†

7.6 Electronic Spectra of Transition Metal Complexes

Perhaps the most striking feature of transition metal complexes is the array of colors they present. Color results when a complex absorbs frequencies in the visible region of the spectrum, causing transitions from the ground electronic state to certain of the excited states of the configuration. Since the electronic states arise from d-electron configurations on the metal ion, the absorptions are said to result from *d–d transitions*. The unabsorbed portion of the spectrum is transmitted, and results in the perceived color. Figure 7.28 shows the visible absorption spectra of $[M(H_2O)_6]^{n+}$ complexes of first-row transition metal ions in aqueous solution. For example, the d^1 ion $[Ti(H_2O)_6]^{3+}$ has maximum absorbance at 20,000 cm^{-1}, corresponding to green light. The transmitted frequencies are red and to a lesser extent blue, which combine to give the characteristic purple color of the ion in solution.

The absorption of a complex ion at a given wavelength follows the Beer-Lambert law

$$A = \log(I/I_0) = \epsilon cb \qquad (7.15)$$

where A is the absorbance, I_0 is the intensity of the incident radiation, I is the transmitted intensity, ϵ is the molar absorptivity, c is the molar concentration, and b is the path length of the light through the sample. With constant concentration and fixed path length it is customary to plot the spectra as ab-

---

*All octahedral terms warrant the use of the subscript g notation. However, since there is no ambiguity, in the Tanabe and Sugano diagrams the g is customarily omitted for clarity.
†A. L. Hormann and C. F. Shaw, *J. Chem. Educ.* **1987,** *64*, 918.

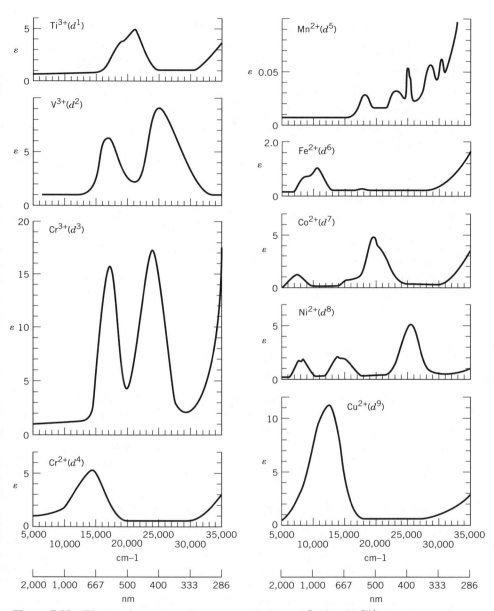

Figure 7.28   Electronic spectra of aqueous solutions of $[M(H_2O)_6]^{n+}$ complexes of first-row transition metals. [Reproduced with permission from B. N. Figgis, *Introduction to Ligand Fields*, Wiley-Interscience, New York, 1966, pp. 221 and 224.]

sorptivity (ϵ) versus frequency or wavelength. As the examples shown in Fig. 7.28 indicate, the molar absorptivities at the maximum absorbing frequencies in the visible region are relatively low for d–d transitions of octahedral complexes (e.g., $\epsilon \approx 5 - 100$). By contrast, other systems, such as organic dyes, show values of molar absorptivities that are often $\geq 10^4$.* The ϵ values are low for visible absorptions of complex ions because the electronic transitions are actually forbidden by either or both of the following two quantum mechanical selection rules:

1. *LaPorte's Rule.* If the system is centrosymmetric, transitions between states with the same inversion symmetry ($g \rightarrow g$, $u \rightarrow u$) are forbidden, but transitions between states of different inversion symmetries ($g \rightarrow u$, $u \rightarrow g$) are allowed.

2. *Spin Multiplicity Rule.* Transitions between states with different spin multiplicities are forbidden.

These selection rules, which would seem to preclude any visible absorption spectra for octahedral complexes, are clearly violated routinely, as evidenced by the colors that are so characteristic of transition metal compounds. We can understand the mechanisms by which these rules break down by considering the symmetry of the transition moments.

For purposes of molecular spectroscopy we routinely assume that the internal energy of the system can be expressed as the sum of rotational, vibrational, and electronic contributions,

$$E_{\text{int}} = E_r + E_v + E_e$$

This implies that the overall wave function is

$$\Psi = \psi_r \psi_v \psi_e$$

for which the individual energies are given by separate Schrödinger equations of the form $\mathcal{H}\psi = E\psi$. By this model, the absorption spectra of metal complexes involve one or more transitions between an electronic ground state ψ_e and some excited state ψ_e'. The transition will be observable as a band in the absorption spectrum if there is a nonzero *transition moment* of the form

$$M_e = \int \psi_e \boldsymbol{\mu} \psi_e' d\tau \tag{7.16}$$

in which $\boldsymbol{\mu}$ is the electronic dipole moment operator, whose components resolve as $\boldsymbol{\mu} = \boldsymbol{\mu}_x + \boldsymbol{\mu}_y + \boldsymbol{\mu}_z$. As we saw in connection with vibrational spectra (cf. Section 6.2), M_e will be nonzero if the symmetry of the transition belongs to the totally symmetric representation, which for O_h is A_{1g}. The product $\psi_e \boldsymbol{\mu} \psi_e'$ in Eq. (7.16) can be totally symmetric only if the product of two of the terms is the same symmetry species as the third. Now in O_h x, y, and z transform as T_{1u}, which is then the symmetry of the electric dipole moment operator. However, we know that all the ψ_e's are *gerade* for an octahedral complex. Thus, the product $\psi_e \boldsymbol{\mu} \psi_e'$ must be *ungerade* ($g \times u \times g = u$) and

---

*In addition to d–d transitions, transition metal complexes typically have charge transfer transitions between the metal ion and the ligands (M→L or M←L). These have very high molar absorptivities, but the absorption usually falls in the ultraviolet region. Here we are only concerned with transitions in the visible region.

cannot be totally symmetric. In principle, then, Eq. (7.16) must always be zero for an octahedral complex, as stipulated by LaPorte's rule.

Transitions do occur because the electronic wave functions are not fully independent of the vibrational functions. This means that the appropriate integral has a form such as

$$\int (\psi_e \psi_v) \boldsymbol{\mu} (\psi_e' \, \psi_v') \, d\tau \qquad (7.17)$$

This *vibrational–electronic* interaction is called *vibronic coupling*. As we saw in Section 6.2, the vibrational ground state is totally symmetric, so $\psi_e \psi_v$ has the same symmetry as ψ_e alone. Thus, inclusion of ψ_v in Eq. (7.17) does not affect the symmetry of the transition from what we had previously. However, ψ_v' may have the symmetry of any of the normal modes of the complex. Regardless of the vibrational spectroscopic activity, if any one of these has the appropriate symmetry so that the direct product $\psi_e \boldsymbol{\mu} \psi_e' \psi_v'$ contains the totally symmetric representation, the integral will not vanish. Regardless of the electronic terms involved, we know that $\psi_e \boldsymbol{\mu} \psi_e'$ is *ungerade*, so ψ_v' must have an identical *ungerade* symmetry in order for the entire product $\psi_e \boldsymbol{\mu} \psi_e' \psi_v'$ to contain the totally symmetric representation. For example, in the case of a d^1 ion in an octahedral field the only possible electronic transition is from the $^2T_{2g}$ ground state to the ^2E_g excited state (cf. Fig. 7.25). The symmetry of $\psi_e \boldsymbol{\mu} \psi_e'$ is given by the direct product $^2T_{2g} \times T_{1u} \times E_g$. Multiplying the characters of these irreducible representations gives the reducible representation

O_h	E	$8C_3$	$6C_2$	$6C_4$	$3C_2$	i	$6S_4$	$8S_6$	$3\sigma_h$	$6\sigma_d$
Γ_e	18	0	0	0	2	-18	0	0	-2	0

which gives $\Gamma_e = A_{1u} + A_{2u} + 2E_u + 2T_{1u} + 2T_{2u}$. As expected, all species are *ungerade*. If any one of these matches with the symmetry of a normal mode, then the direct product for $\psi_e \boldsymbol{\mu} \psi_e' \psi_v'$ will contain the totally symmetric representation, and the integral for the transition moment will not vanish. Now, the $3n-6$ normal modes of an octahedral ML_6 complex (cf. Section 6.1 and Appendix C) are $\Gamma_{3n-6} = A_{1g} + E_g + 2T_{1u} + T_{2g} + T_{2u}$. As we see, within Γ_e and Γ_{3n-6} there are matches of T_{1u} and T_{2u}, which means the $^2T_{2g} \rightarrow {}^2E_g$ transition is *vibronically allowed* through possible coupling with the normal modes $\nu_3(T_{1u})$, $\nu_4(T_{1u})$, and $\nu_6(T_{2u})$. By a similar analysis, it can be shown that the possible electronic transitions for any octahedral complex will be vibronically allowed.

Note that the normal modes responsible for the breakdown of the LaPorte rule are those in which the vibration destroys the center of symmetry. Of course, static loss of centrosymmetry would have the same result, as in the D_3 complex $[Co(en)_3]^{2+}$. However, whether static or dynamic, such perturbations from ideal centrosymmetry are minor, and the resulting molar absorptivities are small. By contrast, the LaPorte rule does not apply to tetrahedral, trigonal bipyramidal, and other noncentrosymmetric complexes. As might be expected, such species tend to have higher molar absorptivities ($\epsilon \approx 100\text{–}200$), but not as high as some other LaPorte-allowed transitions. All complex ions show very high absorbance at higher frequencies, usually in the ultraviolet region, due to metal–ligand charge transfer transitions. These are $g \longleftrightarrow u$, allowed by the LaPorte rule, and accordingly have very high molar absorptivities ($\epsilon \approx 10,000$).

We can understand the origin of the spin multiplicity selection rule if we recognize that ψ_e consists of both orbital and spin contributions; that is, $\psi_e = \psi_o \psi_s$. On this basis we might rewrite Eq. (7.17) as

$$\int (\psi_o \psi_s \psi_v) \boldsymbol{\mu} (\psi_o' \psi_s' \psi_v') \, d\tau \tag{7.18}$$

Building on our previous analysis of the LaPorte rule, this can only be non-vanishing if $\psi_s = \psi_s'$, which means the two wave functions have the same overall spin quantum number, S. Otherwise, the symmetry of the transition, which is totally symmetric on the basis of the orbital and vibrational terms, would change and not be totally symmetric.

Despite the spin multiplicity rule, transitions between different spin states do occur, although with molar absorptivities that are even smaller than those for LaPorte-forbidden transitions. In a spectrum with bands from vibronically-allowed transitions with the same multiplicity, spin-forbidden transitions are likely to be too weak to be observed. However, in d^5 high-spin complexes, the only conceivable d–d transitions are spin forbidden and give rise to bands with molar absorptivities typically 0.01–1. For example, Fig. 7.28 shows the visible spectrum of $[Mn(H_2O)_6]^{2+}$, which has many weak bands with $\epsilon \leq 0.4$. The weak intensity of these bands accounts for the barely perceptible faint pink color of this complex in solution.

Spin-forbidden transitions occur because of spin–orbital coupling. As we have noted (cf. Section 7.4), the Russell–Saunders coupling scheme, which assumes separately definable L and S values, is only an approximation, which becomes less valid with the heavier transition metals. As a result, spin-forbidden transitions are more common among second and third series transition metal complexes.

The observed spectra of octahedral transition metal complexes can be assigned on the basis of the Tanabe and Sugano diagrams. All transitions are presumed to originate from the ground-state term to the various upper-state terms. The absorption spectra of most transition metal complexes, except as noted, consist principally of bands arising from transitions that are LaPorte-forbidden (vibronically allowed) and spin-allowed. Therefore, the most intense bands arise from transitions to excited states with the same spin multiplicity as the ground state term. For example, for a d^7 high-spin octahedral complex, the Tanabe and Sugano diagram (cf. Fig. 7.27, left side) leads us to expect three spin-allowed transitions: $^4T_{1g}(F) \rightarrow {}^4T_{2g}(F)$, $^4T_{1g}(F) \rightarrow {}^4A_{2g}$, $^4T_{1g}(F) \rightarrow {}^4T_{1g}(P)$. For $[Co(H_2O)_6]^{2+}$ these bands are observed at 8000 cm^{-1}, 19,600 cm^{-1}, and 21,600 cm^{-1}, respectively (cf. Fig. 7.28).

In our discussion of correlation diagrams (Section 7.4) we noted that the term splittings are reversed for d^n and d^{10-n} configurations of the same symmetry, and likewise they are reversed for $d^n(O_h)$ and $d^n(T_d)$ cases. These relationships are used to advantage in an abbreviated set of correlation diagrams, originally devised by Orgel,* which may be used to predict the spin-allowed transitions of octahedral and tetrahedral high-spin complexes. Since the Orgel diagrams are only intended for use with spin-allowed transitions, no correlations are shown for states with different spin multiplicities from that of the ground-state term. Consequently, there is no Orgel diagram for d^5, since only spin-forbidden transitions are possible for the high-spin case. Moreover, the Orgel diagrams cannot be used to interpret the spectra of low-spin complexes. These restrictions allow the Orgel diagrams to take advantage of yet another relationship between correlation diagrams, which can be verified by inspecting the Tanabe and Sugano diagrams: *In the same ligand field (O_h or T_d), terms with the same spin multiplicity as the ground state have identical splitting patterns for d^n and $d^{n\pm5}$ configurations.* With this and the previously identified relationships, the Orgel diagrams are related to each other as follows:

1. $d^n(O_h)$ and $d^{n\pm5}$ (O_h) have the same diagram.
2. $d^n(T_d)$ and $d^{n\pm5}$ (T_d) have the same diagram.
3. d^n, $d^{n\pm5}$ (O_h) is the reverse of d^n, $d^{n\pm5}$ (T_d), and vice versa.
4. $d^n(O_h)$ is the reverse of d^{10-n} (O_h), and $d^n(T_d)$ is the reverse of $d^{10-n}(T_d)$.

As a result, we need only two diagrams, both of which are easily committed to memory.

The simpler of the two Orgel diagrams (Fig. 7.29, top) can be generated by extrapolating the lines for the term splitting scheme for the d^1 octahedral case to the left side of the diagram (cf. Fig. 7.25). By the relationships we have seen, the right side becomes the diagram for d^1 and d^6 octahedral and d^4 and d^9 tetrahedral complexes, and the left side becomes the diagram for d^4 and d^9 octahedral and d^1 and d^6 tetrahedral complexes. For these cases, the spectrum is expected to show a single band ($T_2 \to E$ or $E \to T_2$). The separation between the two states is Δ_o or Δ_t as the case may be, so the observed frequency corresponds to the crystal field splitting energy.

The diagram generated from the d^2 octahedral case is more complex but its construction employs the same principles (cf. Fig. 7.29, bottom). This is the diagram for d^2 and d^7 octahedral and d^3 and d^8 tetrahedral complexes on the right, and d^3 and d^8 octahedral and d^2 and d^7 tetrahedral complexes on the left. The correlation lines for the $T_1(F)$ and $T_1(P)$ states curve away from each other as a consequence of the noncrossing rule (cf. Section 7.4). The straight-line projections for these terms (without mutual repulsions) are

*L. E. Orgel, *J. Chem. Phys.* **1955**, *23*, 1004 and 1819.

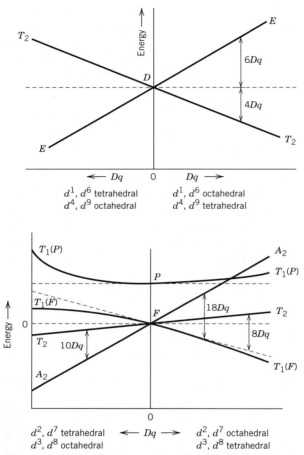

Figure 7.29 Orgel diagrams for interpreting spectra of octahedral and tetrahedral complexes. [Adapted with permission from L. E. Orgel, *J. Chem. Phys.* **1955,** *23,* 1004.]

shown as dashed lines in Fig. 7.29. In general, three absorption bands are expected for complexes covered by this diagram. For complexes treated by the left side of the diagram, the crystal field splitting energy (Δ_o or Δ_t) can be obtained directly from the lowest frequency band in the visible absorption spectrum (assigned as ν_1). For complexes treated by the right side of the diagram, however, the crystal field splitting cannot be obtained directly, owing to the mutual repulsion of the two T_1 states. In these cases it is necessary to carry out a calculation involving estimation of the Racah parameter. Details of this process are given in some advanced inorganic chemistry texts.*

*For example, see B. E. Douglas, D. H. McDaniel, and J. J. Alexander, *Concepts and Models of Inorganic Chemistry, 3rd ed.,* John Wiley & Sons, New York, 1994, p. 252; or G. L. Miessler and D. A. Tarr, *Inorganic Chemistry,* Prentice-Hall, Englewood Cliffs, NJ, 1991, pp. 331–333.

The Orgel diagrams (and also the corresponding Tanabe and Sugano diagrams) predict either one or three bands for the absorption spectra of octahedral complexes. As examination of Fig. 7.28 reveals, actual spectra do not always show this ideal behavior. For example, note that the spectrum of $[Ti(H_2O)_6]^{3+}$ shows evidence of two bands, rather than the expected one band from a $^2T_{2g} \rightarrow ^2E_g$ transition. This is the consequence of the Jahn–Teller effect (cf. Section 7.2). Both the ground-state and excited-state terms are subject to distortion, but the doubly degenerate ^2E_g excited state-term is expected to experience the more pronounced distortion. If we assume a tetragonal distortion leading to D_{4h}, the excited state will be split into two new terms, $^2A_{1g}$ and $^2B_{1g}$. The relative energy ordering of these new terms cannot be predicted, since it will depend on the nature of the distortion (e.g., $^2A_{1g} <$ $^2B_{1g}$ for a stretching distortion along z). Regardless of the ordering of the new terms the observed effect is the same. Ignoring any splitting of the ground-state term, the splitting of the excited state gives rise to two possible transitions (Fig. 7.30), which account for the band shape in the spectrum of $[Ti(H_2O)_6]^{3+}$ shown in Fig. 7.28. Similar evidence of band splitting from Jahn–Teller distortions, either in the ground state or an excited state, can be found in some of the other spectra shown in Fig. 7.28.

While Jahn–Teller effects create additional features, some spectra, such as that of $[V(H_2O)_6]^{3+}$ shown in Fig. 7.28, have only two bands when three are expected from the Orgel diagram. In these cases the highest energy band (ν_3) either falls in the ultraviolet, beyond the region shown, or is obscured by the tail of the very strong LaPorte-allowed charge-transfer band in the ultraviolet. For example, in the case of $[V(H_2O)_6]^{3+}$ the missing band, which corresponds to a $^3T_{1g}(F) \rightarrow ^3A_{2g}$ transition, is estimated to lie at about 36,000 cm^{-1}. Since this transition corresponds to the simultaneous excitation of two electrons, the band is expected to have an extremely low molar absorptivity. This prevents it from being observed in a region where the charge transfer bands are beginning to rise in intensity.*

*B. N. Figgis, *Introduction to Ligand Fields,* Wiley-Interscience, New York, 1966, pp. 218–220.

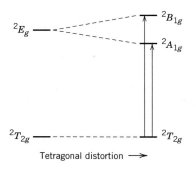

Figure 7.30 Jahn–Teller splitting of the ^2E_g excited state by a tetragonal distortion of a d^1 octahedral complex creates two possible spin-allowed transitions. For simplicity, Jahn–Teller distortion of the $^2T_{2g}$ ground state has been ignored. The ordering shown for the split terms $^2A_{1g}$ and $^2B_{1g}$ presumes a tetragonal distortion involving a stretch along z. In general, neither the nature of the distortion nor the ordering of split terms can be predicted.

Problems

7.1   In the gas-phase synthesis of buckminsterfullerene, C_{60}, it is possible to trap one or more metal atoms inside the "bucky ball." Such *endohedrally doped fullerenes*, as they are called, are given formulas of the type $M_n@C_{60}$. Suppose a single transition metal ion having some d^n- configuration ($n \neq 0$) were trapped in the center of a "bucky ball," forming the species $M@C_{60}$. How would you expect the energies of the d orbitals to be affected by this environment?

7.2   Determine the CFT splitting among d orbitals in ligand fields with the following geometries. In each case, label the levels by Mulliken symbol and specific d orbitals. (a) ML_2—linear; (b) ML_3—trigonal planar; (c) ML_5—trigonal bipyramidal; (d) ML_5—square pyramidal; (e) ML_6—trigonal antiprismatic; (f) ML_7—capped trigonal prismatic (the capped position is above one rectangular face of the trigonal prism); (g) ML_7—pentagonal bipyramidal; (h) ML_8—square antiprismatic; (i) ML_8—cubic; (j) ML_9—tricapped trigonal prismatic.

7.3   For an octahedral ML_6 complex, show the effects on the orbital energies of the t_{2g} and e_g levels from an orthorhombic distortion in which the two positions along z are stretched ($r'_{\pm z} > r_{\pm z}$) and the two positions along x are equally compressed ($r'_{\pm x} < r_{\pm x}$). Is this a possible Jahn–Teller distortion?

7.4   In the manner shown in Section 7.1 for O_h, demonstrate that in the environment of a tetrahedral ML_4 complex (T_d) the operations of C_3 and C_3^2 make the d orbitals of the t_2 set degenerate with each other and make those of the e set degenerate with each other.

7.5   We have noted that no operation of O_h interconverts the t_{2g} and e_g sets of d orbitals. Suppose you could add additional symmetry operations to an octahedron, with the only stipulation being that the new element(s) be coincident with the existing elements of O_h. What symmetry operations would you add to make the t_{2g} and e_g sets of d orbitals equivalent? [*Hint:* It may be useful to consider the $d_{2z^2-x^2-y^2}$ orbital as a combination of wave functions for the unconventional orbitals $d_{z^2-x^2}$ and $d_{z^2-y^2}$:

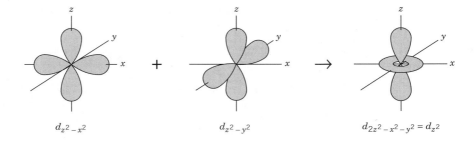

$$d_{z^2-x^2} \quad + \quad d_{z^2-y^2} \quad \rightarrow \quad d_{2z^2-x^2-y^2} = d_{z^2}$$

These wave functions are among six that define orbitals with the familiar four-lobed "cloverleaf" form seen for d_{xy}, d_{yz}, d_{zx}, and $d_{x^2-y^2}$. Like $d_{x^2-y^2}$, the lobes of $d_{z^2-x^2}$ and $d_{z^2-y^2}$ lie along the axes indicated in the subscripts. Owing to restrictions on the quantum number m_l, only five functions can have physical significance, so the $d_{z^2-y^2}$ and $d_{z^2-x^2}$ functions are customarily discarded in favor of a single function, the conventional d_{z^2} wave function (here, $d_{2z^2-x^2-y^2}$), with the form

$$d_{2z^2-x^2-y^2} = (1/\sqrt{2})(d_{z^2-x^2} + d_{z^2-y^2})$$

For the purposes of this exercise, take these two components as if they had independent existence.]

7.6 Using Eqs. (7.2)–(7.6), verify that the five d orbitals of a central metal ion in a complex with the following point group symmetries transform by the irreducible representations indicated in the character tables: (a) D_{3h}, (b) D_{2h}, (c) C_{4v}, (d) I_h (Use the subgroup I and correlate the results to I_h.)

7.7 As in Problem 7.6, determine the irreducible representations by which the seven f orbitals transform in the following point groups: (a) T_d, (b) O_h, (c) D_{4h}, (d) D_{3d}.

7.8 Consider square planar ML_4 transition metal complexes. (a) What configurations of d electrons would cause Jahn–Teller distortions? (b) What kinds of Jahn–Teller distortions might be expected for such complexes? (c) Why are virtually all ML_4 square planar complexes undistorted?

7.9 Consider a trigonal bipyramidal ML_5 transition metal complex.

(a) Construct a simplified sigma-only MO scheme, assuming that only $(n-1)d$ orbitals on the central metal participate in bonding interactions with ligand σ-SALCs (i.e., ignore interactions with metal ns and np AOs). Assume that the energies of the SALCs lie lower than those of the metal AOs.

(b) How does the electron filling in this MO scheme compare with the presumed filling in the d orbitals in the CFT model?

(c) Sketch the LCAOs of the bonding MOs formed between the metal d orbitals and their matching σ-SALCs.

(d) If the metal's ns and np orbitals were included in your model, with which SALCs could they potentially form bonding and antibonding combinations? How might your simplified MO scheme need to be modified to accommodate these additional interactions?

7.10 Consider a square pyramidal ML_5 transition metal complex. Assume that M is slightly above the basal plane, such that θ, the angle between the axial bond and the four basal bonds, falls within the range $90° < \theta < 109.5°$.

(a) Determine the symmetries of σ-SALCs for the five ligands, and determine the symmetries of the $(n-1)d$, ns, and np AOs on M.

(b) Sketch the LCAOs for the bonding combinations between metal AOs and ligand σ-SALCs. Where two metal AOs compete for overlap with the same SALC, decide on the basis of your sketches which one will form the more effective bonding and antibonding interactions, and assume that the other AO is essentially nonbonding.

(c) Construct a qualitative MO scheme. (Do not be concerned about the exact order of bonding MOs filled by ligand electrons or that of their antibonding counterparts.)

(d) How does the filling of electrons in your MO scheme compare with the presumed filling in the d orbitals in the CFT model?

7.11   Consider a linear ML_2 transition metal complex.

(a) Determine the symmetries of both σ-SALCs and π-SALCs.

(b) Assuming that ligand p orbitals are used for both sigma and pi-bonding, sketch the forms and write mathematical expressions for the σ- and π-SALCs.

(c) Identify the symmetries of $(n-1)d$, ns, and np AOs on the central metal. Indicate which AOs can form bonding and antibonding interactions with ligand σ- and π-SALCs. Are any AOs or SALCs nonbonding on the basis of symmetry?

(d) Propose a qualitative MO scheme. Note any simplifying assumptions you have made. Discuss how your scheme would need to be modified if these assumptions proved to be invalid in a specific case.

7.12   Consider a square planar ML_4 transition metal complex.

(a) Determine the symmetries of both σ-SALCs and π-SALCs.

(b) Identify the symmetries of $(n-1)d$, ns, and np AOs on the central metal. Indicate which AOs can form bonding and antibonding interactions with ligand σ- and π-SALCs. Are any AOs or SALCs nonbonding on the basis of symmetry?

(c) Assuming that ligand p orbitals are used for both sigma and pi bonding, write mathematical expressions for the σ- and π-SALCs.

(d) Sketch the LCAOs of bonding and nonbonding MOs.

7.13   Verify that the free-ion terms G, H, and I from d orbital configurations in an octahedral field split into the ligand-field terms shown in Table 7.2. Would the orbitals g, h, and i split in the same ways?

7.14   Given Table 7.2, work out the ligand-field terms corresponding to the free-ion terms S, P, D, F, G, H, I from d orbital configurations in the following point groups: (a) D_{4h}, (b) D_3, (c) D_{2d}, (d) $D_{\infty h}$.

7.15   Consider an octahedral ML_6 transition metal complex with a d^7 low-spin configuration.

(a) The ground-state ligand-field term is ^2E_g. Using "line-and-arrow" notation (e.g., $\uparrow$), show the microstates giving rise to this term.

(b) The ground-state ligand-field term originates from a ^2G free-ion term. What other ligand-field terms originate from ^2G?

(c) How many microstates comprise the ^2G term? Show that the complete set of ligand-field terms originating from ^2G is composed of the same total number of microstates.

7.16   Why do configurations d^n and d^{10-n} give rise to identical ligand-field terms in any given complex ion environment?

7.17   Using a procedure similar to that described for the d^2 octahedral case, construct the correlation diagram for a d^2 ion in a tetrahedral field.

7.18   Develop the correlation diagram for a d^9 ion in a square planar field. Assume that the d orbital energies are ordered as shown on the extreme right side of Fig. 7.9.

7.19 Give explanations for the following:

(a) FeF_6^{3-} is colorless, but $Fe(CN)_6^{3-}$ is red.

(b) Molar absorptivities for absorption bands in the visible spectra of ML_4 tetrahedral complexes tend to be higher than those of ML_6 octahedral complexes.

(c) The visible spectrum of $[Cr(H_2O)_6]^{3+}$ shows two bands (cf. Fig. 7.28).

(d) The visible spectrum of $[Fe(H_2O)_6]^{2+}$ shows one band with two distinct maxima (cf. Fig. 7.28).

(e) CrO_4^{2-} and MnO_4^- are intensely colored ions, despite the d^0 configuration of the central transition metal.

7.20 Consider a d^8 square planar ML_4 transition-metal complex.

(a) Assuming the d orbital energy sequence shown on the extreme right of Fig. 7.9, the singlet states in order of increasing energy correspond to the following configurations: $e_g{}^4 a_{1g}{}^2 b_{2g}{}^2 < e_g{}^4 a_{1g}{}^2 b_{2g}{}^1 b_{1g}{}^1 < e_g{}^4 a_{1g}{}^1 b_{2g}{}^2 b_{1g}{}^1 < e_g{}^3 a_{1g}{}^2 b_{2g}{}^2 b_{1g}{}^1$. Determine the ligand-field term symbols for these four states.

(b) The configurations given in part (a) suggest three possible spin-allowed transitions. Identify them, using state-to-state notation (e.g., for d^1 O_h, the single spin-allowed transition is $^2T_{2g} \rightarrow {}^2E_g$).

(c) Since square planar complexes are centrosymmetric, d–d electronic transitions are LaPorte-forbidden. As with octahedral complexes, such transitions are spectroscopically observable because they are vibronically allowed. Which specific normal modes of square planar ML_4 couple with each of the three spin-allowed electronic transitions to make them vibronically allowed?

APPENDIX A

Point Group Character Tables

1. The Nonaxial Groups

C_1	E
A	1

C_s	E	σ_h		
A'	1	1	x, y, R_z	$x^2, y^2,$ z^2, xy
A''	1	-1	z, R_x, R_y	yz, xz

C_i	E	i		
A_g	1	1	R_x, R_y, R_z	x^2, y^2, z^2 xy, xz, yz
A_u	1	-1	x, y, z	

2. The C_n Groups

C_2	E	C_2		
A	1	1	z, R_z	x^2, y^2, z^2, xy
B	1	-1	x, y, R_x, R_y	yz, xz

C_3	E	C_3	C_3^2		$\epsilon = \exp(2\pi i/3)$
A	1	1	1	z, R_z	$x^2 + y^2, z^2$
E	$\begin{cases} 1 \\ 1 \end{cases}$	$\begin{matrix} \epsilon \\ \epsilon^* \end{matrix}$	$\left.\begin{matrix} \epsilon^* \\ \epsilon \end{matrix}\right\}$	$(x, y), (R_x, R_y)$	$(x^2 - y^2, xy), (yz, xz)$

C_4	E	C_4	C_2	C_4^3		
A	1	1	1	1	z, R_z	$x^2 + y^2, z^2$
B	1	-1	1	-1		$x^2 - y^2, xy$
E	$\begin{cases} 1 \\ 1 \end{cases}$	$\begin{matrix} i \\ -i \end{matrix}$	$\begin{matrix} -1 \\ -1 \end{matrix}$	$\left.\begin{matrix} -i \\ i \end{matrix}\right\}$	$(x, y), (R_x, R_y)$	(yz, xz)

C_5	E	C_5	C_5^2	C_5^3	C_5^4			$\epsilon = \exp(2\pi i/5)$
A	1	1	1	1	1	z, R_z		$x^2 + y^2,\ z^2$
E_1	$\begin{cases}1\\1\end{cases}$	$\begin{matrix}\epsilon\\\epsilon^*\end{matrix}$	$\begin{matrix}\epsilon^2\\\epsilon^{2*}\end{matrix}$	$\begin{matrix}\epsilon^{2*}\\\epsilon^2\end{matrix}$	$\begin{matrix}\epsilon^*\\\epsilon\end{matrix}\Big\}$	$(x, y), (R_x, R_y)$		(yz, xz)
E_2	$\begin{cases}1\\1\end{cases}$	$\begin{matrix}\epsilon^2\\\epsilon^{2*}\end{matrix}$	$\begin{matrix}\epsilon^*\\\epsilon\end{matrix}$	$\begin{matrix}\epsilon\\\epsilon^*\end{matrix}$	$\begin{matrix}\epsilon^{2*}\\\epsilon^2\end{matrix}\Big\}$			$(x^2 - y^2, xy)$

C_6	E	C_6	C_3	C_2	C_3^2	C_6^5			$\epsilon = \exp(2\pi i/6)$
A	1	1	1	1	1	1	z, R_z		$x^2 + y^2,\ z^2$
B	1	-1	1	-1	1	-1			
E_1	$\begin{cases}1\\1\end{cases}$	$\begin{matrix}\epsilon\\\epsilon^*\end{matrix}$	$\begin{matrix}-\epsilon^*\\-\epsilon\end{matrix}$	$\begin{matrix}-1\\-1\end{matrix}$	$\begin{matrix}-\epsilon\\-\epsilon^*\end{matrix}$	$\begin{matrix}\epsilon^*\\\epsilon\end{matrix}\Big\}$	$(x, y),$ (R_x, R_y)		(xz, yz)
E_2	$\begin{cases}1\\1\end{cases}$	$\begin{matrix}-\epsilon^*\\-\epsilon\end{matrix}$	$\begin{matrix}-\epsilon\\-\epsilon^*\end{matrix}$	$\begin{matrix}1\\1\end{matrix}$	$\begin{matrix}-\epsilon^*\\-\epsilon\end{matrix}$	$\begin{matrix}-\epsilon\\-\epsilon^*\end{matrix}\Big\}$			$(x^2 - y^2, xy)$

C_7	E	C_7	C_7^2	C_7^3	C_7^4	C_7^5	C_7^6			$\epsilon = \exp(2\pi i/7)$
A	1	1	1	1	1	1	1	z, R_z		$x^2 + y^2,\ z^2$
E_1	$\begin{cases}1\\1\end{cases}$	$\begin{matrix}\epsilon\\\epsilon^*\end{matrix}$	$\begin{matrix}\epsilon^2\\\epsilon^{2*}\end{matrix}$	$\begin{matrix}\epsilon^3\\\epsilon^{3*}\end{matrix}$	$\begin{matrix}\epsilon^{3*}\\\epsilon^3\end{matrix}$	$\begin{matrix}\epsilon^{2*}\\\epsilon^2\end{matrix}$	$\begin{matrix}\epsilon^*\\\epsilon\end{matrix}\Big\}$	$(x, y),$ (R_x, R_y)		(xz, yz)
E_2	$\begin{cases}1\\1\end{cases}$	$\begin{matrix}\epsilon^2\\\epsilon^{2*}\end{matrix}$	$\begin{matrix}\epsilon^{3*}\\\epsilon^3\end{matrix}$	$\begin{matrix}\epsilon^*\\\epsilon\end{matrix}$	$\begin{matrix}\epsilon\\\epsilon^*\end{matrix}$	$\begin{matrix}\epsilon^3\\\epsilon^{3*}\end{matrix}$	$\begin{matrix}\epsilon^{2*}\\\epsilon^2\end{matrix}\Big\}$			$(x^2 - y^2, xy)$
E_3	$\begin{cases}1\\1\end{cases}$	$\begin{matrix}\epsilon^3\\\epsilon^{3*}\end{matrix}$	$\begin{matrix}\epsilon^*\\\epsilon\end{matrix}$	$\begin{matrix}\epsilon^2\\\epsilon^{2*}\end{matrix}$	$\begin{matrix}\epsilon^{2*}\\\epsilon^2\end{matrix}$	$\begin{matrix}\epsilon\\\epsilon^*\end{matrix}$	$\begin{matrix}\epsilon^{3*}\\\epsilon^3\end{matrix}\Big\}$			

C_8	E	C_8	C_4	C_8^3	C_2	C_8^5	C_4^3	C_8^7			$\epsilon = \exp(2\pi i/8)$
A	1	1	1	1	1	1	1	1	z, R_z		$x^2 + y^2,\ z^2$
B	1	-1	1	-1	1	-1	1	-1			
E_1	$\begin{cases}1\\1\end{cases}$	$\begin{matrix}\epsilon\\\epsilon^*\end{matrix}$	$\begin{matrix}i\\-i\end{matrix}$	$\begin{matrix}-\epsilon^*\\-\epsilon\end{matrix}$	$\begin{matrix}-1\\-1\end{matrix}$	$\begin{matrix}-\epsilon\\-\epsilon^*\end{matrix}$	$\begin{matrix}-i\\i\end{matrix}$	$\begin{matrix}\epsilon^*\\\epsilon\end{matrix}\Big\}$	$(x, y),$ (R_x, R_y)		(xz, yz)
E_2	$\begin{cases}1\\1\end{cases}$	$\begin{matrix}i\\-i\end{matrix}$	$\begin{matrix}-1\\-1\end{matrix}$	$\begin{matrix}-i\\i\end{matrix}$	$\begin{matrix}1\\1\end{matrix}$	$\begin{matrix}i\\-i\end{matrix}$	$\begin{matrix}-1\\-1\end{matrix}$	$\begin{matrix}-i\\i\end{matrix}\Big\}$			$(x^2 - y^2, xy)$
E_3	$\begin{cases}1\\1\end{cases}$	$\begin{matrix}-\epsilon\\-\epsilon^*\end{matrix}$	$\begin{matrix}i\\-i\end{matrix}$	$\begin{matrix}\epsilon^*\\\epsilon\end{matrix}$	$\begin{matrix}-1\\-1\end{matrix}$	$\begin{matrix}\epsilon\\\epsilon^*\end{matrix}$	$\begin{matrix}-i\\i\end{matrix}$	$\begin{matrix}-\epsilon^*\\-\epsilon\end{matrix}\Big\}$			

3. The D_n Groups

D_2	E	$C_2(z)$	$C_2(y)$	$C_2(x)$		
A	1	1	1	1		x^2, y^2, z^2
B_1	1	1	-1	-1	z, R_z	xy
B_2	1	-1	1	-1	y, R_y	xz
B_3	1	-1	-1	1	x, R_x	yz

D_3	E	$2C_3$	$3C_2$		
A_1	1	1	1		$x^2 + y^2, z^2$
A_2	1	1	-1	z, R_z	
E	2	-1	0	$(x, y), (R_x, R_y)$	$(x^2 - y^2, xy), (xz, yz)$

D_4	E	$2C_4$	$C_2 (=C_4^2)$	$2C_2'$	$2C_2''$		
A_1	1	1	1	1	1		$x^2 + y^2, z^2$
A_2	1	1	1	-1	-1	z, R_z	
B_1	1	-1	1	1	-1		$x^2 - y^2$
B_2	1	-1	1	-1	1		xy
E	2	0	-2	0	0	$(x, y), (R_x, R_y)$	(xz, yz)

D_5	E	$2C_5$	$2C_5^2$	$5C_2$		
A_1	1	1	1	1		$x^2 + y^2, z^2$
A_2	1	1	1	-1	z, R_z	
E_1	2	$2 \cos 72°$	$2 \cos 144°$	0	$(x, y), (R_x, R_y)$	(xz, yz)
E_2	2	$2 \cos 144°$	$2 \cos 72°$	0		$(x^2 - y^2, xy)$

D_6	E	$2C_6$	$2C_3$	C_2	$3C_2'$	$3C_2''$		
A_1	1	1	1	1	1	1		$x^2 + y^2, z^2$
A_2	1	1	1	1	-1	-1	z, R_z	
B_1	1	-1	1	-1	1	-1		
B_2	1	-1	1	-1	-1	1		
E_1	2	1	-1	-2	0	0	$(x, y), (R_x, R_y)$	(xz, yz)
E_2	2	-1	-1	2	0	0		$(x^2 - y^2, xy)$

4. The C_{nv} Groups

C_{2v}	E	C_2	$\sigma_v(xz)$	$\sigma_v'(yz)$		
A_1	1	1	1	1	z	x^2, y^2, z^2
A_2	1	1	-1	-1	R_z	xy
B_1	1	-1	1	-1	x, R_y	xz
B_2	1	-1	-1	1	y, R_x	yz

C_{3v}	E	$2C_3$	$3\sigma_v$		
A_1	1	1	1	z	$x^2 + y^2, z^2$
A_2	1	1	-1	R_z	
E	2	-1	0	$(x, y), (R_x, R_y)$	$(x^2 - y^2, xy), (xz, yz)$

Raman

C_{4v}	E	$2C_4$	C_2	$2\sigma_v$	$2\sigma_d$		
A_1	1	1	1	1	1	z	$x^2 + y^2, z^2$
A_2	1	1	1	-1	-1	R_z	
B_1	1	-1	1	1	-1		$x^2 - y^2$
B_2	1	-1	1	-1	1		xy
E	2	0	-2	0	0	$(x, y), (R_x, R_y)$	(xz, yz)

C_{5v}	E	$2C_5$	$2C_5^2$	$5\sigma_v$		
A_1	1	1	1	1	z	$x^2 + y^2, z^2$
A_2	1	1	1	-1	R_z	
E_1	2	$2\cos 72°$	$2\cos 144°$	0	$(x, y), (R_x, R_y)$	(xz, yz)
E_2	2	$2\cos 144°$	$2\cos 72°$	0		$(x^2 - y^2, xy)$

C_{6v}	E	$2C_6$	$2C_3$	C_2	$3\sigma_v$	$3\sigma_d$		
A_1	1	1	1	1	1	1	z	$x^2 + y^2, z^2$
A_2	1	1	1	1	-1	-1	R_z	
B_1	1	-1	1	-1	1	-1		
B_2	1	-1	1	-1	-1	1		
E_1	2	1	-1	-2	0	0	$(x, y), (R_x, R_y)$	(xz, yz)
E_2	2	-1	-1	2	0	0		$(x^2 - y^2, xy)$

5. The C_{nh} Groups

C_{2h}	E	C_2	i	σ_h		
A_g	1	1	1	1	R_z	x^2, y^2, z^2, xy
B_g	1	-1	1	-1	R_x, R_y	xz, yz
A_u	1	1	-1	-1	z	
B_u	1	-1	-1	1	x, y	

C_{3h}	E	C_3	C_3^2	σ_h	S_3	S_3^5		$\epsilon = \exp(2\pi i/3)$
A'	1	1	1	1	1	1	R_z	$x^2 + y^2, z^2$
E'	$\begin{cases} 1 \\ 1 \end{cases}$	$\begin{matrix} \epsilon \\ \epsilon^* \end{matrix}$	$\begin{matrix} \epsilon^* \\ \epsilon \end{matrix}$	$\begin{matrix} 1 \\ 1 \end{matrix}$	$\begin{matrix} \epsilon \\ \epsilon^* \end{matrix}$	$\begin{matrix} \epsilon^* \\ \epsilon \end{matrix}$	(x, y)	$(x^2 - y^2, xy)$
A''	1	1	1	-1	-1	-1	z	
E''	$\begin{cases} 1 \\ 1 \end{cases}$	$\begin{matrix} \epsilon \\ \epsilon^* \end{matrix}$	$\begin{matrix} \epsilon^* \\ \epsilon \end{matrix}$	$\begin{matrix} -1 \\ -1 \end{matrix}$	$\begin{matrix} -\epsilon \\ -\epsilon^* \end{matrix}$	$\begin{matrix} -\epsilon^* \\ -\epsilon \end{matrix}$	(R_x, R_y)	(xz, yz)

C_{4h}	E	C_4	C_2	C_4^3	i	S_4^3	σ_h	S_4		
A_g	1	1	1	1	1	1	1	1	R_z	$x^2+y^2,\ z^2$
B_g	1	-1	1	-1	1	-1	1	-1		$x^2-y^2,\ xy$
E_g	$\left\{\begin{matrix}1\\1\end{matrix}\right.$	$\begin{matrix}i\\-i\end{matrix}$	$\begin{matrix}-1\\-1\end{matrix}$	$\begin{matrix}-i\\i\end{matrix}$	$\begin{matrix}1\\1\end{matrix}$	$\begin{matrix}i\\-i\end{matrix}$	$\begin{matrix}-1\\-1\end{matrix}$	$\left.\begin{matrix}-i\\i\end{matrix}\right\}$	(R_x, R_y)	(xz, yz)
A_u	1	1	1	1	-1	-1	-1	-1	z	
B_u	1	-1	1	-1	-1	1	-1	1		
E_u	$\left\{\begin{matrix}1\\1\end{matrix}\right.$	$\begin{matrix}i\\-i\end{matrix}$	$\begin{matrix}-1\\-1\end{matrix}$	$\begin{matrix}-i\\i\end{matrix}$	$\begin{matrix}-1\\-1\end{matrix}$	$\begin{matrix}-i\\i\end{matrix}$	$\begin{matrix}1\\1\end{matrix}$	$\left.\begin{matrix}i\\-i\end{matrix}\right\}$	(x, y)	

C_{5h}	E	C_5	C_5^2	C_5^3	C_5^4	σ_h	S_5	S_5^7	S_5^3	S_5^9		$\epsilon = \exp(2\pi i/5)$
A'	1	1	1	1	1	1	1	1	1	1	R_z	$x^2+y^2,\ z^2$
E_1'	$\left\{\begin{matrix}1\\1\end{matrix}\right.$	$\begin{matrix}\epsilon\\\epsilon^*\end{matrix}$	$\begin{matrix}\epsilon^2\\\epsilon^{2*}\end{matrix}$	$\begin{matrix}\epsilon^{2*}\\\epsilon^2\end{matrix}$	$\begin{matrix}\epsilon^*\\\epsilon\end{matrix}$	$\begin{matrix}1\\1\end{matrix}$	$\begin{matrix}\epsilon\\\epsilon^*\end{matrix}$	$\begin{matrix}\epsilon^2\\\epsilon^{2*}\end{matrix}$	$\begin{matrix}\epsilon^{2*}\\\epsilon^2\end{matrix}$	$\left.\begin{matrix}\epsilon^*\\\epsilon\end{matrix}\right\}$	(x, y)	
E_2'	$\left\{\begin{matrix}1\\1\end{matrix}\right.$	$\begin{matrix}\epsilon^2\\\epsilon^{2*}\end{matrix}$	$\begin{matrix}\epsilon^*\\\epsilon\end{matrix}$	$\begin{matrix}\epsilon\\\epsilon^*\end{matrix}$	$\begin{matrix}\epsilon^{2*}\\\epsilon^2\end{matrix}$	$\begin{matrix}1\\1\end{matrix}$	$\begin{matrix}\epsilon^2\\\epsilon^{2*}\end{matrix}$	$\begin{matrix}\epsilon^*\\\epsilon\end{matrix}$	$\begin{matrix}\epsilon\\\epsilon^*\end{matrix}$	$\left.\begin{matrix}\epsilon^{2*}\\\epsilon^2\end{matrix}\right\}$		$(x^2-y^2,\ xy)$
A''	1	1	1	1	1	-1	-1	-1	-1	-1	z	
E_1''	$\left\{\begin{matrix}1\\1\end{matrix}\right.$	$\begin{matrix}\epsilon\\\epsilon^*\end{matrix}$	$\begin{matrix}\epsilon^2\\\epsilon^{2*}\end{matrix}$	$\begin{matrix}\epsilon^{2*}\\\epsilon^2\end{matrix}$	$\begin{matrix}\epsilon^*\\\epsilon\end{matrix}$	$\begin{matrix}-1\\-1\end{matrix}$	$\begin{matrix}-\epsilon\\-\epsilon^*\end{matrix}$	$\begin{matrix}-\epsilon^2\\-\epsilon^{2*}\end{matrix}$	$\begin{matrix}-\epsilon^{2*}\\-\epsilon^2\end{matrix}$	$\left.\begin{matrix}-\epsilon^*\\-\epsilon\end{matrix}\right\}$	(R_x, R_y)	(xz, yz)
E_2''	$\left\{\begin{matrix}1\\1\end{matrix}\right.$	$\begin{matrix}\epsilon^2\\\epsilon^{2*}\end{matrix}$	$\begin{matrix}\epsilon^*\\\epsilon\end{matrix}$	$\begin{matrix}\epsilon\\\epsilon^*\end{matrix}$	$\begin{matrix}\epsilon^{2*}\\\epsilon^2\end{matrix}$	$\begin{matrix}-1\\-1\end{matrix}$	$\begin{matrix}-\epsilon^2\\-\epsilon^{2*}\end{matrix}$	$\begin{matrix}-\epsilon^*\\-\epsilon\end{matrix}$	$\begin{matrix}-\epsilon\\-\epsilon^*\end{matrix}$	$\left.\begin{matrix}-\epsilon^{2*}\\-\epsilon^2\end{matrix}\right\}$		

C_{6h}	E	C_6	C_3	C_2	C_3^2	C_6^5	i	S_3^5	S_6^5	σ_h	S_6	S_3		$\epsilon = \exp(2\pi i/6)$
A_g	1	1	1	1	1	1	1	1	1	1	1	1	R_z	$x^2+y^2,\ z^2$
B_g	1	-1	1	-1	1	-1	1	-1	1	-1	1	-1		
E_{1g}	$\left\{\begin{matrix}1\\1\end{matrix}\right.$	$\begin{matrix}\epsilon\\\epsilon^*\end{matrix}$	$\begin{matrix}-\epsilon^*\\-\epsilon\end{matrix}$	$\begin{matrix}-1\\-1\end{matrix}$	$\begin{matrix}-\epsilon\\-\epsilon^*\end{matrix}$	$\begin{matrix}\epsilon^*\\\epsilon\end{matrix}$	$\begin{matrix}1\\1\end{matrix}$	$\begin{matrix}\epsilon\\\epsilon^*\end{matrix}$	$\begin{matrix}-\epsilon^*\\-\epsilon\end{matrix}$	$\begin{matrix}-1\\-1\end{matrix}$	$\begin{matrix}-\epsilon\\-\epsilon^*\end{matrix}$	$\left.\begin{matrix}\epsilon^*\\\epsilon\end{matrix}\right\}$	(R_x, R_y)	(xz, yz)
E_{2g}	$\left\{\begin{matrix}1\\1\end{matrix}\right.$	$\begin{matrix}-\epsilon^*\\-\epsilon\end{matrix}$	$\begin{matrix}-\epsilon\\-\epsilon^*\end{matrix}$	$\begin{matrix}1\\1\end{matrix}$	$\begin{matrix}-\epsilon^*\\-\epsilon\end{matrix}$	$\begin{matrix}-\epsilon\\-\epsilon^*\end{matrix}$	$\begin{matrix}1\\1\end{matrix}$	$\begin{matrix}-\epsilon^*\\-\epsilon\end{matrix}$	$\begin{matrix}-\epsilon\\-\epsilon^*\end{matrix}$	$\begin{matrix}1\\1\end{matrix}$	$\begin{matrix}-\epsilon^*\\-\epsilon\end{matrix}$	$\left.\begin{matrix}-\epsilon\\-\epsilon^*\end{matrix}\right\}$		$(x^2-y^2,\ xy)$
A_u	1	1	1	1	1	1	-1	-1	-1	-1	-1	-1	z	
B_u	1	-1	1	-1	1	-1	-1	1	-1	1	-1	1		
E_{1u}	$\left\{\begin{matrix}1\\1\end{matrix}\right.$	$\begin{matrix}\epsilon\\\epsilon^*\end{matrix}$	$\begin{matrix}-\epsilon^*\\-\epsilon\end{matrix}$	$\begin{matrix}-1\\-1\end{matrix}$	$\begin{matrix}-\epsilon\\-\epsilon^*\end{matrix}$	$\begin{matrix}\epsilon^*\\\epsilon\end{matrix}$	$\begin{matrix}-1\\-1\end{matrix}$	$\begin{matrix}-\epsilon\\-\epsilon^*\end{matrix}$	$\begin{matrix}\epsilon^*\\\epsilon\end{matrix}$	$\begin{matrix}1\\1\end{matrix}$	$\begin{matrix}\epsilon\\\epsilon^*\end{matrix}$	$\left.\begin{matrix}-\epsilon^*\\-\epsilon\end{matrix}\right\}$	(x, y)	
E_{2u}	$\left\{\begin{matrix}1\\1\end{matrix}\right.$	$\begin{matrix}-\epsilon^*\\-\epsilon\end{matrix}$	$\begin{matrix}-\epsilon\\-\epsilon^*\end{matrix}$	$\begin{matrix}1\\1\end{matrix}$	$\begin{matrix}-\epsilon^*\\-\epsilon\end{matrix}$	$\begin{matrix}-\epsilon\\-\epsilon^*\end{matrix}$	$\begin{matrix}-1\\-1\end{matrix}$	$\begin{matrix}\epsilon^*\\\epsilon\end{matrix}$	$\begin{matrix}\epsilon\\\epsilon^*\end{matrix}$	$\begin{matrix}-1\\-1\end{matrix}$	$\begin{matrix}\epsilon^*\\\epsilon\end{matrix}$	$\left.\begin{matrix}\epsilon\\\epsilon^*\end{matrix}\right\}$		

6. The D_{nh} Groups

D_{2h}	E	$C_2(z)$	$C_2(y)$	$C_2(x)$	i	$\sigma(xy)$	$\sigma(xz)$	$\sigma(yz)$		
A_g	1	1	1	1	1	1	1	1		x^2, y^2, z^2
B_{1g}	1	1	-1	-1	1	1	-1	-1	R_z	xy
B_{2g}	1	-1	1	-1	1	-1	1	-1	R_y	xz
B_{3g}	1	-1	-1	1	1	-1	-1	1	R_x	yz
A_u	1	1	1	1	-1	-1	-1	-1		
B_{1u}	1	1	-1	-1	-1	-1	1	1	z	
B_{2u}	1	-1	1	-1	-1	1	-1	1	y	
B_{3u}	1	-1	-1	1	-1	1	1	-1	x	

D_{3h}	E	$2C_3$	$3C_2$	σ_h	$2S_3$	$3\sigma_v$		
A_1'	1	1	1	1	1	1		$x^2 + y^2, z^2$
A_2'	1	1	-1	1	1	-1	R_z	
E'	2	-1	0	2	-1	0	(x, y)	$(x^2 - y^2, xy)$
A_1''	1	1	1	-1	-1	-1		
A_2''	1	1	-1	-1	-1	1	z	
E''	2	-1	0	-2	1	0	(R_x, R_y)	(xz, yz)

D_{4h}	E	$2C_4$	C_2	$2C_2'$	$2C_2''$	i	$2S_4$	σ_h	$2\sigma_v$	$2\sigma_d$		
A_{1g}	1	1	1	1	1	1	1	1	1	1		$x^2 + y^2, z^2$
A_{2g}	1	1	1	-1	-1	1	1	1	-1	-1	R_z	
B_{1g}	1	-1	1	1	-1	1	-1	1	1	-1		$x^2 - y^2$
B_{2g}	1	-1	1	-1	1	1	-1	1	-1	1		xy
E_g	2	0	-2	0	0	2	0	-2	0	0	(R_x, R_y)	(xz, yz)
A_{1u}	1	1	1	1	1	-1	-1	-1	-1	-1		
A_{2u}	1	1	1	-1	-1	-1	-1	-1	1	1	z	
B_{1u}	1	-1	1	1	-1	-1	1	-1	-1	1		
B_{2u}	1	-1	1	-1	1	-1	1	-1	1	-1		
E_u	2	0	-2	0	0	-2	0	2	0	0	(x, y)	

D_{5h}	E	$2C_5$	$2C_5^2$	$5C_2$	σ_h	$2S_5$	$2S_5^3$	$5\sigma_v$		
A_1'	1	1	1	1	1	1	1	1		$x^2 + y^2, z^2$
A_2'	1	1	1	-1	1	1	1	-1	R_z	
E_1'	2	$2\cos 72°$	$2\cos 144°$	0	2	$2\cos 72°$	$2\cos 144°$	0	(x, y)	
E_2'	2	$2\cos 144°$	$2\cos 72°$	0	2	$2\cos 144°$	$2\cos 72°$	0		$(x^2 - y^2, xy)$
A_1''	1	1	1	1	-1	-1	-1	-1		
A_2''	1	1	1	-1	-1	-1	-1	1	z	
E_1''	2	$2\cos 72°$	$2\cos 144°$	0	-2	$-2\cos 72°$	$-2\cos 144°$	0	(R_x, R_y)	(xz, yz)
E_2''	2	$2\cos 144°$	$2\cos 72°$	0	-2	$-2\cos 144°$	$-2\cos 72°$	0		

D_{6h}	E	$2C_6$	$2C_3$	C_2	$3C_2'$	$3C_2''$	i	$2S_3$	$2S_6$	σ_h	$3\sigma_d$	$3\sigma_v$		
A_{1g}	1	1	1	1	1	1	1	1	1	1	1	1		x^2+y^2, z^2
A_{2g}	1	1	1	1	-1	-1	1	1	1	1	-1	-1	R_z	
B_{1g}	1	-1	1	-1	1	-1	1	-1	1	-1	1	-1		
B_{2g}	1	-1	1	-1	-1	1	1	-1	1	-1	-1	1		
E_{1g}	2	1	-1	-2	0	0	2	1	-1	-2	0	0	(R_x, R_y)	(xz, yz)
E_{2g}	2	-1	-1	2	0	0	2	-1	-1	2	0	0		(x^2-y^2, xy)
A_{1u}	1	1	1	1	1	1	-1	-1	-1	-1	-1	-1		
A_{2u}	1	1	1	1	-1	-1	-1	-1	-1	-1	1	1	z	
B_{1u}	1	-1	1	-1	1	-1	-1	1	-1	1	-1	1		
B_{2u}	1	-1	1	-1	-1	1	-1	1	-1	1	1	-1		
E_{1u}	2	1	-1	-2	0	0	-2	-1	1	2	0	0	(x, y)	
E_{2u}	2	-1	-1	2	0	0	-2	1	1	-2	0	0		

D_{8h}	E	$2C_8^3$	$2C_8$	$2C_4$	C_2	$4C_2'$	$4C_2''$	i	$2S_8^3$	$2S_8$	$2S_4$	σ_h	$4\sigma_d$	$4\sigma_v$		
A_{1g}	1	1	1	1	1	1	1	1	1	1	1	1	1	1		x^2+y^2, z^2
A_{2g}	1	1	1	1	1	-1	-1	1	1	1	1	1	-1	-1	R_z	
B_{1g}	1	-1	-1	1	1	1	-1	1	-1	-1	1	1	1	-1		
B_{2g}	1	-1	-1	1	1	-1	1	1	-1	-1	1	1	-1	1		
E_{1g}	2	$\sqrt{2}$	$-\sqrt{2}$	0	-2	0	0	2	$\sqrt{2}$	$-\sqrt{2}$	0	-2	0	0	(R_x, R_y)	(xz, yz)
E_{2g}	2	0	0	-2	2	0	0	2	0	0	-2	2	0	0		(x^2-y^2, xy)
E_{3g}	2	$-\sqrt{2}$	$\sqrt{2}$	0	-2	0	0	2	$-\sqrt{2}$	$\sqrt{2}$	0	-2	0	0		
A_{1u}	1	1	1	1	1	1	1	-1	-1	-1	-1	-1	-1	-1		
A_{2u}	1	1	1	1	1	-1	-1	-1	-1	-1	-1	-1	1	1	z	
B_{1u}	1	-1	-1	1	1	1	-1	-1	1	1	-1	-1	-1	1		
B_{2u}	1	-1	-1	1	1	-1	1	-1	1	1	-1	-1	1	-1		
E_{1u}	2	$\sqrt{2}$	$-\sqrt{2}$	0	-2	0	0	-2	$-\sqrt{2}$	$\sqrt{2}$	0	2	0	0	(x, y)	
E_{2u}	2	0	0	-2	2	0	0	-2	0	0	2	-2	0	0		
E_{3u}	2	$-\sqrt{2}$	$\sqrt{2}$	0	-2	0	0	-2	$\sqrt{2}$	$-\sqrt{2}$	0	2	0	0		

7. The D_{nd} Groups

D_{2d}	E	$2S_4$	C_2	$2C_2'$	$2\sigma_d$		
A_1	1	1	1	1	1		x^2+y^2, z^2
A_2	1	1	1	-1	-1	R_z	
B_1	1	-1	1	1	-1		x^2-y^2
B_2	1	-1	1	-1	1	z	xy
E	2	0	-2	0	0	$(x, y),$ (R_x, R_y)	(xz, yz)

D_{3d}	E	$2C_3$	$3C_2$	i	$2S_6$	$3\sigma_d$		
A_{1g}	1	1	1	1	1	1		x^2+y^2, z^2
A_{2g}	1	1	-1	1	1	-1	R_z	
E_g	2	-1	0	2	-1	0	(R_x, R_y)	$(x^2-y^2, xy),$ (xz, yz)
A_{1u}	1	1	1	-1	-1	-1		
A_{2u}	1	1	-1	-1	-1	1	z	
E_u	2	-1	0	-2	1	0	(x, y)	

D_{4d}	E	$2S_8$	$2C_4$	$2S_8^3$	C_2	$4C_2'$	$4\sigma_d$		
A_1	1	1	1	1	1	1	1		x^2+y^2, z^2
A_2	1	1	1	1	1	-1	-1	R_z	
B_1	1	-1	1	-1	1	1	-1		
B_2	1	-1	1	-1	1	-1	1	z	
E_1	2	$\sqrt{2}$	0	$-\sqrt{2}$	-2	0	0	(x, y)	
E_2	2	0	-2	0	2	0	0		(x^2-y^2, xy)
E_3	2	$-\sqrt{2}$	0	$\sqrt{2}$	-2	0	0	(R_x, R_y)	(xz, yz)

D_{5d}	E	$2C_5$	$2C_5^2$	$5C_2$	i	$2S_{10}^3$	$2S_{10}$	$5\sigma_d$		
A_{1g}	1	1	1	1	1	1	1	1		x^2+y^2, z^2
A_{2g}	1	1	1	-1	1	1	1	-1	R_z	
E_{1g}	2	$2\cos 72°$	$2\cos 144°$	0	2	$2\cos 72°$	$2\cos 144°$	0	(R_x, R_y)	(xz, yz)
E_{2g}	2	$2\cos 144°$	$2\cos 72°$	0	2	$2\cos 144°$	$2\cos 72°$	0		(x^2-y^2, xy)
A_{1u}	1	1	1	1	-1	-1	-1	-1		
A_{2u}	1	1	1	-1	-1	-1	-1	1	z	
E_{1u}	2	$2\cos 72°$	$2\cos 144°$	0	-2	$-2\cos 72°$	$-2\cos 144°$	0	(x, y)	
E_{2u}	2	$2\cos 144°$	$2\cos 72°$	0	-2	$-2\cos 144°$	$-2\cos 72°$	0		

D_{6d}	E	$2S_{12}$	$2C_6$	$2S_4$	$2C_3$	$2S_{12}^5$	C_2	$6C_2'$	$6\sigma_d$		
A_1	1	1	1	1	1	1	1	1	1		x^2+y^2, z^2
A_2	1	1	1	1	1	1	1	-1	-1	R_z	
B_1	1	-1	1	-1	1	-1	1	1	-1		
B_2	1	-1	1	-1	1	-1	1	-1	1	z	
E_1	2	$\sqrt{3}$	1	0	-1	$-\sqrt{3}$	-2	0	0	(x, y)	
E_2	2	1	-1	-2	-1	1	2	0	0		(x^2-y^2, xy)
E_3	2	0	-2	0	2	0	-2	0	0		
E_4	2	-1	-1	2	-1	-1	2	0	0		
E_5	2	$-\sqrt{3}$	1	0	-1	$\sqrt{3}$	-2	0	0	(R_x, R_y)	(xz, yz)

8. The S_n Groups

S_4	E	S_4	C_2	S_4^3		
A	1	1	1	1	R_z	x^2+y^2, z^2
B	1	-1	1	-1	z	x^2-y^2, xy
E	$\begin{cases}1 \\ 1\end{cases}$	$\begin{matrix}i \\ -i\end{matrix}$	$\begin{matrix}-1 \\ -1\end{matrix}$	$\begin{matrix}-i \\ i\end{matrix}$	$(x, y), (R_x, R_y)$	(xz, yz)

S_6	E	C_3	C_3^2	i	S_6^5	S_6			$\epsilon = \exp(2\pi i/3)$
A_g	1	1	1	1	1	1	R_z		$x^2 + y^2,\ z^2$
E_g	$\begin{cases}1\\1\end{cases}$	$\begin{matrix}\epsilon\\\epsilon^*\end{matrix}$	$\begin{matrix}\epsilon^*\\\epsilon\end{matrix}$	$\begin{matrix}1\\1\end{matrix}$	$\begin{matrix}\epsilon\\\epsilon^*\end{matrix}$	$\left.\begin{matrix}\epsilon^*\\\epsilon\end{matrix}\right\}$	(R_x, R_y)		$(x^2-y^2, xy),$ (xz, yz)
A_u	1	1	1	−1	−1	−1	z		
E_u	$\begin{cases}1\\1\end{cases}$	$\begin{matrix}\epsilon\\\epsilon^*\end{matrix}$	$\begin{matrix}\epsilon^*\\\epsilon\end{matrix}$	$\begin{matrix}-1\\-1\end{matrix}$	$\begin{matrix}-\epsilon\\-\epsilon^*\end{matrix}$	$\left.\begin{matrix}-\epsilon^*\\-\epsilon\end{matrix}\right\}$	(x, y)		

S_8	E	S_8	C_4	S_8^3	C_2	S_8^5	C_4^3	S_8^7			$\epsilon = \exp(2\pi i/8)$
A	1	1	1	1	1	1	1	1	R_z		$x^2 + y^2,\ z^2$
B	1	−1	1	−1	1	−1	1	−1	z		
E_1	$\begin{cases}1\\1\end{cases}$	$\begin{matrix}\epsilon\\\epsilon^*\end{matrix}$	$\begin{matrix}i\\-i\end{matrix}$	$\begin{matrix}-\epsilon^*\\-\epsilon\end{matrix}$	$\begin{matrix}-1\\-1\end{matrix}$	$\begin{matrix}-\epsilon\\-\epsilon^*\end{matrix}$	$\begin{matrix}-i\\i\end{matrix}$	$\left.\begin{matrix}\epsilon^*\\\epsilon\end{matrix}\right\}$	$(x, y),$ (R_x, R_y)		
E_2	$\begin{cases}1\\1\end{cases}$	$\begin{matrix}i\\-i\end{matrix}$	$\begin{matrix}-1\\-1\end{matrix}$	$\begin{matrix}-i\\i\end{matrix}$	$\begin{matrix}1\\1\end{matrix}$	$\begin{matrix}i\\-i\end{matrix}$	$\begin{matrix}-1\\-1\end{matrix}$	$\left.\begin{matrix}-i\\i\end{matrix}\right\}$			$(x^2 - y^2, xy)$
E_3	$\begin{cases}1\\1\end{cases}$	$\begin{matrix}-\epsilon^*\\-\epsilon\end{matrix}$	$\begin{matrix}-i\\i\end{matrix}$	$\begin{matrix}\epsilon\\\epsilon^*\end{matrix}$	$\begin{matrix}-1\\-1\end{matrix}$	$\begin{matrix}\epsilon^*\\\epsilon\end{matrix}$	$\begin{matrix}i\\-i\end{matrix}$	$\left.\begin{matrix}-\epsilon\\-\epsilon^*\end{matrix}\right\}$			(xz, yz)

9. The Cubic Groups

T	E	$4C_3$	$4C_3^2$	$3C_2$			$\epsilon = \exp(2\pi i/3)$
A	1	1	1	1			$x^2 + y^2 + z^2$
E	$\begin{cases}1\\1\end{cases}$	$\begin{matrix}\epsilon\\\epsilon^*\end{matrix}$	$\begin{matrix}\epsilon^*\\\epsilon\end{matrix}$	$\begin{matrix}1\\1\end{matrix}$			$(2z^2 - x^2 - y^2,$ $x^2 - y^2)$
T	3	0	0	−1	$(R_x, R_y, R_z),\ (x, y, z)$		(xy, xz, yz)

T_h	E	$4C_3$	$4C_3^2$	$3C_2$	i	$4S_6^5$	$4S_6$	$3\sigma_h$			$\epsilon = \exp(2\pi i/3)$
A_g	1	1	1	1	1	1	1	1			$x^2 + y^2 + z^2$
E_g	$\begin{cases}1\\1\end{cases}$	$\begin{matrix}\epsilon\\\epsilon^*\end{matrix}$	$\begin{matrix}\epsilon^*\\\epsilon\end{matrix}$	$\begin{matrix}1\\1\end{matrix}$	$\begin{matrix}1\\1\end{matrix}$	$\begin{matrix}\epsilon\\\epsilon^*\end{matrix}$	$\begin{matrix}\epsilon^*\\\epsilon\end{matrix}$	$\left.\begin{matrix}1\\1\end{matrix}\right\}$			$(2z^2 - x^2 - y^2,\ x^2 - y^2)$
T_g	3	0	0	−1	3	0	0	−1	(R_x, R_y, R_z)		(xz, yz, xy)
A_u	1	1	1	1	−1	−1	−1	−1			
E_u	$\begin{cases}1\\1\end{cases}$	$\begin{matrix}\epsilon\\\epsilon^*\end{matrix}$	$\begin{matrix}\epsilon^*\\\epsilon\end{matrix}$	$\begin{matrix}1\\1\end{matrix}$	$\begin{matrix}-1\\-1\end{matrix}$	$\begin{matrix}-\epsilon\\-\epsilon^*\end{matrix}$	$\begin{matrix}-\epsilon^*\\-\epsilon\end{matrix}$	$\left.\begin{matrix}-1\\-1\end{matrix}\right\}$			
T_u	3	0	0	−1	−3	0	0	1	(x, y, z)		

T_d	E	$8C_3$	$3C_2$	$6S_4$	$6\sigma_d$		
A_1	1	1	1	1	1		$x^2 + y^2 + z^2$
A_2	1	1	1	−1	−1		
E	2	−1	2	0	0		$(2z^2 - x^2 - y^2, x^2 - y^2)$
T_1	3	0	−1	1	−1	(R_x, R_y, R_z)	
T_2	3	0	−1	−1	1	(x, y, z)	(xy, xz, yz)

O	E	$8C_3$	$3C_2(=C_4^2)$	$6C_4$	$6C_2$		
A_1	1	1	1	1	1		$x^2+y^2+z^2$
A_2	1	1	1	-1	-1		
E	2	-1	2	0	0		$(2z^2-x^2-y^2, x^2-y^2)$
T_1	3	0	-1	1	-1	$(R_x, R_y, R_z), (x, y, z)$	
T_2	3	0	-1	-1	1		(xy, xz, yz)

O_h	E	$8C_3$	$6C_2$	$6C_4$	$3C_2(=C_4^2)$	i	$6S_4$	$8S_6$	$3\sigma_h$	$6\sigma_d$		
A_{1g}	1	1	1	1	1	1	1	1	1	1		$x^2+y^2+z^2$
A_{2g}	1	1	-1	-1	1	1	-1	1	1	-1		
E_g	2	-1	0	0	2	2	0	-1	2	0		$(2z^2-x^2-y^2, x^2-y^2)$
T_{1g}	3	0	-1	1	-1	3	1	0	-1	-1	(R_x, R_y, R_z)	
T_{2g}	3	0	1	-1	-1	3	-1	0	-1	1		(xz, yz, xy)
A_{1u}	1	1	1	1	1	-1	-1	-1	-1	-1		
A_{2u}	1	1	-1	-1	1	-1	1	-1	-1	1		
E_u	2	-1	0	0	2	-2	0	1	-2	0		
T_{1u}	3	0	-1	1	-1	-3	-1	0	1	1	(x, y, z)	
T_{2u}	3	0	1	-1	-1	-3	1	0	1	-1		

Rotational

trace

10. The Groups $C_{\infty v}$ and $D_{\infty h}$ for Linear Molecules

$C_{\infty v}$	E	$2C_\infty^\Phi$	$\cdots$	$\infty \sigma_v$		
$A_1 \equiv \Sigma^+$	1	1	$\cdots$	1	z	x^2+y^2, z^2
$A_2 \equiv \Sigma^-$	1	1	$\cdots$	-1	R_z	
$E_1 \equiv \Pi$	2	$2\cos\Phi$	$\cdots$	0	$(x, y), (R_x, R_y)$	(xz, yz)
$E_2 \equiv \Delta$	2	$2\cos 2\Phi$	$\cdots$	0		(x^2-y^2, xy)
$E_3 \equiv \Phi$	2	$2\cos 3\Phi$	$\cdots$	0		
$\cdots$	$\cdots$	$\cdots$	$\cdots$	$\cdots$		

$D_{\infty h}$	E	$2C_\infty^\Phi$	$\cdots$	$\infty \sigma_v$	i	$2S_\infty^\Phi$	$\cdots$	∞C_2		
Σ_g^+	1	1	$\cdots$	1	1	1	$\cdots$	1		x^2+y^2, z^2
Σ_g^-	1	1	$\cdots$	-1	1	1	$\cdots$	-1	R_z	
Π_g	2	$2\cos\Phi$	$\cdots$	0	2	$-2\cos\Phi$	$\cdots$	0	(R_x, R_y)	(xz, yz)
Δ_g	2	$2\cos 2\Phi$	$\cdots$	0	2	$2\cos 2\Phi$	$\cdots$	0		(x^2-y^2, xy)
$\cdots$	$\cdots$	$\cdots$	$\cdots$	$\cdots$	$\cdots$	$\cdots$	$\cdots$	$\cdots$		
Σ_u^+	1	1	$\cdots$	1	-1	-1	$\cdots$	-1	z	
Σ_u^-	1	1	$\cdots$	-1	-1	-1	$\cdots$	1		
Π_u	2	$2\cos\Phi$	$\cdots$	0	-2	$2\cos\Phi$	$\cdots$	0	(x, y)	
Δ_u	2	$2\cos 2\Phi$	$\cdots$	0	-2	$-2\cos 2\Phi$	$\cdots$	0		
$\cdots$	$\cdots$	$\cdots$	$\cdots$	$\cdots$	$\cdots$	$\cdots$	$\cdots$	$\cdots$		

11. The Icosahedral Groups

I	E	$12C_5$	$12C_5^2$	$20C_3$	$15C_2$		$\eta^{\pm} = \frac{1}{2}(1 \pm 5^{1/2})$
A	1	1	1	1	1		$x^2 + y^2 + z^2$
T_1	3	η^+	η^-	0	-1	$(x, y, z),$	
						(R_x, R_y, R_z)	
T_2	3	η^-	η^+	0	-1		
G	4	-1	-1	1	0		
H	5	0	0	-1	1		$(2z^2 - x^2 - y^2,$
							$x^2 - y^2,$
							$xy, yz, zx)$

I_h	E	$12C_5$	$12C_5^2$	$20C_3$	$15C_2$	i	$12S_{10}$	$12S_{10}^3$	$20S_6$	15σ		$\eta^{\pm} = \frac{1}{2}(1 \pm 5^{1/2})$
A_g	1	1	1	1	1	1	1	1	1	1		$x^2 + y^2 + z^2$
T_{1g}	3	η^+	η^-	0	-1	3	η^-	η^+	0	-1	(R_x, R_y, R_z)	
T_{2g}	3	η^-	η^+	0	-1	3	η^+	η^-	0	-1		
G_g	4	-1	-1	1	0	4	-1	-1	1	0		
H_g	5	0	0	-1	1	5	0	0	-1	1		$(2z^2 - x^2 - y^2,$
												$x^2 - y^2,$
												$xy, yz, zx)$
A_u	1	1	1	1	1	-1	-1	-1	-1	-1		
T_{1u}	3	η^+	η^-	0	-1	-3	$-\eta^-$	$-\eta^+$	0	1	(x, y, z)	
T_{2u}	3	η^-	η^+	0	-1	-3	$-\eta^+$	$-\eta^-$	0	1		
G_u	4	-1	-1	1	0	-4	1	1	-1	0		
H_u	5	0	0	-1	1	-5	0	0	1	-1		

Note: In these groups and others containing C_5, the following relationships may be useful:

$$\eta^+ = 1/2(1 + 5^{1/2}) = 1.61803 \ldots = -2 \cos 144°$$

$$\eta^- = 1/2(1 - 5^{1/2}) = -0.61803 \ldots = -2 \cos 72°$$

$$\eta^+\eta^+ = 1 + \eta^+, \qquad \eta^-\eta^- = 1 + \eta^-, \qquad \eta^+\eta^- = -1$$

APPENDIX B

Correlation Tables

The tables in this appendix show correlations between the species (irreducible representations) of a parent group and many (if not all) of its subgroups. Where the correlation to a particular subgroup depends on which element of the parent group is retained in the subgroup, the retained element is listed under the heading of the subgroup in the table. For example, in the table for C_{2v}, either the $\sigma(xz)$ or the $\sigma(yz)$ plane can be retained in forming the subgroup C_s, resulting in the two different correlations listed. When the identification of the retained element might be ambiguous in the subgroup, its labels in the parent group and subgroup are linked by an arrow. For example, in the correlation table for C_{6v}, the listing $\sigma_v \rightarrow \sigma(xz)$ for C_{2v} means that the σ_v plane of C_{6v} becomes the $\sigma(xz)$ plane of C_{2v}.

The tables for certain large-order parent groups do not list correlations to all of the smaller subgroups. In such cases, carry out the correlation in two steps, using correlations to and from an intermediate group. For example, to correlate O_h with C_{2v}, first use the O_h table to correlate with T_d, and then use the T_d table to complete the correlation to C_{2v}.

Correlations between two smaller groups for which no individual correlation table is shown can be deduced by finding the related groups within a table for a larger group of which they are both subgroups. For example, the correlation between the groups C_4 and C_2 can be found within the correlation table for C_{4v}. Accordingly, in the descent from C_4 to C_2, the correlations are $A \rightarrow A$, $B \rightarrow A$, and $E \rightarrow 2B$.

Paired complex conjugate irreducible representations, which occur in groups C_n, C_{nh}, and S_n with $n \geq 3$ and the cubic groups T and T_h, are indicated by surrounding the Mulliken symbol in braces—for example, $\{E\}$. Correlations of these species to species in another group (either parent group or subgroup) always carry over both complex conjugate irreducible representations to the same species in the related group. If the related group is a subgroup that does not allow degeneracy, the complex conjugate pair become two real-number nondegenerate species in the subgroup.

Correlations between $C_{\infty v}$ and C_{2v} are given in Table 3.8, and those between $D_{\infty h}$ and D_{2h} are given in Table 3.9 (see p. 80).

C_{2v}	C_2	C_s $\sigma(xz)$	C_s $\sigma(yz)$
A_1	A	A'	A'
A_2	A	A''	A''
B_1	B	A'	A''
B_2	B	A''	A'

C_{3v}	C_3	C_s
A_1	A	A'
A_2	A	A''
E	E	$A' + A''$

C_{4v}	C_4	C_{2v} σ_v	C_{2v} σ_d	C_2	C_s σ_v	C_s σ_d
A_1	A	A_1	A_1	A	A'	A'
A_2	A	A_2	A_2	A	A''	A''
B_1	B	A_1	A_2	A	A'	A''
B_2	B	A_2	A_1	A	A''	A'
E	E	$B_1 + B_2$	$B_1 + B_2$	$2B$	$A' + A''$	$A' + A''$

C_{5v}	C_5	C_s
A_1	A	A'
A_2	A	A''
E_1	$\{E_1\}$	$A' + A''$
E_2	$\{E_2\}$	$A' + A''$

C_{6v}	C_6	C_{3v} σ_v	C_{3v} σ_d	C_{2v} $\sigma_v \to \sigma(xz)$	C_3	C_2	C_s σ_v	C_s σ_d
A_1	A	A_1	A_1	A_1	A	A	A'	A'
A_2	A	A_2	A_2	A_2	A	A	A''	A''
B_1	B	A_1	A_2	B_1	A	B	A'	A''
B_2	B	A_2	A_1	B_2	A	B	A''	A'
E_1	$\{E_1\}$	E	E	$B_1 + B_2$	$\{E\}$	$2B$	$A' + A''$	$A' + A''$
E_2	$\{E_2\}$	E	E	$A_1 + A_2$	$\{E\}$	$2A$	$A' + A''$	$A' + A''$

C_{2h}	C_2	C_s	C_i
A_g	A	A'	A_g
B_g	B	A''	A_g
A_u	A	A''	A_u
B_u	B	A'	A_u

C_{3h}	C_3	C_s
A'	A	A'
$\{E'\}$	$\{E\}$	$2A'$
A''	A	A''
$\{E''\}$	$\{E\}$	$2A''$

C_{4h}	C_4	S_4	C_{2h}	C_2	C_s	C_i
A_g	A	A	A_g	A	A'	A_g
B_g	B	B	A_g	A	A'	A_g
$\{E_g\}$	$\{E\}$	$\{E\}$	$2B_g$	$2B$	$2A''$	$2A_g$
A_u	A	B	A_u	A	A''	A_u
B_u	B	A	A_u	A	A''	A_u
$\{E_u\}$	$\{E\}$	$\{E\}$	$2B_u$	$2B$	$2A'$	$2A_u$

C_{5h}	C_5	C_s
A'	A	A'
$\{E_1'\}$	$\{E_1\}$	$2A'$
$\{E_2'\}$	$\{E_2\}$	$2A'$
A''	A	A''
$\{E_1''\}$	$\{E_1\}$	$2A''$
$\{E_2''\}$	$\{E_2\}$	$2A''$

C_{6h}	C_6	C_{3h}	S_6	C_{2h}	C_3	C_2	C_s	C_i
A_g	A	A'	A_g	A_g	A	A	A'	A_g
B_g	B	A''	A_g	B_g	A	B	A''	A_g
$\{E_{1g}\}$	$\{E_1\}$	$\{E''\}$	$\{E_g\}$	$2B_g$	$\{E\}$	$2B$	$2A''$	$2A_g$
$\{E_{2g}\}$	$\{E_2\}$	$\{E'\}$	$\{E_g\}$	$2A_g$	$\{E\}$	$2A$	$2A'$	$2A_g$
A_u	A	A''	A_u	A_u	A	A	A''	A_u
B_u	B	A'	A_u	B_u	A	B	A'	A_u
$\{E_{1u}\}$	$\{E_1\}$	$\{E'\}$	$\{E_u\}$	$2B_u$	$\{E\}$	$2B$	$2A'$	$2A_u$
$\{E_{2u}\}$	$\{E_2\}$	$\{E''\}$	$\{E_u\}$	$2A_u$	$\{E\}$	$2A$	$2A''$	$2A_u$

D_{2h}	D_2	C_{2v} $C_2(z)$	C_{2v} $C_2(y)$	C_{2v} $C_2(x)$	C_{2h} $C_2(z)$	C_{2h} $C_2(y)$	C_{2h} $C_2(x)$	C_2 $C_2(z)$	C_2 $C_2(y)$	C_2 $C_2(x)$	C_s $\sigma(xy)$	C_s $\sigma(xz)$	C_s $\sigma(yz)$
A_g	A	A_1	A_1	A_1	A_g	A_g	A_g	A	A	A	A'	A'	A'
B_{1g}	B_1	A_2	B_2	B_1	A_g	B_g	B_g	A	B	B	A'	A''	A''
B_{2g}	B_2	B_1	A_2	B_2	B_g	A_g	B_g	B	A	B	A''	A'	A''
B_{3g}	B_3	B_2	B_1	A_2	B_g	B_g	A_g	B	B	A	A''	A''	A'
A_u	A	A_2	A_2	A_2	A_u	A_u	A_u	A	A	A	A''	A''	A''
B_{1u}	B_1	A_1	B_1	B_2	A_u	B_u	B_u	A	B	B	A''	A'	A'
B_{2u}	B_2	B_2	A_1	B_1	B_u	A_u	B_u	B	A	B	A'	A''	A'
B_{3u}	B_3	B_1	B_2	A_1	B_u	B_u	A_u	B	B	A	A'	A'	A''

D_{3h}	D_3	C_{3v}	C_{3h}	C_{2v} $\sigma_h \to \sigma_v(yz)$	C_s σ_h	C_s σ_v
A_1'	A_1	A_1	A'	A_1	A'	A'
A_2'	A_2	A_2	A'	B_2	A'	A''
E'	E	E	$\{E'\}$	$A_1 + B_2$	$2A'$	$A' + A''$
A_1''	A_1	A_2	A''	A_2	A''	A''
A_2''	A_2	A_1	A''	B_1	A''	A'
E''	E	E	$\{E''\}$	$A_2 + B_1$	$2A''$	$A' + A''$

Other subgroups: C_3, C_2

D_{4h}	D_4	C_{4v}	C_{4h}	C_4	D_{2h} C_2'	D_{2h} C_2''	D_{2d} $C_2 \to C_2'$	D_{2d} $C_2'' \to C_2'$
A_{1g}	A_1	A_1	A_g	A	A_g	A_g	A_1	A_1
A_{2g}	A_2	A_2	A_g	A	B_{1g}	B_{1g}	A_2	A_2
B_{1g}	B_1	B_1	B_g	B	A_g	B_{1g}	B_1	B_2
B_{2g}	B_2	B_2	B_g	B	B_{1g}	A_g	B_2	B_1
E_g	E	E	$\{E_g\}$	$\{E\}$	$B_{2g} + B_{3g}$	$B_{2g} + B_{3g}$	E	E
A_{1u}	A_1	A_2	A_u	A	A_u	A_u	B_1	B_1
A_{2u}	A_2	A_1	A_u	A	B_{1u}	B_{1u}	B_2	B_2
B_{1u}	B_1	B_2	B_u	B	A_u	B_{1u}	A_1	A_2
B_{2u}	B_2	B_1	B_u	B	B_{1u}	A_u	A_2	A_1
E_u	E	E	$\{E_u\}$	$\{E\}$	$B_{2u} + B_{3u}$	$B_{2u} + B_{3u}$	E	E

See the next table for additional subgroups.

D_{4h} (cont.)	S_4	D_2 C_2'	D_2 C_2''	C_{2v} C_2, σ_v	C_{2v} C_2, σ_d	C_{2v} C_2'	C_{2v} C_2''
A_{1g}	A	A	A	A_1	A_1	A_1	A_1
A_{2g}	A	B_1	B_1	A_2	A_2	B_1	B_1
B_{1g}	B	A	B_1	A_1	A_2	A_1	B_1
B_{2g}	B	B_1	A	A_2	A_1	B_1	A_1
E_g	$\{E\}$	$B_2 + B_3$	$B_2 + B_3$	$B_1 + B_2$	$B_1 + B_2$	$A_2 + B_2$	$A_2 + B_2$
A_{1u}	B	A	A	A_2	A_2	A_2	A_2
A_{2u}	B	B_1	B_1	A_1	A_1	B_2	B_2
B_{1u}	A	A	B_1	A_2	A_1	A_2	B_2
B_{2u}	A	B_1	A	A_1	A_2	B_2	A_2
E_u	$\{E\}$	$B_2 + B_3$	$B_2 + B_3$	$B_1 + B_2$	$B_1 + B_2$	$A_1 + B_1$	$A_1 + B_1$

See the next table for additional subgroups.

D_{4h} (cont.)	C_{2h} C_2	C_{2h} C_2'	C_{2h} C_2''	C_s σ_h	C_s σ_v	C_s σ_d
A_{1g}	A_g	A_g	A_g	A'	A'	A'
A_{2g}	A_g	B_g	B_g	A'	A''	A''
B_{1g}	A_g	A_g	B_g	A'	A'	A''
B_{2g}	A_g	B_g	A_g	A'	A''	A'
E_g	$2B_g$	$A_g + B_g$	$A_g + B_g$	$2A''$	$A' + A''$	$A' + A''$
A_{1u}	A_u	A_u	A_u	A''	A''	A''
A_{2u}	A_u	B_u	B_u	A''	A'	A'
B_{1u}	A_u	A_u	B_u	A''	A''	A'
B_{2u}	A_u	B_u	A_u	A''	A'	A''
E_u	$2B_u$	$A_u + B_u$	$A_u + B_u$	$2A'$	$A' + A''$	$A' + A''$

Other subgroups: $3C_2$, C_i

D_{5h}	D_5	C_{5v}	C_{5h}	C_5	C_{2v} $\sigma_h \rightarrow \sigma(xz)$
A_1'	A_1	A_1	A'	A	A_1
A_2'	A_2	A_2	A'	A	B_1
E_1'	E_1	E_1	$\{E_1'\}$	$\{E_1\}$	$A_1 + B_1$
E_2'	E_2	E_2	$\{E_2'\}$	$\{E_2\}$	$A_1 + B_1$
A_1''	A_1	A_2	A''	A	A_2
A_2''	A_2	A_1	A''	A	B_2
E_1''	E_1	E_1	$\{E_1''\}$	$\{E_1\}$	$A_2 + B_2$
E_2''	E_2	E_2	$\{E_2''\}$	$\{E_2\}$	$A_2 + B_2$

Other subgroups: C_2, $2C_s$

D_{6h}	D_6	C_{6v}	C_{6h}	C_6	D_{3h} C_2'	D_{3h} C_2''	D_{3d} C_2'	D_{3d} C_2''	D_{2h} $\sigma_h \rightarrow \sigma(xy)$ $\sigma_v \rightarrow \sigma(yz)$
A_{1g}	A_1	A_1	A_g	A	A_1'	A_1'	A_{1g}	A_{1g}	A_g
A_{2g}	A_2	A_2	A_g	A	A_2'	A_2'	A_{2g}	A_{2g}	B_{1g}
B_{1g}	B_1	B_2	B_g	B	A_1''	A_2''	A_{1g}	A_{2g}	B_{2g}
B_{2g}	B_2	B_1	B_g	B	A_2''	A_1''	A_{2g}	A_{1g}	B_{3g}
E_{1g}	E_1	E_1	$\{E_{1g}\}$	$\{E_1\}$	E''	E''	E_g	E_g	$B_{2g} + B_{3g}$
E_{2g}	E_2	E_2	$\{E_{2g}\}$	$\{E_2\}$	E'	E'	E_g	E_g	$A_g + B_{1g}$
A_{1u}	A_1	A_2	A_u	A	A_1''	A_1''	A_{1u}	A_{1u}	A_u
A_{2u}	A_2	A_1	A_u	A	A_2''	A_2''	A_{2u}	A_{2u}	B_{1u}
B_{1u}	B_1	B_1	B_u	B	A_1'	A_2'	A_{1u}	A_{2u}	B_{2u}
B_{2u}	B_2	B_2	B_u	B	A_2'	A_1'	A_{2u}	A_{1u}	B_{3u}
E_{1u}	E_1	E_1	$\{E_{1u}\}$	$\{E_1\}$	E'	E'	E_u	E_u	$B_{2u} + B_{3u}$
E_{2u}	E_2	E_2	$\{E_{2u}\}$	$\{E_2\}$	E''	E''	E_u	E_u	$A_u + B_{1u}$

Other subgroups: $2D_3$, $2C_{3v}$, C_{3h}, S_6, D_2, $2C_{2v}$, $3C_{2h}$, $3C_2$, $3C_s$, C_i

D_{2d}	S_4	D_2 $C_2 \to C_2(z)$	C_{2v}
A_1	A	A	A_1
A_2	A	B_1	A_2
B_1	B	A	A_2
B_2	B	B_1	A_1
E	$\{E\}$	$B_2 + B_3$	$B_1 + B_2$

Other subgroups: $2C_2$, C_s

D_{3d}	D_3	S_6	C_{3v}	C_3	C_{2h}
A_{1g}	A_1	A_g	A_1	A	A_g
A_{2g}	A_2	A_g	A_2	A	B_g
E_g	E	$\{E_g\}$	E	$\{E\}$	$A_g + B_g$
A_{1u}	A_1	A_u	A_2	A	A_u
A_{2u}	A_2	A_u	A_1	A	B_u
E_u	E	$\{E_u\}$	E	$\{E\}$	$A_u + B_u$

Other subgroups: C_2, C_s, C_i

D_{4d}	D_4	S_8	C_{4v}	C_4	C_{2v}	C_2 $C_2(z)$	C_2 C_2'	C_s
A_1	A_1	A	A_1	A	A_1	A	A	A'
A_2	A_2	A	A_2	A	A_2	A	B	A''
B_1	A_1	B	A_2	A	A_2	A	A	A''
B_2	A_2	B	A_1	A	A_1	A	B	A'
E_1	E	$\{E_1\}$	E	$\{E\}$	$B_1 + B_2$	$2B$	$A + B$	$A' + A''$
E_2	$B_1 + B_2$	$\{E_2\}$	$B_1 + B_2$	$2B$	$A_1 + A_2$	$2A$	$A + B$	$A' + A''$
E_3	E	$\{E_3\}$	E	$\{E\}$	$B_1 + B_2$	$2B$	$A + B$	$A' + A''$

D_{5d}	D_5	C_{5v}	C_5	C_2
A_{1g}	A_1	A_1	A	A
A_{2g}	A_2	A_2	A	B
E_{1g}	E_1	E_1	$\{E_1\}$	$A+B$
E_{2g}	E_2	E_2	$\{E_2\}$	$A+B$
A_{1u}	A_1	A_2	A	A
A_{2u}	A_2	A_1	A	B
E_{1u}	E_1	E_1	$\{E_1\}$	$A+B$
E_{2u}	E_2	E_2	$\{E_2\}$	$A+B$

Other subgroups: C_s, C_i

D_{6d}	D_6	C_{6v}	D_3	D_{2d}	S_4	C_2 $C_2(z)$	C_2 C_2'
A_1	A_1	A_1	A_1	A_1	A	A	A
A_2	A_2	A_2	A_2	A_2	A	A	B
B_1	A_1	A_2	A_1	B_1	B	A	A
B_2	A_2	A_1	A_2	B_2	B	A	B
E_1	E_1	E_1	E	E	$\{E\}$	$2B$	$A+B$
E_2	E_2	E_2	E	B_1+B_2	$2B$	$2A$	$A+B$
E_3	B_1+B_2	B_1+B_2	A_1+A_2	E	$\{E\}$	$2B$	$A+B$
E_4	E_2	E_2	E	A_1+A_2	$2A$	$2A$	$A+B$
E_5	E_1	E_1	E	E	$\{E\}$	$2B$	$A+B$

Other subgroups: C_6, C_{3v}, C_3, D_2, C_{2v}, C_s

T	C_3	D_2	C_2
A	A	A	A
$\{E\}$	$\{E\}$	$2A$	$2A$
T	$A+\{E\}$	$B_1+B_2+B_3$	$A+2B$

T_h	T	S_6	D_{2h}	D_2
A_g	A	A_g	A_g	A
$\{E_g\}$	$\{E\}$	$\{E_g\}$	$2A_g$	$2A$
T_g	T	$A_g + \{E_g\}$	$B_{1g} + B_{2g} + B_{3g}$	$B_1 + B_2 + B_3$
A_u	A	A_u	A_u	A
$\{E_u\}$	$\{E\}$	$\{E_u\}$	$2A_u$	$2A$
T_u	T	$A_u + \{E_u\}$	$B_{1u} + B_{2u} + B_{3u}$	$B_1 + B_2 + B_3$

Other subgroups: C_{2v}, C_{2h}, C_3, C_2, C_s, C_i

T_d	T	C_{3v}	C_{2v}	D_{2d}
A_1	A	A_1	A_1	A_1
A_2	A	A_2	A_2	B_1
E	$\{E\}$	E	$A_1 + A_2$	$A_1 + B_1$
T_1	T	$A_2 + E$	$A_2 + B_1 + B_2$	$A_2 + E$
T_2	T	$A_1 + E$	$A_1 + B_1 + B_2$	$B_2 + E$

Other subgroups: S_4, D_2, C_3, C_2, C_s

O	T	D_4	C_4	D_3	D_2 $3C_2$	D_2 $C_2, 2C_2'$	C_3	C_2 C_2	C_2 C_2'
A_1	A	A_1	A	A_1	A	A	A	A	A
A_2	A	B_1	B	A_2	A	B_1	A	A	B
E	$\{E\}$	$A_1 + B_1$	$A + B$	E	$2A$	$A + B_1$	$\{E\}$	$2A$	$A + B$
T_1	T	$A_2 + E$	$A + E$	$A_2 + E$	$B_1 + B_2 + B_3$	$B_1 + B_2 + B_3$	$A + \{E\}$	$A + 2B$	$A + 2B$
T_2	T	$B_2 + E$	$B + E$	$A_1 + E$	$B_1 + B_2 + B_3$	$A + B_2 + B_3$	$A + \{E\}$	$A + 2B$	$2A + B$

O_h	O	T_d	T_h	D_{4h}	D_{3d}
A_{1g}	A_1	A_1	A_g	A_{1g}	A_{1g}
A_{2g}	A_2	A_2	A_g	B_{1g}	A_{2g}
E_g	E	E	$\{E_g\}$	$A_{1g} + B_{1g}$	E_g
T_{1g}	T_1	T_1	T_g	$A_{2g} + E_g$	$A_{2g} + E_g$
T_{2g}	T_2	T_2	T_g	$B_{2g} + E_g$	$A_{1g} + E_g$
A_{1u}	A_1	A_2	A_u	A_{1u}	A_{1u}
A_{2u}	A_2	A_1	A_u	B_{1u}	A_{2u}
E_u	E	E	$\{E_u\}$	$A_{1u} + B_{1u}$	E_u
T_{1u}	T_1	T_2	T_u	$A_{2u} + E_u$	$A_{2u} + E_u$
T_{2u}	T_2	T_1	T_u	$B_{2u} + E_u$	$A_{1u} + E_u$

Other subgroups: T, D_4, C_{4v}, C_{4h}, C_4, D_3, S_6, C_{3v}, C_3, $2D_{2h}$, D_{2d}, $2D_2$, S_4, $3C_{2v}$, $2C_{2h}$, $2C_2$, C_i, C_s

I	T	D_5	C_5	D_3	C_3	D_2	C_2
A	A	A_1	A	A_1	A	A	A
T_1	T	$A_2 + E_1$	$A + \{E_1\}$	$A_2 + E$	$A + \{E\}$	$B_1 + B_2 + B_3$	$A + 2B$
T_2	T	$A_2 + E_2$	$A + \{E_2\}$	$A_2 + E$	$A + \{E\}$	$B_1 + B_2 + B_3$	$A + 2B$
G	$A + T$	$E_1 + E_2$	$\{E_1\} + \{E_2\}$	$A_1 + A_2 + E$	$2A + \{E\}$	$A + B_1 + B_2 + B_3$	$2A + 2B$
H	$\{E\} + T$	$A_1 + E_1 + E_2$	$A + \{E_1\} + \{E_2\}$	$A_1 + 2E$	$A + 2\{E\}$	$2A + B_1 + B_2 + B_3$	$3A + 2B$

R_3	O	D_4	D_3
S	A_1	A_1	A_1
P	T_1	$A_2 + E$	$A_2 + E$
D	$E + T_2$	$A_1 + B_1 + B_2 + E$	$A_1 + 2E$
F	$A_2 + T_1 + T_2$	$A_2 + B_1 + B_2 + 2E$	$A_1 + 2A_2 + 2E$
G	$A_1 + E + T_1 + T_2$	$2A_1 + A_2 + B_1 + B_2 + 2E$	$2A_1 + A_2 + 3E$
H	$E + 2T_1 + T_2$	$A_1 + 2A_2 + B_1 + B_2 + 3E$	$A_1 + 2A_2 + 4E$

APPENDIX C

Normal Modes of Vibration of Some Common Structures

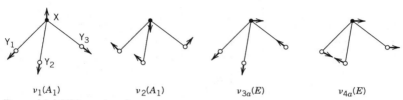

$\nu_1(A_1)$ $\nu_2(A_1)$ $\nu_{3a}(E)$ $\nu_{4a}(E)$

Pyramidal XY$_3$ molecules.

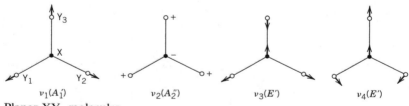

$\nu_1(A_1')$ $\nu_2(A_2'')$ $\nu_3(E')$ $\nu_4(E')$

Planar XY$_3$ molecules.

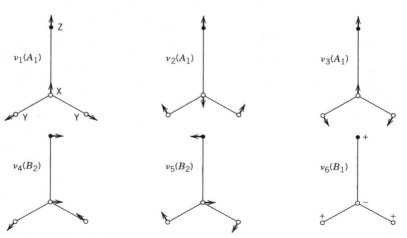

Planar ZXY$_2$ molecules.

284

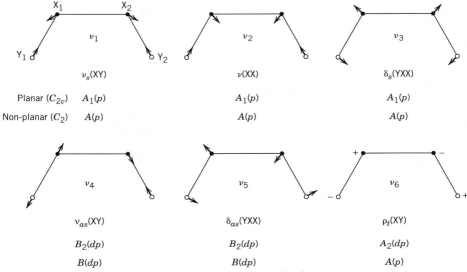

Nonlinear X_2Y_2 molecules (p, polarized; dp, depolarized).

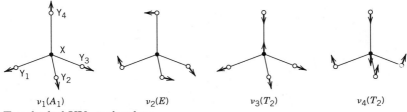

Tetrahedral XY_4 molecules.

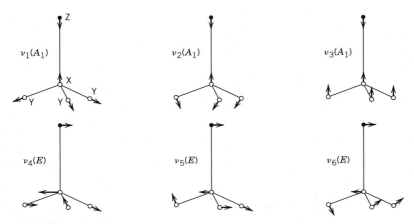

C_{3v} ZXY_3 molecules.

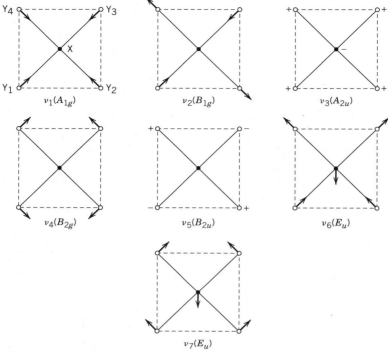

Square planar XY_4 molecules.

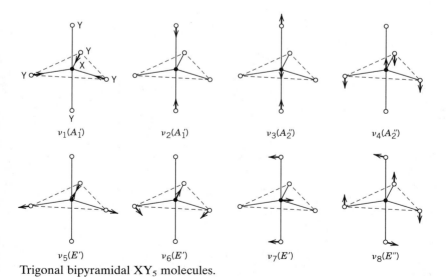

Trigonal bipyramidal XY_5 molecules.

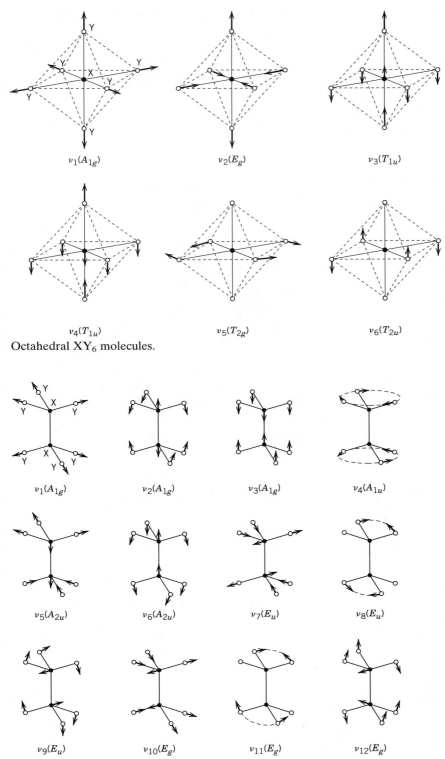

$\nu_1(A_{1g})$ $\nu_2(E_g)$ $\nu_3(T_{1u})$

$\nu_4(T_{1u})$ $\nu_5(T_{2g})$ $\nu_6(T_{2u})$

Octahedral XY_6 molecules.

$\nu_1(A_{1g})$ $\nu_2(A_{1g})$ $\nu_3(A_{1g})$ $\nu_4(A_{1u})$

$\nu_5(A_{2u})$ $\nu_6(A_{2u})$ $\nu_7(E_u)$ $\nu_8(E_u)$

$\nu_9(E_u)$ $\nu_{10}(E_g)$ $\nu_{11}(E_g)$ $\nu_{12}(E_g)$

Ethane-type X_2Y_6 molecules.

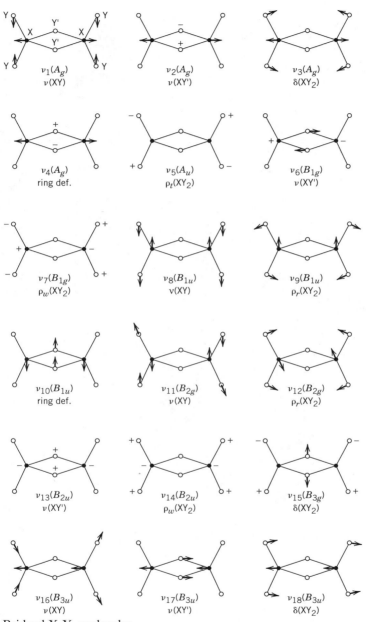

$\nu_1(A_g)$
$\nu(XY)$

$\nu_2(A_g)$
$\nu(XY')$

$\nu_3(A_g)$
$\delta(XY_2)$

$\nu_4(A_g)$
ring def.

$\nu_5(A_u)$
$\rho_t(XY_2)$

$\nu_6(B_{1g})$
$\nu(XY')$

$\nu_7(B_{1g})$
$\rho_w(XY_2)$

$\nu_8(B_{1u})$
$\nu(XY)$

$\nu_9(B_{1u})$
$\rho_r(XY_2)$

$\nu_{10}(B_{1u})$
ring def.

$\nu_{11}(B_{2g})$
$\nu(XY)$

$\nu_{12}(B_{2g})$
$\rho_r(XY_2)$

$\nu_{13}(B_{2u})$
$\nu(XY')$

$\nu_{14}(B_{2u})$
$\rho_w(XY_2)$

$\nu_{15}(B_{3g})$
$\delta(XY_2)$

$\nu_{16}(B_{3u})$
$\nu(XY)$

$\nu_{17}(B_{3u})$
$\nu(XY')$

$\nu_{18}(B_{3u})$
$\delta(XY_2)$

Bridged X_2Y_6 molecules.

APPENDIX D

*Tanabe and Sugano Diagrams**

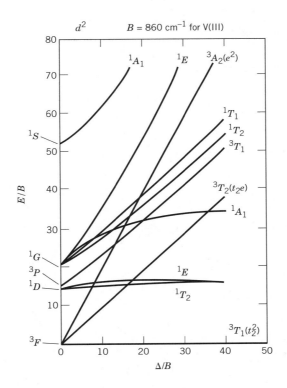

*Adapted with permission from T. Tanabe and S. Sugano, *J. Phys. Soc. Japan* **1954,** *9,* 766.

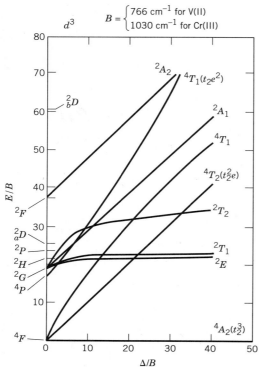

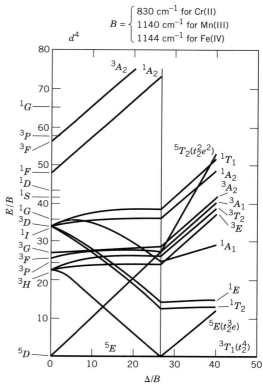

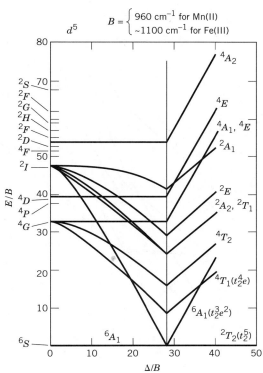

d^5

$B = \begin{cases} 960 \text{ cm}^{-1} \text{ for Mn(II)} \\ \sim 1100 \text{ cm}^{-1} \text{ for Fe(III)} \end{cases}$

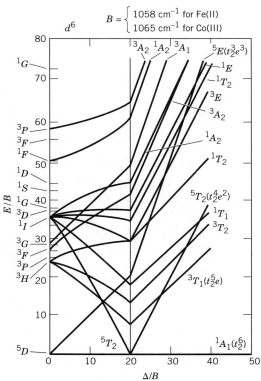

d^6

$B = \begin{cases} 1058 \text{ cm}^{-1} \text{ for Fe(II)} \\ 1065 \text{ cm}^{-1} \text{ for Co(III)} \end{cases}$

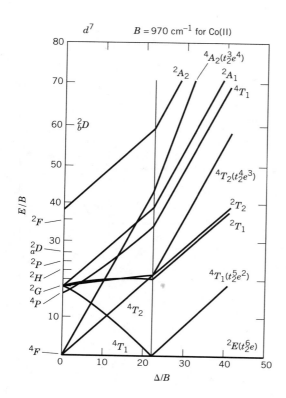

d^7 $B = 970\ \text{cm}^{-1}$ for Co(II)

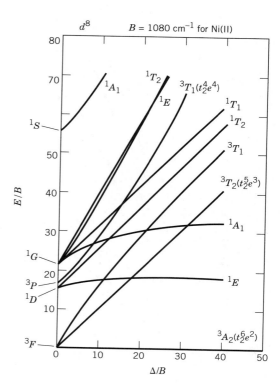

d^8 $B = 1080\ \text{cm}^{-1}$ for Ni(II)

INDEX

Entries for symbols customarily set in capital letter italics (e.g., Schönflies notations, Mulliken symbols, overall quantum numbers) are given before those for words, phrases, formulas, and orbital notations, which are intermingled in alphabetical order.

in $[Mn(H_2O]^{2+}$, 252
spin multiplicity rule for, 250,
  252
vibronically allowed, 251, 252
Enantiomers, 33
Equivalent electrons, defined, 228

F irreducible representation
  notation, 60
Fermi resonance, 195–196
Flow chart for point group
  classification, 28–29
Force constant, 165

Great Orthogonality Theorem,
  relationships from, 61–64
Group, requirements of, 18–20. *See
  also* Point Group(s)
Group-subgroup relationships, 20,
  73–79. *See also* Appendix B

Hamiltonian operator, 88
Harmonic oscillator, 166
Hermann-Mauguin notation, 21
High spin configurations:
  in octahedral complexes,
    207–209
  in tetrahedral complexes, 209
H_2O_2, lack of chirality, 34
Hole formalism, 244–245
Hybrid orbitals, 95–104
  AO composition by group
    theory, 97–104
  sp^3 wave functions, 95–96,
    154–156
  wave functions by projection
    operators, 153–156
  wave functions for *tbp* MX_5, 159

I, notation for identity, 2
I_h, symmetry of, 27
Icosahedron, symmetry of, 27
Identity, 2–3, 19
Improper axis, *see* Axis (axes),
  improper

Improper rotation, *see* Rotation-
  reflection
Infrared spectra:
  activity of normal modes in,
    180–181, 183–184
  of CCl_4 Fermi doublet, 196
Internal energy, 250
Inverse, 20
Inversion, 2, 8–9
Inversion center, 2, 8
Isotopic splitting, 196–197

J quantum number, 229
Jahn-Teller distortion, 210–214
  effect on electronic spectra, 255
Jahn-Teller theorem, 210

Kronecker delta function, 64, 140,
  155 (f. n.)

L quantum number, 229–230
LaPorte's rule, 250–252
  break down of, 251, 252
LCAO method, 104–105
Ligand field theory, 202, 214
Low spin configurations:
  in octahedral complexes,
    207–209
  in square planar complexes,
    213–214
  in tetrahedral complexes, 209

M_L quantum number, 230–231
M_S quantum number, 231–232
Matrix (matrices):
  character of, 50
  conformability, 47
  defined, 47
  dimensions of, 47
  elements, defined, 47
  identity, 155 (f. n.)
  indexing system, 47–48
  inverse, 155
  multiplication of, 48
  operator, 16, 44, 174